Advances in Experimental Medicine and Biology

Volume 752

For further volumes:
http://www.springer.com/series/5584

G. Cliff Lamb • Nicolas DiLorenzo
Editors

Current and Future Reproductive Technologies and World Food Production

Springer

Editors
G. Cliff Lamb
North Florida Research
and Education Centre
University of Florida
Marianna, FL, USA

Nicolas DiLorenzo
North Florida Research
and Education Centre
University of Florida
Marianna, FL, USA

ISSN 0065-2598 ISSN 2214-8019 (electronic)
ISBN 978-1-4614-8886-6 ISBN 978-1-4614-8887-3 (eBook)
DOI 10.1007/978-1-4614-8887-3
Springer New York Heidelberg Dordrecht London

Library of Congress Control Number: 2013950986

Printed on acid-free paper

Springer is part of Springer Science+Business Media (www.springer.com)

Preface

The title for the 2011 Society for the Study of Reproduction Annual Meeting in Portland Oregon was "Reproduction and the World's Future." One of the mini-symposia during the conference was titled "Reproduction Research and Its Impact on Feeding the World's Hungry" and was chaired by Dr. Robert Cushman (ARS/USDA, Clay Center, NE). The mini-symposium covered the broad impacts of reproduction technologies on our current livestock, poultry, and fish industries. Shortly after completing the mini-symposium the speakers and chair convened and discussed the possibility of generating a publication that may expand on the presentations of the mini-symposium. After some discussion, the result is this comprehensive book with contributions from experts throughout the world. These contributions provide various points of view on author perceptions, based on data, of how reproductive technologies have changed animal, poultry, and fish production. In addition, the book provides an insight on the crucial role that future reproductive technologies may play in production systems to ensure a sustainable food supply.

It is commonly known that estimations of world population growth indicate that by the year 2050 there will be nine billion habitants on earth. These estimates impose a tremendous challenge in the current agricultural systems as food supply will need to increase by 100 % in the next 40 years (Food and Agriculture Organization, 2009). This could be achieved by two means: increasing productive resources (land, livestock, and crops) or increasing the productivity of the existing resources. However, a further expansion of the agricultural frontier is not likely to happen to a great extent, and certainly not without incurring an environmental cost. Thus, the refinement of current technologies and development of new technologies aimed at increasing the productivity of resources while minimizing negative environmental impacts will be critical in order to meet the global food demand in the near future.

During the past 50 years assisted reproductive technologies have been developed and refined to increase the number and quality of offspring from genetically superior farm animal livestock species. Artificial insemination (AI), estrous synchronization and fixed-time AI, semen and embryo cryopreservation, multiple ovulation and embryo transfer (MOET), in vitro fertilization, sex determination of sperm or embryos, and nuclear transfer are technologies that are used to enhance the production efficiency of species used for food production. In many cases, the development of these technologies is responsible for significant changes to traditional production practices. This is a single publication that addresses the impacts of current and future reproductive technologies to our world food production that has never been compiled before. The impacts of reproductive technologies (both traditional and advanced) will shape future world food supply. As editors, we are proud of the work that the authors have done to provide a comprehensive overview and hope that readers find value in some if not all of the chapters.

Marianna, FL, USA

G. Cliff Lamb
Nicolas DiLorenzo

Contents

Contributors

Vishwajit S. Chowdhury Division for Arts and Science, Faculty of Arts and Science, Kyushu University, Fukuoka, Japan

Reinaldo F. Cooke Eastern Oregon Agricultural Research Center, Oregon State University, Burns, OR, USA

Robert A. Cushman Reproduction Research Unit, U.S. Meat Animal Research Center, Clay Center, NE, USA

Carl Dahlen Department of Animal Sciences, North Dakota State University, Fargo, ND, USA

Ocilon Gomes de Sá Filho Departamento de Produção Animal, Faculdade de Medicina Veterinária e Zootecnia—Universidade Estadual Paulista, Botucatu, São Paulo, Brazil

Mitsuhiro Furuse Faculty of Agriculture, Laboratory of Regulation in Metabolism and Behavior, Department of Bioresource Sciences, Kyushu University, Fukuoka, Japan

Peter J. Hansen Department of Animal Sciences, D.H. Barron Reproductive and Perinatal Biology Research Program, and Genetics Institute, University of Florida, Gainesville, FL, USA

Robert V. Knox Department of Animal Sciences, University of Illinois, Urbana, IL, USA

Larry A. Kuehn USDA-ARS, U.S. Meat Animal Research Center, Clay Center, NE, USA

G. Cliff Lamb North Florida Research and Education Center, University of Florida, Marianna, FL, USA

Jamie Larson Department of Animal and Dairy Sciences, Mississippi State University, Mississippi State, MS, USA

Cheng-Sheng Lee Center for Tropical and Subtropical Aquaculture, c/o The Oceanic Institute, Waimanalo, HI, USA

Graeme B. Martin UWA Institute of Agriculture M082, The University of Western Australia, Crawley, WA, Australia

Tara G. McDaneld USDA-ARS, U.S. Meat Animal Research Center, Clay Center, NE, USA

Dan Nonneman USDA-ARS, U.S. Meat Animal Research Center, Clay Center, NE, USA

Takeshi Osawa Faculty of Agriculture, Laboratory of Theriogenology, Department of Veterinary Sciences, University of Miyazaki, Miyazaki, Japan

Ramesh Ramachandran Department of Animal Science, Center for Reproductive Biology and Health, The Pennsylvania State University, University Park, PA, USA

George E. Seidel Jr. Animal Reproduction and Biotechnology Laboratory, Department of Biomedical Sciences, Colorado State University, Fort Collins, CO, USA

Warren M. Snelling USDA-ARS, U.S. Meat Animal Research Center, Clay Center, NE, USA

Jeffrey S. Stevenson Department of Animal Sciences and Industry, Kansas State University, Manhattan, KS, USA

Halima Sultana Laboratory of Animal Science, North Florida Research and Education Center (NFREC), University of Florida, Marianna, FL, USA

Esté van Marle-Köster Department of Animal and Wildlife Sciences, University of Pretoria, Hatfield, South Africa

José Luiz Moraes Vasconcelos Departamento de Produção Animal, Faculdade de Medicina Veterinária e Zootecnia—Universidade Estadual Paulista, Botucatu, São Paulo, Brazil

Edward C. Webb Department of Animal and Wildlife Sciences, University of Pretoria, Hatfield, South Africa

Gregory M. Weber National Center for Cool and Coldwater Aquaculture, ARS/USDA, Kearneysville, WV, USA

1 Current and Future Assisted Reproductive Technologies for Mammalian Farm Animals

Peter J. Hansen

Abstract

Reproduction in domestic animals is under control by man and the technologies developed to facilitate that control have a major impact on the efficiency of food production. Reproduction is an energy-intensive process. In beef cattle, for example, over 50 % of the total feed consumption required to produce a unit of meat protein is consumed by the dam of the meat animal (Anim Prod 27:367–379, 1978). Sows are responsible for about 20 % of the total feed needed to produce animals for slaughter (Adv Pork Prod 19:223–237, 2008). Accordingly, energy input to produce food from animal sources is reduced by increasing number of offspring per unit time a breeding female is in the herd. Using beef cattle as an example again, life-cycle efficiency for production of weaned calves is positively related to early age at puberty and short calving intervals (J Anim Sci 57:852–866, 1983). Reproductive technologies also dictate the strategies that can be used to select animals genetically for traits that improve production. Of critical importance has been artificial insemination (AI) (Anim Reprod Sci 62:143–172, 2000; Stud Hist Philos Biol Biomed Sci 38:411–441, 2007; Reprod Domest Anim 43:379–385, 2008; J Dairy Sci 92:5814–5833, 2009) and, as will be outlined in this chapter, emerging technologies offer additional opportunities for improvements in genetic selection. Given the central role of reproduction as a determinant of production efficiency and in genetic selection, improvements in reproductive technologies will be crucial to meeting the challenges created by the anticipated increases in world population (from seven billion people in 2011 to an anticipated nine billion by 2050; World population prospects: the 2010 revision, highlights and advance tables. Working Paper No. ESA/P/WP.220,

P.J. Hansen (✉)
Department of Animal Sciences, D.H. Barron Reproductive and Perinatal Biology Research Program, and Genetics Institute, University of Florida, PO Box 110910, Gainesville, FL 32611-0910, USA
e-mail: Hansen@animal.ufl.edu

G.C. Lamb and N. DiLorenzo (eds.), *Current and Future Reproductive Technologies and World Food Production*, Advances in Experimental Medicine and Biology 752,
DOI 10.1007/978-1-4614-8887-3_1, © Springer Science+Business Media New York 2014

New York) and by difficulties in livestock production wrought by climate change (SAT eJournal 4:1–23, 2007).

The purpose of this chapter will be to highlight current and emerging reproductive technologies that have the potential to improve efficiency of livestock production. The focus will be on technologies that manipulate male and female gametes as well as the stem cells from which they are derived and the preimplantation embryo. While technology is crucial to other interventions in the reproductive process like control of seasonal breeding, hormonal regulation of ovulation, estrous cyclicity and pregnancy establishment, feeding to optimize reproduction, minimizing environmental stress, and selection of genes controlling reproduction, these will not be considered here. Rather the reader is directed to other chapters in this volume as well as some reviews on other aspects of artificial manipulation of reproduction (Reprod Fertil Dev 24:258–266, 2011; Reprod Domest Anim 43:40–47, 2008; Reprod Domest Anim 43:122–128, 2008; Soc Reprod Fertil Suppl 66:87–102, 2009; Comprehensive biotechnology, Amsterdam, pp 477–485; Dairy production medicine, Chichester, pp 153–163; Theriogenology 76:1619–1631, 2011; Theriogenology 76:1568–1582, 2011; Theriogenology 77:1–11, 2012). Given the large number of mammalian species used for production of products useful for man and the diversity in their biology and management, the review will not be comprehensive but instead will use results from species that are most illustrative of the opportunities generated by assisted reproductive technologies.

Keywords
Reproduction • Mammals • Livestock • Semen preservation • Artificial insemination • In vitro fertilization

Introduction: Reproduction as a Centrally Important Component of Animal Production

Reproduction in domestic animals is under control by man and the technologies developed to facilitate that control have a major impact on the efficiency of food production. Reproduction is an energy-intensive process. In beef cattle, for example, over 50 % of the total feed consumption required to produce a unit of meat protein is consumed by the dam of the meat animal (Dickerson 1978). Sows are responsible for about 20 % of the total feed needed to produce animals for slaughter (Ball et al. 2008). Accordingly, energy input to produce food from animal sources is reduced by increasing number of offspring per unit time a breeding female is in the herd. Using beef cattle as an example again, life-cycle efficiency for production of weaned calves is positively related to early age at puberty and short calving intervals (Davis et al. 1983). Reproductive technologies also dictate the strategies that can be used to select animals genetically for traits that improve production. Of critical importance has been artificial insemination (AI) (Johnson et al. 2000; Wilmot 2007; Leboeuf et al. 2008; Carta et al. 2009) and, as will be outlined in this chapter, emerging technologies offer additional opportunities for improvements in genetic selection. Given the central role of reproduction as a determinant of production efficiency and in genetic selection, improvements in reproductive technologies will be crucial to meeting the challenges created by the anticipated increases in world population (from seven billion people in 2011 to an anticipated nine billion by 2050; United Nations et al. 2011) and by difficulties in livestock production wrought by climate change (Thornton et al. 2007).

The purpose of this chapter will be to highlight current and emerging reproductive technologies that have the potential to improve efficiency of livestock production. The focus will be on technologies that manipulate male and female gametes as well as the stem cells from which they are derived and the preimplantation embryo. While technology is crucial to other interventions in the reproductive process like control of seasonal breeding, hormonal regulation of ovulation, estrous cyclicity and pregnancy establishment, feeding to optimize reproduction, minimizing environmental stress, and selection of genes controlling reproduction, these will not be considered here. Rather the reader is directed to other chapters in this volume as well as some reviews on other aspects of artificial manipulation of reproduction (Bisinotto and Santos 2011; Chemineau et al. 2008; Notter 2008; Onteru et al. 2009; Hansen 2011a, b; Thatcher et al. 2011; Wiltbank et al. 2011; De Rensis et al. 2012). Given the large number of mammalian species used for production of products useful for man and the diversity in their biology and management, the review will not be comprehensive but instead will use results from species that are most illustrative of the opportunities generated by assisted reproductive technologies.

Technologies Focused on Male Reproduction

Semen Preservation

The reproductive technology that has the greatest impact on livestock production has been AI. In dairy cattle, for example, the development of AI as the predominant breeding method, combined with extensive record-keeping and formulation of genetic models to estimate estimated breeding values for production traits, has resulted in large increases in milk yield per cow. In US Holsteins, the estimated breeding value for milk yield has increased by 3,670 kg from 1960 to 2010 (http://aipl.arsusda.gov/eval/summary/trend.cfm).

One reason why AI has had a large impact on genetic merit for production is that development of methods for storage of sperm in liquid and frozen form has made it possible to distribute semen widely both geographically and temporally. In cattle, pregnancy rates achieved by AI using frozen semen is similar to that achieved after natural mating (Landivar et al. 1985; Lima et al. 2009) while in other species, including the boar (Johnson et al. 2000), small ruminants (Barbas and Mascarenhas 2009; Cseh et al. 2012), and donkey (Vidament et al. 2009), fertility with frozen semen remains suboptimal. In some species where frozen semen gives poor results, storage of semen in the liquid state can result in high pregnancy rates; the boar (Flowers and Alhusen 1992) and donkey (Vidament et al. 2009) are examples. In the stallion, pregnancy rates after insemination with frozen semen can be as high (Loomis 2001) or higher (Crowe et al. 2008) than with chilled semen although there is variability between individual males (Loomis and Graham 2008).

Breakthroughs in sperm cryopreservation technologies for species in which semen is difficult to freeze have proven elusive. In part, lack of progress reflects inadequate funding as well as the paucity of studies in which semen is used to inseminate females. Most experiments evaluating alterations in cryopreservation protocols use morphological, biochemical, or motility changes in the sperm cell as an indication of cryopreservation success but treatments that improve these endpoints do not necessarily result in higher pregnancy rate after AI (Spizziri et al. 2010). One treatment that appears promising for improving the fertilizing capability of cryopreserved semen is addition of antioxidants to semen extenders (Zhang et al. 2012). Addition of glutathione to extender improved post-thaw sperm motility and acrosomal integrity in the bull and resulted in increased conception rate when semen was used in AI (Perumal et al. 2011). For sows inseminated with frozen-thawed semen, addition of β-mercaptoethanol to the thawing solution improved litter size but not conception rate (Yamaguchi and Funahashi 2012).

It has also been long known that components of seminal plasma interact with sperm in ways that can compromise sperm function (Caballero et al. 2012). There may also be beneficial components of seminal plasma that are lost when seminal plasma is discarded or diluted during the storage process. Addition of seminal plasma before freezing has been reported to improve sperm survival in the bull, boar, ram, and stallion (Robinson et al. 2011; Caballero et al. 2012) and there are reports that addition of seminal plasma after thawing can improve pregnancy rates when ewes are cervically inseminated with frozen-thawed semen using the transcervical (Robinson et al. 2011). Specific components of seminal plasma may also interact with sperm cells or in the reproductive tract to enhance pregnancy rates. Addition of the heterodimer of porcine seminal protein I (PSP1) and PSPII to sex-sorted boar sperm improved fertilization rate when sperm were placed in the oviduct laparoscopically (García et al. 2007). There have also been inconsistent reports in pigs and cattle that addition of transforming growth factor-β to semen improves fertility following AI (Hansen 2011c).

Sexed Semen

In the First Century A.D., Pliny the Elder expressed the belief that gender could be selected by controlling whether semen originated from the left or right testicle. "If the right testicle is tied up, the ram will generate females, and if the left, males" (*Naturalis Historiae*, translated by Bostock and Riley 1890). The long-held goal of animal breeders to control gender became a reality with the development in the 1980s of flow cytometric techniques to separate X-bearing sperm from Y-bearing sperm based on slight differences in DNA content (about 4.2 % in the bull; Moruzzi 1979). In 1989, it was reported that 94 % of offspring from rabbit does inseminated with X-sorted sperm were female and 81 % of offspring from does inseminated with Y-based sperm were male (Johnson et al. 1989). Since that time, flow cytometric techniques have been improved and offspring have been born as a result of fertilization with sex-sorted sperm in many mammalian species including pigs, horses, elk, humans, and dolphins (Garner and Seidel 2008; O'Brien et al. 2009). First introduced in the United States for cattle AI on a wide scale in 2005 (DeJarnette et al. 2009), by 2008 sexed semen was used on 17.8 % of the recorded breeding of Holsteins heifers in the United States and for 0.4 % of the recorded breeding of cows (Norman et al. 2010).

The ability to control gender through use of sexed semen has increased the breeding strategies available to livestock producers. In dairy cattle, for example, De Vries et al. (2008) have pointed out that producers could choose to use sexed semen on the top half of the herd to produce replacement females. The bottom half of the herd could be bred by conventional semen using dairy semen and heifers sold as replacements or bred with conventional or Y-sorted semen from beef culls and calves used for beef production. The availability of genomic testing makes the accuracy of selection of females to receive sexed semen more accurate than previously the case (Weigel et al. 2012).

The process of sperm sorting is very effective in terms of separating X and Y sperm. Large-scale studies in cattle indicate that the sex ratio of calves born from insemination with X-sorted semen varies from 89 to 91 % (DeJarnette et al. 2009; Norman et al. 2010). There are, however, problems with the process associated with sperm damage and yield that limit applications of the technology. Sperm damage results from the dilution of sperm during the sorting process (eliminating possible protective effects of seminal plasma), possible effects of interactions of nuclear dyes with chromatin, damage from the laser used by flow cytometry, electric forces placed on the sperm as part of the sorting process, and sperm aging during sorting (Rath and Johnson 2008). As a result, sorting causes reductions in fertilizing ability of sperm and competence of the resultant embryos to develop to the blastocyst stage (Lu et al. 1999; Wheeler et al. 2006; Rasmussen et al. 2013). Yield of sperm is

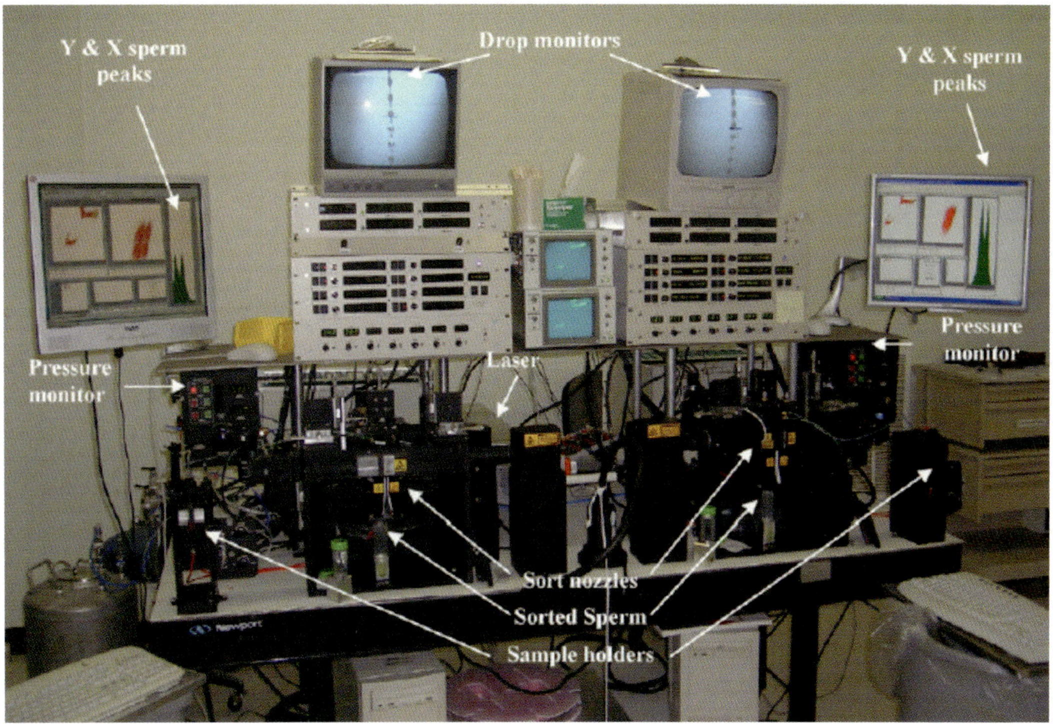

Fig. 1.1 A flow cytometer used for commercial sexing of bovine sperm located at Sexing Technologies, Navasota, TX, USA. The cytometer is a customized Dako MoFlo® model. The image is reproduced from Garner and Seidel (2008) with permission

limited by the fact that sperm unidentified by the flow cytometer as being X or Y are voided to waste. Garner and Seidel (2008) reported, for example, that the MoFlo SX™ flow cytometer sorted 20,000 sperm per second to yield ~6,000 X-sorted and ~6,000 Y-sorted sperm (Fig. 1.1).

It is possible to sort frozen sperm and achieve pregnancies using in vitro fertilization (IVF) or AI, even when the sorted sperm are stored frozen a second time after the sorting process. Preliminary data are indicative that fertility to insemination is low (Underwood et al. 2010a) but blastocyst yield after IVF was similar to that achieved with non-sorted sperm (Underwood et al. 2010b).

Given the currently achieved sorting speeds and loss of sperm inherent in the process, it is not practical to perform AI using sex-sorted sperm with sperm numbers used for conventional AI. In the bovine, for example, straws of sex-sorted sperm contain about two million sperm as compared to 15–30 million sperm for conventional semen (Rath and Johnson 2008). For species requiring large insemination doses such as the pig, insemination with sexed semen has largely been limited to procedures based on deposition of sperm into the reproductive tract via laparoscopy or catheter-mediated uterine insemination (Rath and Johnson 2008). Another effective use of sex-sorted sperm is for IVF. In this way, one straw of sperm can be used to produce multiple embryos. Evidence from cattle indicates that the competence of an embryo produced in vitro with sex-sorted sperm to establish pregnancy is equivalent to that of an embryo produced in vitro with conventional semen (Xu et al. 2006; Rasmussen et al. 2013). Use of sex-sorted sperm for superovulated cattle has usually resulted in lower embryo yields than use of conventional semen (Sartori et al. 2004; Hayakawa et al. 2009; Larson et al. 2010). A recent report in which deep uterine insemination was utilized indicated similar embryo production as for use of conventional semen (An et al. 2010).

Table 1.1 Incidence of stillbirths in male Holstein calves born as a result of insemination with sexed or conventional semen

Female	Type of breeding	Number of calves	Percent stillbirths	Study
Heifer	Conventional semen	793	12.9	DeJarnette et al. (2009)
Heifer	Sexed semen	1,318	19.9	
Heifer	Conventional semen	a	10.8	Norman et al. (2010)
Heifer	Sexed semen		15.6	
Cow	Conventional semen	b	3.6	Norman et al. (2010)
Cow	Sexed semen		2.6	

[a]Based on a total data set of 319,720 calvings of which 48.5 % of calvings from conventional semen and 8.9 % of calvings from sexed semen were single males
[b]Based on a total data set of 1,036,298 of which 49.2 % of calvings from conventional semen and 10.2 % of calvings from sexed semen were single males

The combination of reduced number of sperm in an insemination dose and reduced competence of the sperm to fertilize oocytes and produce developmentally competent embryos means that fertility for females inseminated with sexed sperm is lower than that for females inseminated with conventional semen. Two large-scale studies in cattle indicate the magnitude of the reduction. DeJarnette et al. (2009) analyzed records from 93,481 Holstein heifers and found that conception rate for Holstein heifers was 56 % for those bred with conventional semen vs. 45 % for those bred with sexed sperm. Thus conception rate for sexed sperm was only 80 % of that with conventional semen. Norman et al. (2010) analyzed 1.3 million breedings of Holstein heifers and 10.8 million breedings of Holstein cows. For heifers, the conception rate was 56 % for conventional semen vs. 39 % for sexed sperm (i.e., conception rates for sexed sperm were 70 % of that for conventional semen). For cows, the conception rate was 30 % for conventional semen vs. 25 % for sexed sperm (i.e., conception rates for sexed sperm were 83 % of that for conventional semen).

Calves born as a result of insemination with sexed sperm appear normal except for a small proportion of bull calves born to females inseminated with X-sorted sperm, i.e., when the expected gender was female. The incidence of stillbirths among male calves produced with X-sorted sperm was greater than that for male calves produced with conventional semen in heifers although not in cows (Table 1.1). It has been speculated that Y-bearing sperm identified as X-bearing sperm are more likely to contain excess DNA because of aneuploidy or a large translocation that would predispose calves to stillbirth (DeJarnette et al. 2009).

Resolution of problems limiting the use of sexed semen will depend upon either improving the process of sperm sorting (to reduce sperm damage, increase sorting speed, and reduce the number of sperm that cannot be successfully identified as X or Y) or developing methods for sexing sperm that do not depend on flow cytometers. Much work is underway to improve sorting technology including potential protective effects of antioxidants (Klinc and Rath 2007) and seminal proteins (García et al. 2007; Leahy et al. 2009). Rath et al. (2009) have reported that fertility of heifers inseminated with sexed semen was similar to that inseminated with conventional semen when sorting was performed using a procedure called Sexcell® that involves addition of antioxidants to the flow medium, inhibition of sperm motility by fluoride during sorting and a three-step cooling process.

The impending lapse of the original patent for sex sorting of sperm may lead to innovations in flow cytometric approaches to sperm sorting. What has remained elusive so far are technologies for sperm sexing that do not depend upon flow cytometry. Development of such technologies will depend on identification of either non-cytometric methods to separate sperm on the basis of DNA content or on identification of other differences between X- and

Y-containing sperm that can be exploited in a separation protocol. There are many differences at the genomic level between the two types of sperm but separation of sperm on the basis of these differences will involve identifying probes that can hybridize with sperm DNA while maintaining cell viability or identifying proteins encoded for by sex chromosomes that are synthesized during spermatogenesis. The likelihood of identifying proteins distinctive for X- and Y-containing sperm is very small, however, because the sex chromosomes become condensed and transcriptionally inactive during meiosis (De Vries et al. 2012).

Sperm Fertility Measurements

Fertility varies between males and between ejaculates of individual males, even when sperm numbers used for insemination are adjusted to reduce variation in fertility. In certain species like the stallion, differences between males in freezing are a particular problem (Loomis and Graham 2008). Identification of biological markers of sperm-fertilizing ability could result in more precise processing of semen for AI and improvements in female fertility.

Definitive markers have not yet been found but results are promising. In boars, for example, concentrations of PSP-I in semen were negatively correlated with number of piglets born while concentrations of GPX5 were positively correlated (Novak et al. 2010a). Concentrations of CRISP3 in stallion seminal plasma were positively related to conception rate while concentrations of CLU, KLK1E2, and seminal proteins SP1 and SP2 were negatively correlated with conception rate (Novak et al. 2010b). Similarly, there were differences in amounts of eight proteins in sperm that were related to bull fertility (Park et al. 2012).

Proteins are relatively difficult to measure compared to mRNA. Recently, Kasimanickam et al. (2012) found that 97 % of the variation in sire conception rate between bulls could be explained by measuring the mRNA in sperm for five genes (*AK1*, *IB5*, *TIMP2*, *SNRPN*, and *PLCZ1*).

Technologies Focused on Female Reproduction

Artificial Insemination

A successful AI program depends not only upon a satisfactory process for preservation of semen but also on ensuring deposition of adequate numbers of competent spermatozoa in the female reproductive tract at a time close enough to ovulation to allow fertilization. In many species, incorporation of AI into breeding strategies has been compromised both by anatomical barriers to semen deposition and inadequacy of methods of predicting ovulation. Progress in bypassing anatomical issues has been incremental but progress in predicting ovulation has been revolutionary.

For optimal AI, semen must be deposited directly in the uterus. This is a relatively simple task in some species like cattle, buffalo, and the horse, where the large size of females makes it feasible to use rectal palpation of the reproductive tract to facilitate movement of the AI pipette through the cervix. In smaller animals, rectal palpation is not possible and the alternatives reduce the desirability of AI. In the pig, for example, the most common method of AI involves cervical deposition of 2–3 billion sperm (Roca et al. 2011). This compares to the 15–30 million sperm used for AI in cattle. As a result, the number of inseminations per ejaculate is low and use of frozen or sexed semen becomes problematic.

Intrauterine insemination in pigs is now used commercially. Deposition of semen in utero is achieved by a catheter extended from the corkscrew-shaped part of the insemination pipette placed in the cervix (Fig. 1.2). Recently, it was reported that pregnancy rates and litter sizes similar to those after cervical insemination with three billion sperm per insemination could be achieved by a double uterine insemination technique (see Fig. 1.2) using 750 million sperm per insemination (Mozo-Martín et al. 2012). For use of sexed semen in pigs, laparoscopic methods remain the optimal method for insemination (Roca et al. 2011).

Efforts in the sheep to achieve transcervical insemination have met with mixed results.

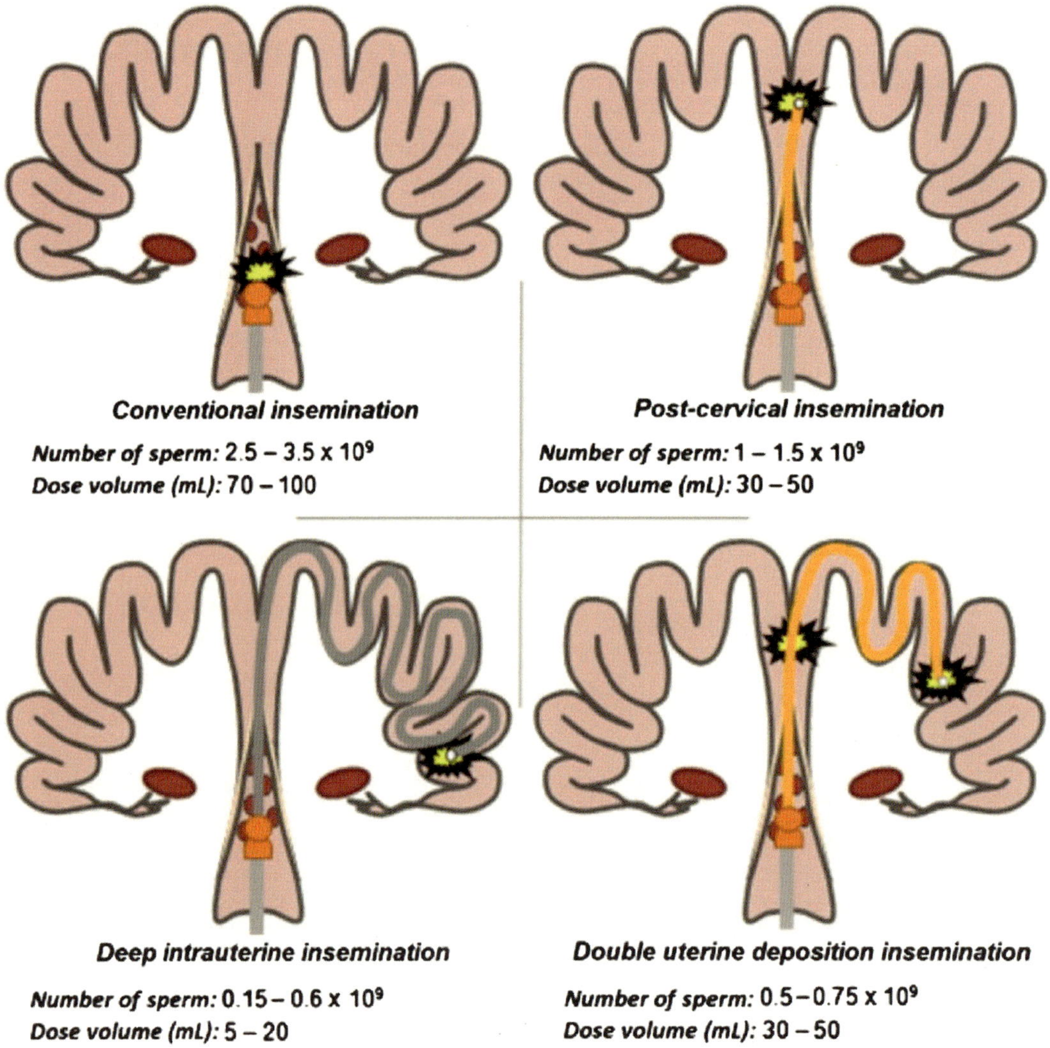

Fig. 1.2 Diagram of various artificial insemination systems in pigs. In conventional insemination, semen is deposited into the cervix using a corkscrew-tipped pipette. For the other methods a tubular extension of the pipette is threaded into the uterus. Figure reproduced from Mozo-Martín et al. (2012) with permission

A system for transcervical insemination has been developed at the University of Guelph (Fig. 1.3) that is based on visualization of the os cervix with a lighted speculum, manipulation of the cervix with a forceps, and manipulation of a pipette through the cervix and into the uterus (Buckrell et al. 1994). Some authors have reported excellent results whereas results were disappointing in other studies (Robinson et al. 2011). Efforts to improve the system have relied on hormonal regimens to dilate the cervix such as estradiol, prostaglandin E2, and oxytocin (Robinson et al. 2011). More information about the biology of cervical relaxation may lead to more consistent results.

For most of its history, timing of AI was based on identifying estrus. The degree to the effectiveness of this approach for ensuring semen is deposited at a time coincident with release of the oocyte into the oviduct varies between species. Indeed, problems exist in many species including lack of overt symptoms of estrus in the absence of a male (the ewe), a prolonged estrus (mare), lack of physical access to females under extensive

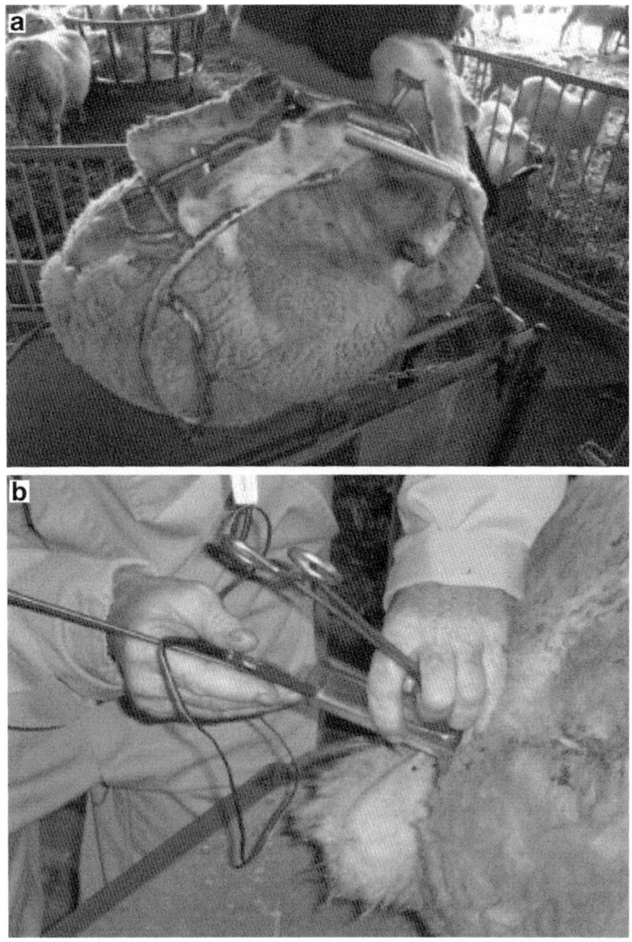

Fig. 1.3 The Guelph transcervical artificial insemination system. Ewes are placed in a dorsally recumbent position (**a**). A lighted plexiglass speculum is inserted into the vagina and used to visualize the cervix. Tissue near the os cervix is grabbed with a forceps to manipulate the cervix and allow manipulation of an insemination pipette into the uterus (**b**). Figure is reproduced from Candappa et al. (2009) with permission

management (beef cattle and sheep), and poor expression of estrus (the dairy cow). In a large study, estrus in dairy cows involved an average of 8.5 standing events over an average of 7.1 h (Dransfield et al. 1998). Some of the problems related to estrus detection have been reduced through development of estrous synchronization programs. More recently, ovulation synchronization programs based on regulation of luteolysis, follicular growth, and ovulation have eliminated the need for estrous detection completely. Most well developed in cattle, the best timed artificial insemination schemes can result in pregnancy rates for subfertile cows that are higher than those following artificial insemination after detected estrus (Bisinotto and Santos 2011).

Embryo Transfer

Embryo transfer (ET) has long been seen as the female equivalent of AI—a method to increase genetic selection by increasing the number of offspring produced by genetically elite females as well as to preserve germplasm in a cryopreserved state. Although the situation is changing somewhat, ET is but a poor cousin to AI as a genetic selection tool. The numbers of offspring that can be produced by AI from a single bull are in the hundreds of thousands while the numbers that can be produced from a single cow using the most intensive ET programs are in the hundreds. Moreover, up until recently, the accuracy of identifying genetically-elite animals was much lower

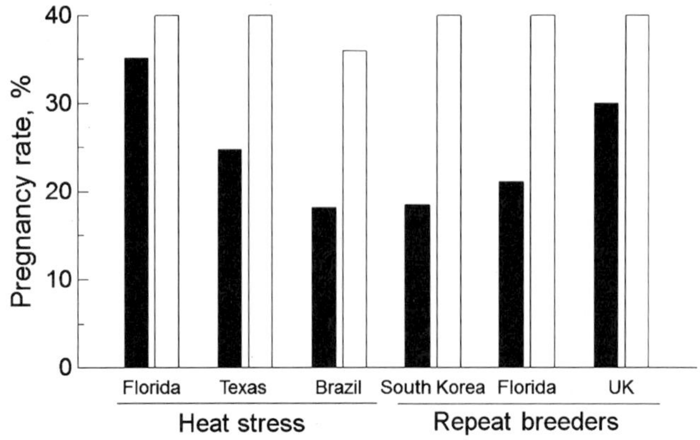

Fig. 1.4 Examples of studies in lactating dairy cows in which pregnancy rates were improved by embryo transfer (*open bar*) as compared to artificial insemination (*filled bar*). Data from heat stress experiments are from Block et al. (2010) (Florida), Stewart et al. (2011) (Texas) and Vasconcelos et al. (2011) (Brazil). Data from repeat breeder studies are from Son et al. (2007) (South Korea), Block et al. (2010) (Florida) and Canu et al. (2010) (UK)

for females than males because of the disparity in number of performance records from offspring.

Nonetheless, changes in technology are improving the usefulness of embryo transfer as a genetic selection tool. The use of genotyping to make genomic predictions of genetically important traits has greatly improved the accuracy of selection of females. In a study in cattle, the reliability of genetic estimates for a variety of traits averaged 50 % using genomic data from the Bovine SNP50 gene array whereas reliability averaged 27 % for the same traits estimated using parent averages (VanRaden et al. 2009). In addition, it is now possible to biopsy a few cells of an embryo, perform whole genome amplification and genotype individual embryos by polymerase chain reaction for gender (Lopes et al. 2001), other specific genes (Polisseni et al. 2010) or for genome-wide analysis (Moghaddaszadeh-Ahrabi et al. 2012). There are commercial services in Canada and Europe offering this latter service. Thus, genetic selection can be performed in the first few days after conception to dramatically shorten generation interval.

It is unlikely that embryo transfer is an economically practical approach for improving genetic merit in commercial populations of livestock although the situation is likely to be different for elite herds where offspring of desirable animals have great financial value. Weigel et al. (2012) estimated the change in the breeding value of dairy cattle for lifetime net merit (an estimate of genetic merit in terms of lifetime economic return) when heifers were subjected to genomic testing using the Illumina 3K genomic chip. Under the best scenario, when 20 % of females were used to produce all replacements and no information was available on the parents of the heifers, the improvement in net merit after subtracting costs of genomic testing was $259. This is much less than the current costs to produce a heifer by ET. Ribeiro et al. (2012) estimated the cost of a female pregnancy produced by ET varied from $261 to $2,418, depending on the procedure used to produce the embryo and expected pregnancy rate. Under the most likely scenario, using oocyte pickup procedures and IVF with sexed semen, the estimated cost of a female pregnancy was $739 when pregnancy rate was 35 % and $634 when pregnancy rate was 40 %. Such costs are acceptable when the offspring are worth thousands of dollars as purebred animals but are too high to make ET a profitable approach to increasing genetic selection in commercial herds. Increased efficiencies in ET are needed before the technology becomes used widely for genetic improvement.

Another potential use for ET is to improve fertility. The rationale is that an embryo available for transfer has already successfully overcome many obstacles to pregnancy including ovulation and fertilization failure and pre-blastocyst embryonic mortality. Use of ET for fertility enhancement has been examined most thoroughly in lactating dairy cows. As shown in Fig. 1.4, there

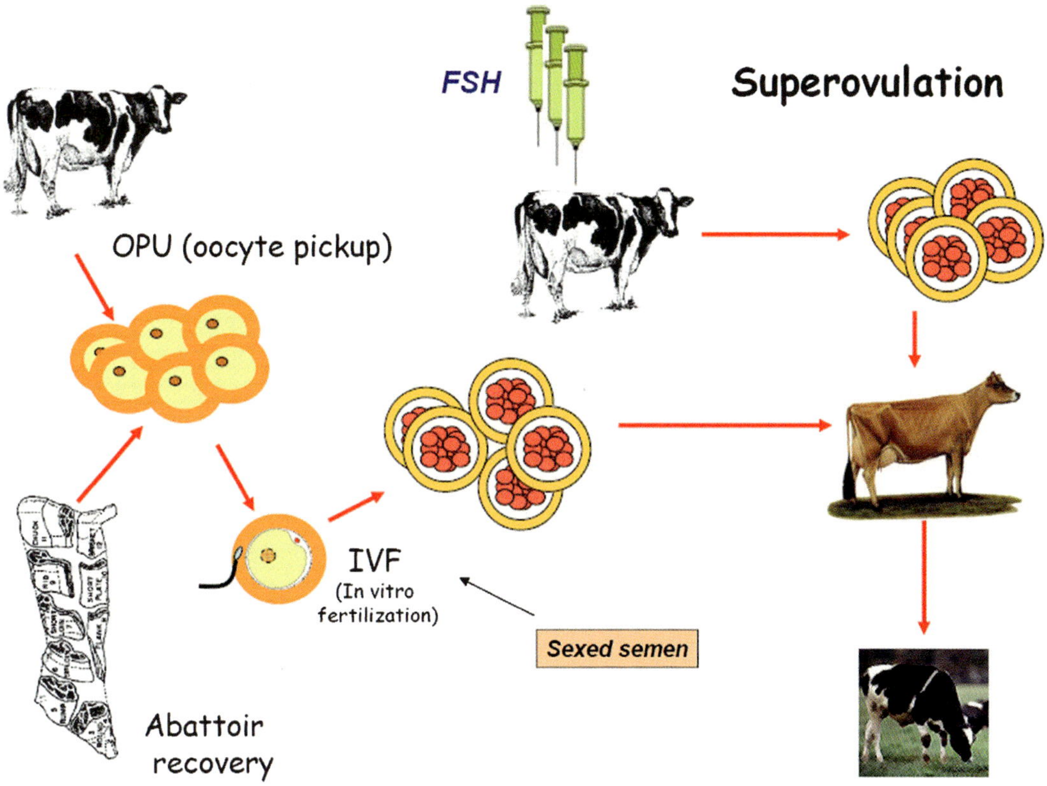

Fig. 1.5 Assisted reproductive technologies for production of embryos in the cow. For superovulation, multiple follicles are recruited to grow by administration of FSH. Following ovulation and AI, embryos are recovered by flushing the uterus. For IVF, oocytes obtained either from excised ovaries (usually from the abattoir) or from living cows (by OPU) are matured in vitro, fertilized with conventional or sexed semen, and then cultured until transfer into recipients. The figure is from Hansen (2013)

is a clear fertility-promoting effect of embryo transfer when fertility is low, as during periods of heat stress (Block et al. 2010; Stewart et al. 2011; Vasconcelos et al. 2011) and in repeat breeder cows (i.e., greater than 2–3 previous unsuccessful inseminations) (Son et al. 2007; Block et al. 2010; Canu et al. 2010). In the absence of low fertility, results have been inconsistent with no improvement in fertility as compared to AI in two studies (Sartori et al. 2006; Rasmussen et al. 2013) whereas, in others, pregnancy rates following ET were higher than following AI (Demetrio et al. 2007; Vasconcelos et al. 2006, 2011).

There are three major means to produce embryos for transfer (Fig. 1.5). The first is to recover embryos from the reproductive tract of females, often after females have been treated hormonally to induce ovulate of a large number of follicles (i.e., superovulation). This method is still widely used today. Global numbers of reported embryo transfer using in vivo-derived embryos in 2010 were 590,561 for bovine, 29,078 for sheep, 1,633 for goat, 28,824 for equine, and 84 for deer (Stroud 2011). The second method is to produce embryos in vitro through IVF. In 2010, the global numbers of reported embryo transfer using in vitro-derived cattle embryos was 339,685 (Stroud 2011). Numbers for other species are not available. The third method, somatic cell nuclear cloning, will be discussed in a later section.

The biggest limitations to in vivo production of embryos are characteristics of the superovulation response, recovery of embryos from the reproductive tract (requiring surgical procedures in pigs, sheep, and goats), and, as discussed earlier, inadequacy of sexed semen in species in which semen is deposited transcervically. The most important of these are issues related to superovulation. In some species, response to superovulation is very poor. In mares (Roser and Meyers-Brown 2012) and water buffalo (Drost 2007), for instance, embryo recovery after superovulation is often less than after natural ovulation. In the pig, in which the high natural ovulation rate limits the need for superovulation, embryonic viability can sometimes be reduced by superovulation (Youngs 2011). In other species, responses to superovulation can be good but variable. In one study in cattle, the number of recovered embryos ranged from 1 to 27 (Rico et al. 2012). An important source of variation in superovulation response is timing of gonadotropin stimulation relative to emergence of a follicular wave (Baruselli et al. 2006; Amiridis and Cseh 2012). There are also inherent differences among females in the size of the antral follicle pool that determine superovulation response in cattle and which can be assessed by measuring concentrations of the granulosa-cell-derived hormone antimullerian hormone (AMH) (Rico et al. 2012).

Another problem with superovulation protocols is the need to give multiple injections of gonadotropins. Superovulation programs based on a single administration of follicle stimulating hormone (FSH) are now being developed in cattle (Tríbulo et al. 2011). The advent of recombinant gonadotropins should eventually lead to the elimination of naturally derived gonadotropins to remove concerns over batch-to-batch variability and pathogen contamination.

The major advantage of systems for producing embryos in vitro is cost (when embryos are produced using excised ovaries recovered at an abattoir) and total yield of embryos per unit time (when oocytes are collected via transvaginal aspiration or surgically). In cattle, for example, superovulation can typically be performed at 60-day intervals. However, oocytes for in vitro embryo production can be harvested as often as twice a week (Merton et al. 2003) and during periods when ovulation is inhibited including the prepuberium, pregnancy and nonbreeding season (Merton et al. 2009; Cseh et al. 2012). In addition, sexed semen can be used very efficiently for IVF and the pregnancy rates following transfer into recipients is similar between cows receiving embryos produced with sexed semen and those receiving embryos produced with conventional semen (Xu et al. 2006; Rasmussen et al. 2013).

Production of embryos by IVF involves incubation of oocytes that completed nuclear maturation (i.e., are in metaphase II) with spermatozoa that have been successfully capacitated. Capacitation of sperm is achieved by a variety of molecules including heparin or caffeine and there is significant male-to-male variation in conditions for optimal capacitation (Gil et al. 2010). In most species, immature oocytes can be collected from antral follicles and then matured in vitro before fertilization. The notable exception is for the mare, where methods to allow in vitro maturation have remained elusive (Deleuze et al. 2010) and oocytes are often allowed to mature in vivo before harvesting (Hinrichs 2010). Depending upon the species, oocytes can be harvested from follicles using excised ovaries (typically from an abattoir), following laparotomy or laparoscopy to visualize the ovary, or transvaginally using a procedure called transvaginal, ultrasound-guided oocyte recovery or oocyte pickup (OPU). Oocyte yields from live animals can be improved by administration of gonadotropins (Merton et al. 2003; Cseh et al. 2012) and, in anestrous sheep, by administration of melatonin (Tsiligianni et al. 2009). Like for superovulation, oocyte yield in cattle can be predicted by AMH (Rico et al. 2012) and single nucleotide polymorphisms in *GDF9*, *FGF8*, *BMRP2*, and *LHCGR* have also been reported to be related to oocyte yield in this species (Santos-Biase et al. 2012). There is a large breed effect in cattle on yield of oocytes from OPU procedures with *Bos indicus* breeds being greater than *B. taurus* (Pontes et al. 2010).

A successful in vitro production system can be defined as one in which a high proportion of oocytes develops to a transferrable stage, the resultant pregnancy rate after transfer to recipients is high, and the offspring born are normal. Two important components to achieving this success are the conditions for oocyte maturation and embryo culture. A variety of modifications of culture media have been evaluated for optimizing oocyte maturation and embryonic development (see reviews by Hinrichs 2010; Vajta et al. 2010; Yoshioka 2011; Block et al. 2011). Data in cattle suggest that a period of withdrawal of FSH support for the follicle (termed FSH coasting) is beneficial for subsequent in vitro maturation (Nivet et al. 2012). Nonetheless, culture systems for most species result in embryos that are abnormal in some respects from embryos produced in vivo. In cattle, transfer of embryos produced in vitro into recipients results in lower pregnancy rates than transfer of embryos produced in vivo (Farin et al. 1999) and there can be increased incidence of abnormalities in the offspring, including, most noticeably, an increase in birth weight (Farin et al. 2006).

Another characteristic of embryos produced in vitro is poor survival to cryopreservation as compared to embryos produced in vivo (Hansen and Block 2004). The reason for the increased cryosensitivity is not completely understood but one cause is the increase in intracellular lipid that accumulates in embryos produced in vitro. Survival of pig embryos can be improved by mechanical delipidation (Men et al. 2012). In vitro survival of bovine embryos can be increased by addition of various agents that inhibit lipid metabolism (Barceló-Fimbres and Seidel 2007; Pereira and Marques 2008). The method of cooling is also important, with vitrification often superior to other methods (Pereira and Marques 2008; Saragusty and Arav 2011).

Intracytoplasmic sperm injection (ICSI) is a technique that can be incorporated into in vitro embryo production systems. In ICSI, a single sperm cell is injected directly into an oocyte using micromanipulators. Since only one sperm is needed and it can be damaged or dead, ICSI can be used to obtain offspring from oligospermic males or males with a high degree of sperm defects. ICSI has also been proposed as a method for eliminating the problem of polyspermy in pigs, which affects 30–50 % of inseminated oocytes in that species (Gil et al. 2010). Live offspring have been produced using ICSI in the bovine, pig, sheep, and horse but results are generally poor (García-Roselló et al. 2009).

Cell-Based Technologies

Novel biotechnologies are resulting from development of methods to control a cell's differentiation status (through manipulation of intracellular and extracellular molecules that regulate pluripotency and differentiation) and genetic composition (through transgenesis and targeted mutagenesis). For assisted reproduction, cell-based technologies have expanded the limits of the possible. Somatic cell nuclear cloning has been used to produce offspring in at least 20 species (Oback 2008; Rodriguez-Osorio et al. 2012), is a commercial reality in cattle, horses and pigs (see www.viagen.com), and has been widely used to produce transgenic pigs for research (Whyte and Prather 2011; Luo et al. 2012). Breakthroughs in stem cell research may soon make it possible to use stem cells to produce gametes of either sex.

Somatic Cell Nuclear Cloning

The process of somatic cell nuclear cloning (SCNT) involves fusion of a somatic cell nucleus (usually encased in the entire cell) with an enucleated nucleus, activation of the newly formed embryo to initiate cell proliferation and culture of the embryo until transfer into recipients (Fig. 1.6). The major agricultural use of SCNT at present is the genetic duplication of elite animals. This use is only practical when the individual to be cloned has a high degree of financial or emotional value because of inefficiencies in the cloning process (see below). Theoretically, nuclear cloning could improve rates of genetic selection by virtue of an increase in selection intensity (only a few sires and dams need be produced) and because the accuracy of selection can be improved by recording performance of specific genotypes under a variety of environments (Dematawewa and Berger 1998). Unless the efficiency of SCNT is increased, however, the increase in genetic merit achieved by incorporation of cloning in selection programs will not be enough to offset the costs.

SCNT is likely to have the greatest impact on livestock production in the future by facilitating the

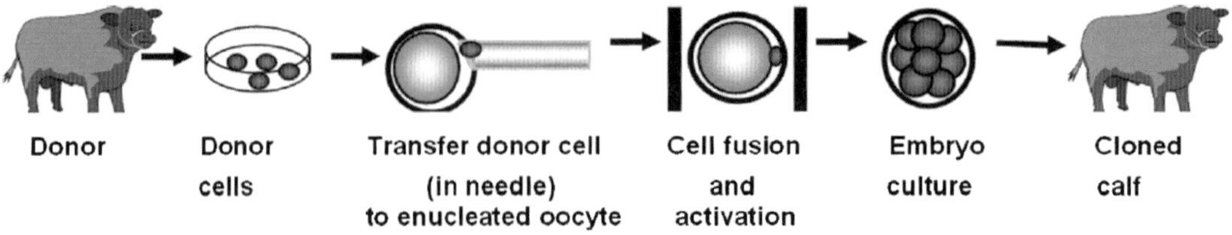

Fig. 1.6 Schematic diagram illustrating the procedures for production of calves by SCNT. Donor cells are grown in culture, fused with an enucleated oocyte, activated to initiate growth, allowed to develop in culture and then transferred to a recipient female. The figure is from Moore and Thatcher (2006) with permission

production of genetically modified animals. Stable transfection of mammalian cells with foreign DNA is an inefficient process and only a small percent of cells that are exposed to a transgene (through microinjection, electroporation, liposome-mediated delivery, retroviral integration, or other technique) are stably incorporated into the genome. Even if only a few out of thousands of cells exposed to foreign DNA become transgenic, transfected cells can be expanded in culture using a selective culture medium that allows only transgenic cells to grow, and the resultant cells can be used for SCNT. This approach is now the predominant one for producing transgenic pigs, the farm animal species most widely used for transgenesis (Whyte and Prather 2011).

The inefficiencies that limit application of cloning are numerous. Each step of the process is liable to a high rate of failure including those leading to the formation of a transferrable embryo and those determining whether a transferred embryo results in a live and healthy neonate (Keefer 2008; Oback 2008; Whitworth and Prather 2010). The percent of transferred embryos resulting in an offspring range from 4 to 29 % of transferred embryos in cattle, 0.5–3 % in pigs, 3–30 % in goats and 0.1–26 % in horses (Galli et al. 2008; Keefer 2008; Whitworth and Prather 2010). Perhaps most disturbing is the high incidence of placental and fetal defects that occur in cloned conceptuses resulting in increased rates of abortion and offspring that are abnormally large (example, cattle) or small (pigs) and that experience a variety of physical malformations (Edwards et al. 2003; Meirelles et al. 2010; Whitworth and Prather 2010; Chavatte-Palmer et al. 2012). Examples of abnormal development of extraembryonic membranes in cloned pig conceptuses are shown in Fig. 1.7. Hydroallantois in cloned cattle and sheep conceptuses is common and can contribute to the death of the recipient or neonate (Edwards et al. 2003; Loi et al. 2006).

Somatic cell nuclear cloning is dependent upon the successful epigenetic reprogramming of the donor nucleus. Reprogramming is incomplete and causes epigenetic alterations (Chavatte-Palmer et al. 2012; Smith et al. 2012) and much effort is being focused on optimizing reprogramming. Among the approaches being evaluated are addition of inhibitors of DNA methylation, histone deacetylases or proteasomes (Whitworth and Prather 2010; Rodriguez-Osorio et al. 2012) and improving oocyte maturation conditions so that, following enucleation, the cytoplast can facilitate donor nucleus reprogramming (Oback 2008). There is also a great deal of interest in generating embryonic stem cells (ES cells) and induced pluripotent stem cells (iPS cells) for use in SCNT because the cells have undergone limited differentiation and, at least for ES cells, can reprogram somatic cells upon fusion (Oback 2009; Rodriguez-Osorio et al. 2012). Experimental evidence (Oback 2009) indicates that less differentiated cells do not necessarily make better candidates for cloning and more knowledge of the molecular basis for cellular reprogramming is needed to optimize cloning. There was also been great improvements in the technical requirements for performing SCNT through the advent of "handmade cloning" procedures that do not depend upon does micromanipulation of cells (Vajta and Callesen 2012).

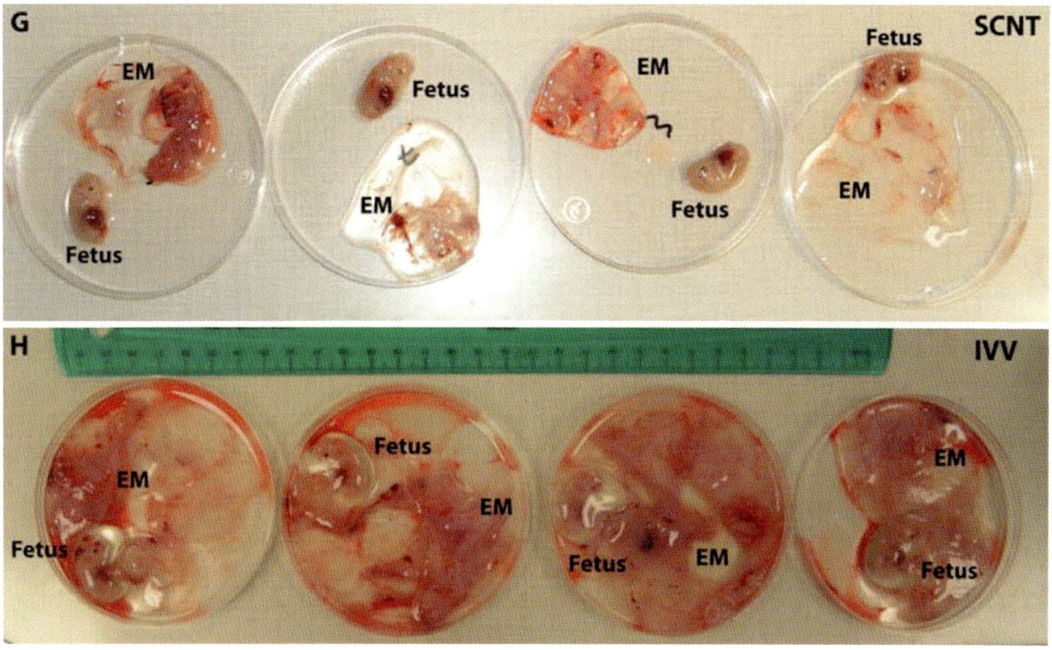

Fig. 1.7 Differences in growth and vasculature of the placenta at Day 30 of gestation between conceptuses produced by somatic cell nuclear cloning (G) (SCNT) or that were produced in vivo (H) (IVV). Note that extraembryonic membranes (EM) are more developed and vascularized for conceptuses produced in vivo. The image is from Whitworth and Prather (2010) and is reproduced with permission

Stem Cells

Stem cells are self-renewing cells that can differentiate into specialized phenotypes. Technologies based on the production and manipulation of stem cells not only has implications for SCNT (discussed above) but also may prove useful for developing novel methods for manipulation of male and female gametes.

Pluripotent stem cells are those that can give rise to virtually any cell type (exclusive of extraembryonic membranes) when placed in specific conditions. Examples include embryonic stem cells, derived from the preimplantation embryo (usually the inner cell mass of blastocysts), embryonic germ cells (derived from primordial germ cells), and iPS cells. This latter category are somatic cells that have been genetically transformed through addition of specific transcription factor genes involved in pluripotency (for example, *OCT4*, *SOX2*, *NANOG*, and *LIN28* for human cells; Yu et al. 2007). Cells with characteristics of iPS cells have been reported in pigs, cattle, sheep, and goats (Nowak-Imialek et al. 2011; Ezashi et al. 2012). While pluripotent, iPS cells may be less undifferentiated than embryonic stem cells because there is evidence that they retain epigenetic modifications from the cell type from which they are derived (Drews et al. 2012).

Stem cells with a more limited differentiation repertoire exist throughout the body and participate in maintaining integrity of regenerating tissues. Of particular pertinence here are spermatogonial stem cells (SSC) and oogonial stem cells (OSC). SSC are well-characterized cells in the testis that give rise to spermatogonia and which can be maintained in culture (Oatley and Brinster 2012). In contrast, the existence of OSC has been controversial (Hutt and Albertini 2006). Nonetheless, there is increasing evidence that, rather than the oocyte pool in adults being incapable of self-renewal, formation of oocytes from progenitor cells can occur to some extent in adult females (Woods and Tilly 2012).

Importantly for assisted reproduction, SSC can be transplanted into the testis, become established in the seminiferous tubules and give rise to spermatozoa. Presence of sperm from transplanted testicular cells in semen has been demonstrated in goats (Honaramooz et al. 2003), pigs (Mikkola et al. 2006) and cattle (Stockwell et al. 2009) and live offspring have been produced from sperm derived from a testicular cell transplant in goats (Honaramooz et al. 2003). Transplantation in livestock is most commonly performed with impure populations of testicular cells that are infused into the testes of prepubertal males via infusion through the rete testis. Unlike for rodents, successful transplantation can occur in immunocompetent recipients that are allogeneic to the SSCs.

Transplantation of germ cells, particularly of SSC maintained and expanded in culture, could have important uses for assisted reproductive technologies. The number of males producing spermatozoa of a specific genotype could be expanded, either to maximize production of semen for artificial insemination, to maintain production of sperm of a given genotype when the original male ages or dies, and to increase distribution of sperm from genetically valuable males in natural mating systems. In addition, SSC transplantation could be an important vehicle for introducing transgenes into populations of livestock, as has been demonstrated in mice (Nagano et al. 2001).

For testicular cell or SSC transplantation to become a routine procedure, additional research is necessary to produce adequate numbers of cells for transplantation and to increase the efficiency of transplantation. Culture conditions for purification of SSCs from the testis and their expansion in culture have still not been made optimal (Oatley 2010). Additionally, the rate of successful transplantation is low and the transplanted cells can decline in number with age. Stockwell et al. (2009) performed transplantation of testicular cells into six prepubertal bulls. When examined 52–56 week later, only one recipient produced spermatozoa of the recipient genotype. At 53 week, 25 % of spermatozoa present in semen were derived from the transplant but this percentage declined to 1 % at 98 week after transplantation.

Recently, OCS have been purified from ovaries of neonatal mice (Zou et al. 2009) and reproductive-age women (White et al. 2012) using fluorescence-activated cell sorting. These cells can be maintained in long-term culture and differentiate into oocytes. In mice, offspring have been produced from oocytes derived from OSC following transplantation into ovaries of infertile mice (Zou et al. 2009). In humans, OSC injected into biopsies of human ovarian cortex survived and were incorporated into follicles following transplantation into immunocompromised mice (White et al. 2012). If these technologies can be developed in farm animals, OCS could be used to greatly expand the number of offspring that can be produced from females using IVF.

Another approach for making oocytes from stem cells is to cause differentiation of embryonic stem cells or iPS cells into oocytes. Such cells might be easier to obtain than OSC. While oocyte-like cells have been generated from embryonic stem cells, the cells cannot progress through meiosis and no offspring have been produced to date from such cells (Woods and Tilly 2012).

Stem cell technologies may also prove useful in modifying males so that all spermatozoa carry the X chromosome. This could be achieved if SSCs derived from genetic females (ES cells, iPS cells, or OSC) were transplanted into the testis.

Where Do We Go from Here?

There are two historical trends in the development of animal production systems that are pertinent to animal physiologists. The first is the increase in efficiency in production of food caused by application of scientific knowledge to animal management. The second is a corresponding growth of unease or opposition from consumers towards the changes in animal management that are the outcome of application of science. Scientists concerned over contemporary opposition to transgenics and SCNT by large segments of the public should realize that there was opposition to AI in the 1920s and 1930s engendered by ethical concerns, apprehension about the fate of purebred breeders and fears of developmental abnormalities in the offspring (Foote 2002; Wilmot 2007). Public opposition, when grounded in unreason, should not be a deterrent to the application of science to animal agriculture.

Many of the techniques described in this chapter are too inefficient at the current level of technology for application to commercial agriculture. This was once true for AI. After being shown a demonstration of AI in April 1923, the Shorthorn Society declined to make policy about the procedure because of the belief it was too difficult for wide adoption (Wilmot 2007). Procedures for AI are still being refined more than 75 years after the procedure first became used on a commercial basis in the 1930s. However, scientific advances occur at a faster pace today than when AI was developed. The biotechnology growth curve is still in the exponential phase, propelled forward by revolutionary developments in molecular biology, cell biology, and genetics. Most of the assisted reproduction techniques that are likely to have a major impact on livestock production in the next few decades are very new. The first calf produced by IVF (Virgil) was born in 1981 (Brackett et al. 1982), the first offspring from sexed semen (a rabbit) was reported in 1989 (Johnson et al. 1989), the first mammal born following somatic cell nuclear cloning of an adult cell (Dolly the sheep) was born in 1996 (Wilmut et al. 1997), the first offspring born as a result of testicular cell transplantation in a livestock species (a goat) was reported in 2003 (Honaramooz et al. 2003), and the first offspring born from OSC transplantation in a mammal (a mouse) was reported in 2009 (Zou et al. 2009). Continued investments in these technologies will almost assuredly result in scientific discoveries that improve their utility.

As stated in the Introduction, the human population faces serious global problems caused by the need to produce more food for a growing world population in the face of climate change. To meet this challenge, while also raising animals in a humane and sustainable manner, we must be bold in following our predecessors and continue to capture the benefits of science for animal production.

References

Amiridis GS, Cseh S (2012) Assisted reproductive technologies in the reproductive management of small ruminants. Anim Reprod Sci 130:152–161

An L, Wu ZH, Wu YF, Zhang XL, Liu X, Zhu YB, Cheng WM, Gao HM, Guo M, Tian JH (2010) Fertility in single-ovulating and superovulated dairy heifers after insemination with low dose sex-sorted sperm. Reprod Domest Anim 45:e344–e350

Ball RO, Samuel RS, Moehn S (2008) Nutrient requirements of prolific sows. Adv Pork Prod 19:223–237

Barbas JP, Mascarenhas RD (2009) Cryopreservation of domestic animal sperm cells. Cell Tissue Bank 10: 49–62

Barceló-Fimbres M, Seidel GE Jr (2007) Effects of fetal calf serum, phenazine ethosulfate and either glucose or fructose during in vitro culture of bovine embryos on embryonic development after cryopreservation. Mol Reprod Dev 74:1395–1405

Baruselli PS, de Sá Filho MF, Martins CM, Nasser LF, Nogueira MF, Barros CM, Bó GA (2006) Superovulation and embryo transfer in Bos indicus cattle. Theriogenology 65:77–88

Bisinotto RS, Santos JE (2011) The use of endocrine treatments to improve pregnancy rates in cattle. Reprod Fertil Dev 24:258–266

Block J, Bonilla L, Hansen PJ (2010) Efficacy of in vitro embryo transfer in lactating dairy cows using fresh or vitrified embryos produced in a novel embryo culture medium. J Dairy Sci 93:5234–2542

Block J, Hansen PJ, Loureiro B, Bonilla L (2011) Improving post-transfer survival of bovine embryos produced in vitro: actions of insulin-like growth factor-1, colony stimulating factor-2 and hyaluronan. Theriogenology 76:1602–1609

Bostock J, Riley H (1890) The natural history of pliny, vol II. George Bell, London

Brackett BG, Bousquet D, Boice ML, Donawick WJ, Evans JF, Dressel MA (1982) Normal development following in vitro fertilization in the cow. Biol Reprod 27:147–158

Buckrell BC, Buschbeck C, Gartley CJ, Kroetsch T, McCutcheon W, Martin J, Penner WK, Walton JS (1994) Further development of a transcervical technique for artificial insemination in sheep using previously frozen semen. Theriogenology 42:601–611

Caballero I, Parrilla I, Almiñana C, del Olmo D, Roca J, Martínez EA, Vázquez JM (2012) Seminal plasma proteins as modulators of the sperm function and their application in sperm biotechnologies. Reprod Domest Anim 47(Suppl 3):12–21

Candappa IB, Bainbridge HC, Price NT, Hourigan KR, Bartlewski PM (2009) A preliminary study on the suitability of Cervidil to induce cervical dilation for artificial insemination in ewes. Res Vet Sci 87:204–206

Canu S, Boland M, Lloyd GM, Newman M, Christie MF, May PJ, Christley RM, Smith RF, Dobson H (2010) Predisposition to repeat breeding in UK cattle and success

of artificial insemination alone or in combination with embryo transfer. Vet Rec 167:44–51
Carta A, Casu S, Salaris S (2009) Invited review: current state of genetic improvement in dairy sheep. J Dairy Sci 92:5814–5833
Chavatte-Palmer P, Camous S, Jammes H, Le Cleac'h N, Guillomot M, Lee RS (2012) Review: placental perturbations induce the developmental abnormalities often observed in bovine somatic cell nuclear transfer. Placenta 33(Suppl):S99–S104
Chemineau P, Guillaume D, Migaud M, Thiéry JC, Pellicer-Rubio MT, Malpaux B (2008) Seasonality of reproduction in mammals: intimate regulatory mechanisms and practical implications. Reprod Domest Anim 43(Suppl 2):40–47
Crowe CA, Ravenhill PJ, Hepburn RJ, Shepherd CH (2008) A retrospective study of artificial insemination of 251 mares using chilled and fixed time frozen-thawed semen. Equine Vet J 40:572–576
Cseh S, Faigl V, Amiridis GS (2012) Semen processing and artificial insemination in health management of small ruminants. Anim Reprod Sci 130:187–192
Davis ME, Rutledge JJ, Cundiff LV, Hauser ER (1983) Life cycle efficiency of beef production: II. Relationship of cow efficiency ratios to traits of the dam and progeny weaned. J Anim Sci 57:852–866
De Rensis F, Saleri R, Tummaruk P, Techakumphu M, Kirkwood RN (2012) Prostaglandin F2α and control of reproduction in female swine: a review. Theriogenology 77:1–11
De Vries A, Overton M, Fetrow J, Leslie K, Eicker S, Rogers G (2008) Exploring the impact of sexed semen on the structure of the dairy industry. J Dairy Sci 91:847–456
de Vries M, Vosters S, Merkx G, D'Hauwers K, Wansink DG, Ramos L, de Boer P (2012) Human male meiotic sex chromosome inactivation. PLoS One 7:e31485
DeJarnette JM, Nebel RL, Marshall CE (2009) Evaluating the success of sex-sorted semen in US dairy herds from on farm records. Theriogenology 71:49–58
Deleuze S, Dubois CS, Caillaud M, Bruneau B, Goudet G, Duchamp G (2010) Influence of cysteamine on in vitro maturation, in vitro and in vivo fertilization of equine oocytes. Reprod Domest Anim 45:1–7
Dematawewa CM, Berger PJ (1998) Break-even cost of cloning in genetic improvement of dairy cattle. J Dairy Sci 81:1136–1147
Demetrio DG, Santos RM, Demetrio CG, Vasconcelos JL (2007) Factors affecting conception rates following artificial insemination or embryo transfer in lactating Holstein cows. J Dairy Sci 90:5073–5082
Dickerson GE (1978) Animal size and efficiency: basic concepts. Anim Prod 27:367–379
Dransfield MB, Nebel RL, Pearson RE, Warnick LD (1998) Timing of insemination for dairy cows identified in estrus by a radiotelemetric estrus detection system. J Dairy Sci 81:1874–1882
Drews K, Jozefczuk J, Prigione A, Adjaye J (2012) Human induced pluripotent stem cells—from mechanisms to clinical applications. J Mol Med (Berl) 90: 735–745
Drost M (2007) Advanced reproductive technology in the water buffalo. Theriogenology 68:450–453
Edwards JL, Schrick FN, McCracken MD, van Amstel SR, Hopkins FM, Welborn MG, Davies CJ (2003) Cloning adult farm animals: a review of the possibilities and problems associated with somatic cell nuclear transfer. Am J Reprod Immunol 50:113–123
Ezashi T, Telugu BP, Roberts RM (2012) Induced pluripotent stem cells from pigs and other ungulate species: an alternative to embryonic stem cells? Reprod Domest Anim 47(Suppl 4):92–97
Farin PW, Slenning BD, Britt JH (1999) Estimates of pregnancy outcomes based on selection of bovine embryos produced in vivo or in vitro. Theriogenology 52:659–670
Farin PW, Piedrahita JA, Farin CE (2006) Errors in development of fetuses and placentas from in vitro-produced bovine embryos. Theriogenology 65:178–191
Flowers WL, Alhusen HD (1992) Reproductive performance and estimates of labor requirements associated with combinations of artificial insemination and natural service in swine. J Anim Sci 70:615–621
Foote RH (2002) The history of artificial insemination: selected notes and notables. J Anim Sci 80(E. Suppl):E22–E32
Galli C, Lagutina I, Duchi R, Colleoni S, Lazzari G (2008) Somatic cell nuclear transfer in horses. Reprod Domest Anim 43(Suppl 2):331–337
García EM, Vázquez JM, Parrilla I, Calvete JJ, Sanz L, Caballero I, Roca J, Vazquez JL, Martínez EA (2007) Improving the fertilizing ability of sex sorted boar spermatozoa. Theriogenology 68:771–778
García-Roselló E, García-Mengual E, Coy P, Alfonso J, Silvestre MA (2009) Intracytoplasmic sperm injection in livestock species: an update. Reprod Domest Anim 44:143–151
Garner DL, Seidel GE Jr (2008) History of commercializing sexed semen for cattle. Theriogenology 69:886–895
Gil MA, Cuello C, Parrilla I, Vazquez JM, Roca J, Martinez EA (2010) Advances in swine in vitro embryo production technologies. Reprod Domest Anim 45(Suppl 2):40–48
Hansen PJ (2011a) Heat stress and climate change. In: Moo-Young M (ed) Comprehensive biotechnology, vol 4, 2nd edn. Elsevier, Amsterdam, pp 477–485
Hansen PJ (2011b) Managing reproduction during heat stress in dairy cows. In: Risco CA, Melendez-Retamel P (eds) Dairy production medicine. Wiley-Blackwell, Chichester, pp 153–163
Hansen PJ (2011c) The immunology of early pregnancy in farm animals. Reprod Domest Anim 46(Suppl 3):18–30
Hansen PJ (2013) Prospects for use of embryo transfer for genetic selection and fertility improvement in cattle. Cattle Pract 21:30–34
Hansen PJ, Block J (2004) Towards an embryocentric world: the current and potential uses of embryo technologies in dairy production. Reprod Fertil Dev 16:1–14
Hayakawa H, Hirai T, Takimoto A, Ideta A, Aoyagi Y (2009) Superovulation and embryo transfer in Holstein cattle using sexed sperm. Theriogenology 71:68–73

Hinrichs K (2010) In vitro production of equine embryos: state of the art. Reprod Domest Anim 45(Suppl 2):3–8
Honaramooz A, Behboodi E, Megee SO, Overton SA, Galantino-Homer H, Echelard Y, Dobrinski I (2003) Fertility and germline transmission of donor haplotype following germ cell transplantation in immunocompetent goats. Biol Reprod 69:1260–1264
Hutt KJ, Albertini DF (2006) Clinical applications and limitations of current ovarian stem cell research: a review. J Exp Clin Assist Reprod 3:6
Johnson LA, Flook JP, Hawk HW (1989) Sex preselection in rabbits: live births from X and Y sperm separated by DNA and cell sorting. Biol Reprod 41:199–203
Johnson LA, Weitze KF, Fiser P, Maxwell WM (2000) Storage of boar semen. Anim Reprod Sci 62:143–172
Kasimanickam V, Kasimanickam R, Arangasamy A, Saberivand A, Stevenson JS, Kastelic JP (2012) Association between mRNA abundance of functional sperm function proteins and fertility of Holstein bulls. Theriogenology 78:2007–2019
Keefer CL (2008) Lessons learned from nuclear transfer (cloning). Theriogenology 69:48–54
Klinc P, Rath D (2007) Reduction of oxidative stress in bovine spermatozoa during flow cytometric sorting. Reprod Domest Anim 42:63–67
Landivar C, Galina CS, Duchateau A, Navarro-Fierro R (1985) Fertility trial in Zebu cattle after a natural or controlled estrus with prostaglandin F2 alpha, comparing natural mating with artificial insemination. Theriogenology 23:421–429
Larson JE, Lamb GC, Funnell BJ, Bird S, Martins A, Rodgers JC (2010) Embryo production in superovulated Angus cows inseminated four times with sexed-sorted or conventional, frozen-thawed semen. Theriogenology 73:698–703
Leahy T, Marti JI, Evans G, Maxwell WM (2009) Seminal plasma proteins protect flow-sorted ram spermatozoa from freeze-thaw damage. Reprod Fertil Dev 21: 571–578
Leboeuf B, Delgadillo JA, Manfredi E, Piacère A, Clément V, Martin P, Pellicer M, Boué P, de Cremoux R (2008) Management of goat reproduction and insemination for genetic improvement in France. Reprod Domest Anim 43(Suppl 2):379–385
Lima FS, Risco CA, Thatcher MJ, Benzaquen ME, Archbald LF, Santos JE, Thatcher WW (2009) Comparison of reproductive performance in lactating dairy cows bred by natural service or timed artificial insemination. J Dairy Sci 92:5456–5466
Loi P, Clinton M, Vackova I, Fulka J Jr, Feil R, Palmieri C, Della Salda L, Ptak G (2006) Placental abnormalities associated with post-natal mortality in sheep somatic cell clones. Theriogenology 65:1110–1121
Loomis PR (2001) The equine frozen semen industry. Anim Reprod Sci 68:191–200
Loomis PR, Graham JK (2008) Commercial semen freezing: individual male variation in cryosurvival and response of stallion sperm to customized freezing protocols. Anim Reprod Sci 105:119–128
Lopes RF, Forell F, Oliveira AT, Rodrigues JL (2001) Splitting and biopsy for bovine embryo sexing under field conditions. Theriogenology 56:1383–1392
Lu K, Cran DG, Seidel GE Jr (1999) In vitro fertilization with flow-cytometrically-sorted bovine sperm. Theriogenology 52:1393–1405
Luo Y, Lin L, Bolund L, Jensen TG, Sørensen CB (2012) Genetically modified pigs for biomedical research. J Inherit Metab Dis 35:695–713
Meirelles FV, Birgel EH, Perecin F, Bertolini M, Traldi AS, Pimentel JR, Komninou ER, Sangalli JR, Neto PF, Nunes MT, Pogliani FC, Meirelles FD, Kubrusly FS, Vannucchi CI, Silva LC (2010) Delivery of cloned offspring: experience in Zebu cattle (*Bos indicus*). Reprod Fertil Dev 22:88–97
Men H, Walters EM, Nagashima H, Prather RS (2012) Emerging applications of sperm, embryo and somatic cell cryopreservation in maintenance, relocation and rederivation of swine genetics. Theriogenology 78: 1720–1729
Merton JS, de Roos AP, Mullaart E, de Ruigh L, Kaal L, Vos PL, Dieleman SJ (2003) Factors affecting oocyte quality and quantity in commercial application of embryo technologies in the cattle breeding industry. Theriogenology 59:651–674
Merton JS, Ask B, Onkundi DC, Mullaart E, Colenbrander B, Nielen M (2009) Genetic parameters for oocyte number and embryo production within a bovine ovum pick-up-in vitro production embryo-production program. Theriogenology 72:885–893
Mikkola M, Sironen A, Kopp C, Taponen J, Sukura A, Vilkki J, Katila T, Andersson M (2006) Transplantation of normal boar testicular cells resulted in complete focal spermatogenesis in a boar affected by the immotile short-tail sperm defect. Reprod Domest Anim 41: 124–128
Moghaddaszadeh-Ahrabi S, Farajnia S, Rahimi-Mianji G, Nejati-Javaremi A (2012) A short and simple improved-primer extension preamplification (I-PEP) procedure for whole genome amplification (WGA) of bovine cells. Anim Biotechnol 23:24–42
Moore K, Thatcher WW (2006) Major advances associated with reproduction in dairy cattle. J Dairy Sci 89:1254–1266
Moruzzi JF (1979) Selecting a mammalian species for the determination of X- and Y-chromosome-bearing sperm. J Reprod Fertil 57:319–323
Mozo-Martín R, Gil L, Gómez-Rincón CF, Dahmani Y, García-Tomás M, Úbeda JL, Grandía J (2012) Use of a novel double uterine deposition artificial insemination technique using low concentrations of sperm in pigs. Vet J 193:251–256
Nagano M, Brinster CJ, Orwig KE, Ryu BY, Avarbock MR, Brinster RL (2001) Transgenic mice produced by retroviral transduction of male germ-line stem cells. Proc Natl Acad Sci U S A 98:13090–13095
Nivet al, Bunel A, Labrecque R, Belanger J, Vigneault C, Blondin P, Sirard MA (2012) FSH withdrawal improves developmental competence of oocytes in the bovine model. Reproduction 143:165–171

Norman HD, Hutchison JL, Miller RH (2010) Use of sexed semen and its effect on conception rate, calf sex, dystocia, and stillbirth of Holsteins in the United States. J Dairy Sci 93:3880–3890
Notter DR (2008) Genetic aspects of reproduction in sheep. Reprod Domest Anim 43(Suppl 2):122–128
Novak S, Ruiz-Sánchez A, Dixon WT, Foxcroft GR, Dyck MK (2010a) Seminal plasma proteins as potential markers of relative fertility in boars. J Androl 31:188–200
Novak S, Smith TA, Paradis F, Burwash L, Dyck MK, Foxcroft GR, Dixon WT (2010b) Biomarkers of in vivo fertility in sperm and seminal plasma of fertile stallions. Theriogenology 74:956–967
Nowak-Imialek M, Kues W, Carnwath JW, Niemann H (2011) Pluripotent stem cells and reprogrammed cells in farm animals. Microsc Microanal 17:474–497
O'Brien JK, Steinman KJ, Robeck TR (2009) Application of sperm sorting and associated reproductive technology for wildlife management and conservation. Theriogenology 71:98–107
Oatley JM (2010) Spermatogonial stem cell biology in the bull: development of isolation, culture, and transplantation methodologies and their potential impacts on cattle production. Soc Reprod Fertil Suppl 67:133–143
Oatley JM, Brinster RL (2012) The germline stem cell niche unit in mammalian testes. Physiol Rev 92:577–595
Oback B (2008) Climbing mount efficiency—small steps, not giant leaps towards higher cloning success in farm animals. Reprod Domest Anim 43(Suppl 2):407–416
Oback B (2009) Cloning from stem cells: different lineages, different species, same story. Reprod Fertil Dev 21:83–94
Onteru SK, Ross JW, Rothschild MF (2009) The role of gene discovery, QTL analyses and gene expression in reproductive traits in the pig. Soc Reprod Fertil Suppl 66:87–102
Park YJ, Kwon WS, Oh SA, Pang MG (2012) Fertility-related proteomic profiling bull spermatozoa separated by percoll. J Proteome Res 11:4162–4168
Pereira RM, Marques CC (2008) Animal oocyte and embryo cryopreservation. Cell Tissue Bank 9:267–277
Perumal P, Selvaraju S, Selvakumar S, Barik AK, Mohanty DN, Das S, Das RK, Mishra PC (2011) Effect of pre-freeze addition of cysteine hydrochloride and reduced glutathione in semen of crossbred Jersey bulls on sperm parameters and conception rates. Reprod Domest Anim 46:636–641
Polisseni J, Sá WF, Guerra Mde O, Machado MA, Serapião RV, Carvalho BC, Camargo LS, Peters VM (2010) Post-biopsy bovine embryo viability and whole genome amplification in preimplantation genetic diagnosis. Fertil Steril 93:783–788
Pontes JH, Silva KC, Basso AC, Rigo AG, Ferreira CR, Santos GM, Sanches BV, Porcionato JP, Vieira PH, Faifer FS, Sterza FA, Schenk JL, Seneda MM (2010) Large-scale in vitro embryo production and pregnancy rates *from Bos taurus*, *Bos indicus*, and *indicus-taurus* dairy cows using sexed sperm. Theriogenology 74: 1349–1355
Rasmussen S, Block J, Seidel GE, Brink Z, McSweeney K, Farin PW, Bonilla L, Hansen PJ (2013) Pregnancy rates of lactating cows after transfer of in vitro produced embryos using X-sorted sperm. Theriogenology 79(3):453–461
Rath D, Johnson LA (2008) Application and commercialization of flow cytometrically sex-sorted semen. Reprod Domest Anim 43(Suppl 2):338–346
Rath D, Moench-Tegeder G, Taylor U, Johnson LA (2009) Improved quality of sex-sorted sperm: a prerequisite for wider commercial application. Theriogenology 71:22–29
Ribeiro ES, Galvão KN, Thatcher WW, Santos JEP (2012) Economic aspects of applying reproductive technologies to dairy herds. Anim Reprod 9:370–387
Rico C, Drouilhet L, Salvetti P, Dalbiès-Tran R, Jarrier P, Touzé JL, Pillet E, Ponsart C, Fabre S, Monniaux D (2012) Determination of anti-Müllerian hormone concentrations in blood as a tool to select Holstein donor cows for embryo production: from the laboratory to the farm. Reprod Fertil Dev 24:932–944
Robinson JJ, McKelvey WA, King ME, Mitchell SE, Mylne MJ, McEvoy TG, Dingwall WS, Williams LM (2011) Traversing the ovine cervix—a challenge for cryopreserved semen and creative science. Animal 5:1791–1804
Roca J, Parrilla I, Rodriguez-Martinez H, Gil MA, Cuello C, Vazquez JM, Martinez EA (2011) Approaches towards efficient use of boar semen in the pig industry. Reprod Domest Anim 46(Suppl 2):79–83
Rodriguez-Osorio N, Urrego R, Cibelli JB, Eilertsen K, Memili E (2012) Reprogramming mammalian somatic cells. Theriogenology 78:1869–1886
Roser JF, Meyers-Brown G (2012) Superovulation in the mare: a work in progress. J Equine Vet Sci 32: 376–386
Santos-Biase WK, Biase FH, Buratini J Jr, Balieiro J, Watanabe YF, Accorsi MF, Ferreira CR, Stranieri P, Caetano AR, Meirelles FV (2012) Single nucleotide polymorphisms in the bovine genome are associated with the number of oocytes collected during ovum pick up. Anim Reprod Sci 134:141–149
Saragusty J, Arav A (2011) Current progress in oocyte and embryo cryopreservation by slow freezing and vitrification. Reproduction 141:1–19
Sartori R, Souza AH, Guenther JN, Carviello DZ, Geiger LN, Schenk JL, Wiltbank MC (2004) Fertilization rate and embryo quality in superovulated Holstein heifers artificially inseminated with X-sorted or unsorted sperm. Anim Reprod 1:86–90
Sartori R, Gümen A, Guenther JN, Souza AH, Caraviello DZ, Wiltbank MC (2006) Comparison of artificial insemination versus embryo transfer in lactating dairy cows. Theriogenology 65:1311–1321
Smith LC, Suzuki J Jr, Goff AK, Filion F, Therrien J, Murphy BD, Kohan-Ghadr HR, Lefebvre R, Brisville AC, Buczinski S, Fecteau G, Perecin F, Meirelles FV (2012) Developmental and epigenetic anomalies in cloned cattle. Reprod Domest Anim 47(Suppl 4): 107–114

Son DS, Choe CY, Cho SR, Choi SH, Kim HJ, Hur TY, Jung YG, Kang HG, Kim IM (2007) A CIDR-based timed embryo transfer protocol increases the pregnancy rate of lactating repeat breeder dairy cows. J Reprod Dev 53:1313–1318
Spizziri BE, Fox MH, Bruemmer JE, Squires EL, Graham JK (2010) Cholesterol-loaded-cyclodextrins and fertility potential of stallions spermatozoa. Anim Reprod Sci 118:255–264
Stewart BM, Block J, Morelli P, Navarette AE, Amstalden M, Bonilla L, Hansen PJ, Bilby TR (2011) Efficacy of embryo transfer in lactating dairy cows during summer using fresh or vitrified embryos produced in vitro with sex-sorted semen. J Dairy Sci 94:3437–3445
Stockwell S, Herrid M, Davey R, Brownlee A, Hutton K, Hill JR (2009) Microsatellite detection of donor-derived sperm DNA following germ cell transplantation in cattle. Reprod Fertil Dev 21:462–468
Stroud B (2011) IETS 2011 Statistics and Data Retrieval Committee Report. Embryo Transf Newsl 29(4): 14–23
Thatcher WW, Santos JE, Staples CR (2011) Dietary manipulations to improve embryonic survival in cattle. Theriogenology 76:1619–1631
Thornton P, Herrero M, Freeman A, Mwai O, Rege E, Jones P, McDermott J (2007) Vulnerability, climate change and livestock—research opportunities and challenges for poverty alleviation. SAT eJournal 4:1–23
Tríbulo A, Rogan D, Tribulo H, Tribulo R, Alasino RV, Beltramo D, Bianco I, Mapletoft RJ, Bó GA (2011) Superstimulation of ovarian follicular development in beef cattle with a single intramuscular injection of Folltropin-V. Anim Reprod Sci 129:7–13
Tsiligianni T, Valasi I, Cseh S, Vainas E, Faigl V, Samartzi F, Papanikolaou T, Dovolou E, Amiridis G (2009) Effects of melatonin treatment on follicular development and oocyte quality in Chios ewes—short communication. Acta Vet Hung 57:331–335
Underwood SL, Bathgate R, Maxwell WM, Evans G (2010a) Birth of offspring after artificial insemination of heifers with frozen-thawed, sex-sorted, re-frozen-thawed bull sperm. Anim Reprod Sci 118:171–175
Underwood SL, Bathgate R, Pereira DC, Castro A, Thomson PC, Maxwell WM, Evans G (2010b) Embryo production after in vitro fertilization with frozen-thawed, sex-sorted, re-frozen-thawed bull sperm. Theriogenology 73:97–102
United Nations, Department of Economic and Social Affairs, Population Division (2011) World population prospects: the 2010 revision, highlights and advance tables. Working Paper No. ESA/P/WP.220, New York
Vajta G, Callesen H (2012) Establishment of an efficient somatic cell nuclear transfer system for production of transgenic pigs. Theriogenology 77:1263–1274
Vajta G, Rienzi L, Cobo A, Yovich J (2010) Embryo culture: can we perform better than nature? Reprod Biomed Online 20:453–469
VanRaden PM, Van Tassell CP, Wiggans GR, Sonstegard TS, Schnabel RD, Taylor JF, Schenkel FS (2009) Invited review: reliability of genomic predictions for North American Holstein bulls. J Dairy Sci 92:16–24
Vasconcelos JL, Demétrio DG, Santos RM, Chiari JR, Rodrigues CA, Sá Filho OG (2006) Factors potentially affecting fertility of lactating dairy cow recipients. Theriogenology 65:192–200
Vasconcelos JL, Jardina DT, Sá Filho OG, Aragon FL, Veras MB (2011) Comparison of progesterone-based protocols with gonadotropin-releasing hormone or estradiol benzoate for timed artificial insemination or embryo transfer in lactating dairy cows. Theriogenology 75:1153–1160
Vidament M, Vincent P, Martin FX, Magistrini M, Blesbois E (2009) Differences in ability of jennies and mares to conceive with cooled and frozen semen containing glycerol or not. Anim Reprod Sci 112:22–35
Weigel KA, Hoffman PC, Herring W, Lawlor TJ Jr (2012) Potential gains in lifetime net merit from genomic testing of cows, heifers, and calves on commercial dairy farms. J Dairy Sci 95:2215–2225
Wheeler MB, Rutledge JJ, Fischer-Brown A, VanEtten T, Malusky S, Beebe DJ (2006) Application of sexed semen technology to in vitro embryo production in cattle. Theriogenology 65:219–227
White YA, Woods DC, Takai Y, Ishihara O, Seki H, Tilly JL (2012) Oocyte formation by mitotically active germ cells purified from ovaries of reproductive-age women. Nat Med 18:413–421
Whitworth KM, Prather RS (2010) Somatic cell nuclear transfer efficiency: how can it be improved through nuclear remodeling and reprogramming? Mol Reprod Dev 77:1001–1015
Whyte JJ, Prather RS (2011) Genetic modifications of pigs for medicine and agriculture. Mol Reprod Dev 78:879–891
Wilmot S (2007) From 'public service' to artificial insemination: animal breeding science and reproductive research in early twentieth-century Britain. Stud Hist Philos Biol Biomed Sci 38:411–441
Wilmut I, Schnieke AE, McWhir J, Kind AJ, Campbell KH (1997) Viable offspring derived from fetal and adult mammalian cells. Nature 385:810–813
Wiltbank MC, Sartori R, Herlihy MM, Vasconcelos JL, Nascimento AB, Souza AH, Ayres H, Cunha AP, Keskin A, Guenther JN, Gumen A (2011) Managing the dominant follicle in lactating dairy cows. Theriogenology 76:1568–1582
Woods DC, Tilly JL (2012) The next (re)generation of ovarian biology and fertility in women: is current science tomorrow's practice? Fertil Steril 98:3–10
Xu J, Guo Z, Su L, Nedambale TL, Zhang J, Schenk J, Moreno JF, Dinnyés A, Ji W, Tian XC, Yang X, Du F (2006) Developmental potential of vitrified Holstein cattle embryos fertilized in vitro with sex-sorted sperm. J Dairy Sci 89:2510–2518
Yamaguchi S, Funahashi H (2012) Effect of the addition of beta-mercaptoethanol to a thawing solution supplemented with caffeine on the function of frozen-thawed boar sperm and on the fertility of sows after artificial insemination. Theriogenology 77:926–932
Yoshioka K (2011) Development and application of a chemically defined medium for the in vitro production of porcine embryos. J Reprod Dev 57:9–16

challenges and future prospects for improving poultry production from the view of a reproductive biologist are discussed in this chapter.

Reproductive Dysfunction in Broiler Breeder Chickens

Approximately 8.5 billion broiler chickens are reared in the USA annually for meat production. Broiler breeder chickens are genetically selected for faster growth, higher feed intake, and greater muscle yield in their progenies. Just like their progenies, the parental line of broiler breeder chickens also displays hyperphagia that leads to reproductive problems (Robinson et al. 2007). Consequently, broiler breeder hens have the poorest reproductive efficiency of all commercial avian species. Current management practices involve cumbersome and often imprecise feed restriction methods to limit body growth in an effort to increase egg production. Despite adopting laborious methods, egg production remains suboptimal due to excessive follicular recruitment that often leads to internal or double (nonviable) ovulations. It is not clear, however, what factor(s) promotes this excessive follicular recruitment in the broiler breeder hen ovary. The following review covers some of the significant research areas that hold promise for improving reproductive efficiency in broiler breeder chickens.

Broiler Breeder Hens Have Excessive Visceral Adiposity and Multiple Ovarian Follicular Hierarchies

Broiler breeder female chickens are hyperphagic. When allowed unrestricted access to feed, they gain twice as much weight as feed-restricted chickens. A major part of this excess body weight is due to the increased deposition of visceral adipose tissue (abdominal fat pad, mesenteric fat, and fat around visceral organs). Accumulation of excessive visceral adipose tissue due to unrestricted feeding, as seen in Fig. 2.1, often leads to hypertrophy of adipocytes with excess triglycerides. Coincidentally, the ovary of the ad libitum-fed broiler breeder hen also develops multiple follicular hierarchies. A normal ovary has a typical hierarchy of 4–6 preovulatory follicles (F1–F4 in Fig. 2.2) that are greater than 10 mm in diameter. The largest follicle (F1) ovulates every 26–28 h and the preovulatory follicular hierarchy is maintained by sequential recruitment of prehierarchical follicles. However, in broiler breeder hens that had unrestricted access to feed, the normal follicular hierarchy is disrupted by selection and growth of more than one follicle resulting in multiple hierarchy, multiple ovulations, and internal ovulations. This is one of the main reasons for poor reproductive efficiency of broiler breeder hens. As in females, male broiler breeder chickens that are fed ad libitum were found to have reduced duration of fertility, possibly contributing

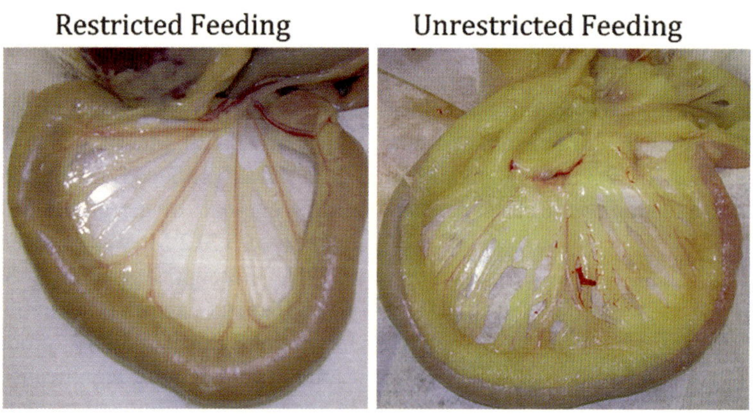

Fig. 2.1 Photographs of a part of small intestine in situ in a broiler breeder hen (18 weeks-old) showing excessive mesenteric adipose tissue accumulation due to unrestricted feeding (*right*) compared to lesser level of adipose tissue following restricted feeding (*left*) (Ramachandran, unpublished data)

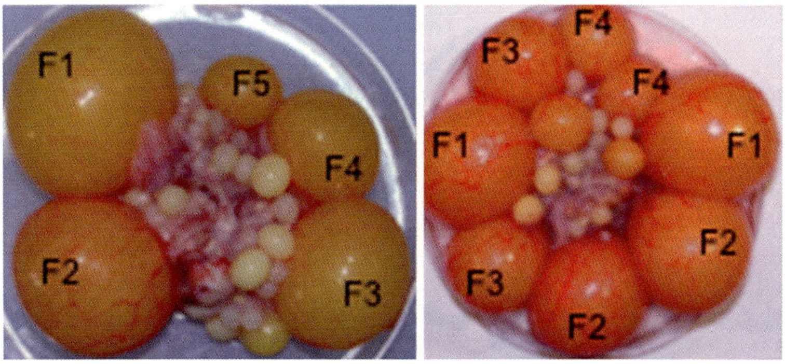

Fig. 2.2 Ovarian follicular hierarchy in leghorn (*left*) and broiler breeder hen (*right*). F1–F5 denote preovulatory follicles. Note double hierarchy of preovulatory follicles in broiler breeder hen ovary (picture courtesy: Dr. Alan Johnson, Department of Animal Science, Pennsylvania State University)

to a reduced fertility in artificially inseminated and naturally mated flocks (Goerzen et al. 1996).

Gonadotropins

Gonadotropins are critical for ovarian follicular development as well as egg production, initiation, and maintenance. Several studies have attempted to determine if gonadotropin secretion was altered in broiler breeder chickens in response to feed restriction. One of the studies suggests that plasma LH and FSH concentrations in Shaver Starbro broiler breeder pullets were significantly higher in ad libitum-fed chickens compared with feed-restricted hens (Renema et al. 1999). Similarly, ad libitum-fed broiler breeder pullets showed the highest responsiveness to ovarian hormones and to cLHRH-I in releasing FSH prior to sexual maturity compared with feed-restricted pullets, suggesting that feeding regimen can modify pituitary sensitivity to cLHRH-I and to gonadal hormones (Bruggeman et al. 1998b). Feed restriction of Hybro G broiler breeder pullets between 7 and 15 weeks-of-age followed by ad libitum feeding led to improved reproductive performance, although pituitary and plasma LH and FSH concentrations, and median eminence levels of cLHRH-I, were not different compared with pullets fed ad libitum (Bruggeman et al. 1998a). This raises the possibility that FSH responsiveness in the pre-hierarchical follicles is increased in response to overfeeding in broiler breeder chickens. FSHR were found to be expressed in both theca and granulosa cells of the developing ovarian follicle (You et al. 1996) but altered FSH signaling in broiler breeder ovaries in response to overfeeding has not been investigated. While it is unequivocally clear that ad libitum feeding reduces reproductive efficiency in broiler breeder hens, the underlying mechanisms involving hypothalamic–pituitary–ovarian axis and sensitivity to FSH at the ovarian level remain to be elucidated.

Metabolic Hormones

The root-cause(s) for the ovarian dysfunction in broiler breeder hens most likely lies within the ovarian follicles and visceral adipose tissue that tend to accumulate excessive triacylglycerol and fatty acids as a result of overeating (Chen et al. 2006). There are evidences to suggest that ad libitum-fed broiler breeder hens suffer with lipotoxicity leading to upregulation of proinflammatory cytokines expression in the liver and an increase in circulating levels of ceramide and sphingomyelin (Pan et al. 2012). In another study, ad libitum feeding of broiler breeder hens was associated with an increased apoptosis of granulosa cell and suppressed Akt activation (Xie et al. 2012). Furthermore, treatment of granulosa cells with palmitic acid, a saturated fatty acid, was found to activate apoptotic machinery in the granulosa cells. Leptin is an adipocytokine hormone that affects various metabolic and reproductive functions mediated through the hypothalamic–pituitary–gonadal axis in mammals (Barash et al. 1996; Bluher and Mantzoros 2007). Although existence of leptin in avian species is debatable (Sharp et al. 2008; Simon et al. 2009), leptin receptor is unequivocally expressed in various tissues including the thecal layer of the ovarian follicles in chickens (Cassy et al. 2004; Ohkubo et al. 2000). Injection of leptin-like substance to fasted laying hens was found to delay cessation of egg laying, attenuate regression of yellow hierarchical follicles, altered ovarian steroidogenesis (Paczoska-Eliasiewicz et al. 2003). Discovery of chicken leptin or endogenous ligand(s) for leptin receptor will improve our understanding on the role of leptin in ovarian dysfunction in broiler breeder hens.

The role of IGF on excessive adipose tissue deposition and ovarian dysfunction in broiler breeder hens has been investigated. Systemic levels of IGF-I and IGF-II were found to be elevated in broiler breeder pullets in response to feed restriction (Bruggeman et al. 1997; Hocking et al. 1994). In another study, the proportion of carcass fat in ad libitum-fed chickens was found to be positively correlated with plasma glucagon, IGF-II, and 17β-estradiol but negatively correlated with plasma insulin, insulin/glucagon ratio, IGF-I, thyroxine, and triiodothyronine suggesting that ad libitum feeding favors fat deposition (Sun et al. 2006). Consequently, excessive accumulation of carcass fat is likely to be detrimental to overall metabolism and in particular, to the reproductive system. Treatment of granulosa cells isolated from F1, F2, and F3 preovulatory follicles of broiler breeder hens with IGF-I alone or in combination with LH significantly increased granulosa cell proliferation in birds fed ad libitum more than feed-restricted hens suggesting that IGF-I may play an important role in accelerating the rate of maturation of follicles (Onagbesan et al. 1999). The precise role of IGF, GH, or insulin on ovarian follicular development in broiler breeder hen ovaries remains to be elucidated.

Inhibin/Activin

Inhibin, a hormone secreted predominantly by the granulosa cells of the ovarian follicle, acts as a negative feedback regulator of pituitary FSH secretion (Johnson et al. 1993; Vanmontfort et al. 1992, 1995). Expression of the inhibin α-subunit and inhibin/activin βA and βB subunits, as well the activin type II receptor have been documented in the developing follicles of broiler breeder hen ovaries suggesting a paracrine role for inhibin and activin within the ovary (Slappey and Davis 2003). Plasma inhibin levels were negatively correlated with FSH and positively correlated with progesterone levels in female chickens (Lovell et al. 2001; Vanmontfort et al. 1992). A practical application exists in modifying inhibin action to improve egg production in chickens. Active immunization of chickens against inhibin in broiler breeder hens was found to increase cumulative number of eggs produced by 9.5 % at the end of week 40 (Satterlee et al. 2002). Immunoneutralization of inhibin in chickens is likely to control the entry of ovarian follicles into preovulatory hierarchy (Lovell et al. 2001). Further studies are required to determine whether inhibin signaling can be altered in commercial settings to improve egg reproduction efficiency.

Anti-Müllerian Hormone

Anti-Müllerian hormone is predominantly secreted by the granulosa cells of the ovarian follicle in adult female chickens (Wojtusik and Johnson 2012). Recently, a possible role for AMH in excessive follicular recruitment in broiler breeder hens has been reported. As expected, AMH gene expression was found to be significantly higher in broiler breeder hen ovaries compared to Leghorn chicken ovaries (Johnson et al. 2009). Similarly, AMH gene expression was higher in the ovaries of fully fed broiler breeder hens compared with feed-restricted hens. AMH was postulated to enhance granulosa cell proliferation in an autocrine or paracrine mechanism but excessive AMH is likely to inhibit follicle development (Johnson et al. 2009). The ovarian AMH gene expression appears to be susceptible to vitamin D levels since a dose-dependent decrease in AMH mRNA levels was detected to vitamin D treatment (Wojtusik and Johnson 2012). Increased serum levels of AMH are associated with polycystic ovarian syndrome in women (Cook et al. 2002), a condition that resembles excessive follicular recruitment as occurring in broiler breeder hens that are fed ad libitum. At the present time, methods to quantify circulating levels of AMH in chickens are not available and therefore, a correlation between plasma AMH levels and egg production are not known. Future studies are required to determine if the manipulation of ovarian AMH levels leads to normalizing ovarian follicular hierarchy and higher egg production in broiler breeder hens.

Artificial Insemination

Comprehensive reviews on the history, methods, and challenges of AI in commercially important avian species can be found elsewhere (Blesbois 2007; Donoghue and Wishart 2000; Long 2006). AI technology is critical to turkey meat production as AI is almost exclusively used for turkey breeding. This is due to a disparity in the sizes of toms and hens as toms often exceed 33 kg in body weight while the hens are only 9 kg thus rendering mating challenging (Donoghue and Wishart 2000). In contrast, AI is not commonly used in chickens as broiler breeder hens are typically reared in floor pens instead of cages and due to low fertility of cryopreserved chicken semen (Donoghue and Wishart 2000). However, AI may become essential and practically relevant if future genetic selection of broilers favor a body conformation that limits physical mating. AI technology in turkeys utilizes fresh liquid semen (Donoghue and Wishart 2000) as storage of liquid semen greater than 6–24 h greatly reduces fertility. Poor fertilizing ability of the frozen/thawed avian spermatozoa can be attributed to several factors including greater sensitivity to the freezing/thawing process, deleterious effects of the cryoprotectant on survival, and the ability to withstand longer storage/selection in the SST of the female reproductive tract.

Recent studies have focused on mitochondrial function (Froman and Feltmann 2010) and the composition of the plasma membrane (Long 2006) with a view to develop cryopreservation methods that maintain the integrity of spermatozoa upon freezing/thawing and longevity once inside the SST. Mitochondria provide energy for sperm mobility and survival in the female reproductive tract and as such, conservation of mitochondrial integrity and function are critical for successful sperm cryopreservation. The function of chicken spermatozoa mitochondria can be temporarily inactivated using a calcium ion chelator prior to cooling to 10 °C and can be reactivated within a 5-h period (Froman and Feltmann 2010). In this study, a fertility rate of 88 % was achieved when the spermatozoa stored for 3 h were reactivated and used for insemination. A mass spectrometric analysis of proteins extracted from chicken spermatozoa revealed that expression levels of proteins related to ATP metabolism and glycolysis differ in high- versus low-sperm-mobility New Hampshire chicken lines (Froman et al. 2011). This suggests that mitochondrial function and energy levels are critical for sperm mobility.

Carbohydrates on the spermatozoa plasma membrane are found to be significantly altered during cryopreservation, and the degree of such modification was influenced by the type of cryoprotectant and freezing–thawing rates (Pelaez et al. 2011). In this regard, the type of cryoprotectant and freezing process was also found to alter the ability of chicken spermatozoa to undergo acrosome reaction (Moce et al. 2010). Fluidity of the avian spermatozoa plasma membrane was found to be affected with a significant decrease in cholesterol/phospholipid ratio following cryopreservation (Blesbois et al. 2005). Membrane fluidity is also one of the predictors for the success rate of semen cryopreservation in the chicken (Blesbois et al. 2008). Cryopreservation of spermatozoa of turkeys and sandhill cranes (*Grus canadensis*), using dimethylacetamine as the cryoprotectant resulted in greater viability for the frozen/thawed crane semen compared with turkey semen emphasizing the need for optimization of protocol suitable for each species (Blanco et al. 2012).

More studies are required to develop an appropriate cryopreservation method taking into consideration the anatomy and physiology unique to avian spermatozoa.

Sex Selection

Male chicks in egg producing flocks are not useful for the egg production and therefore, approximately, one-half of the chicks hatched in the poultry egg industry are typically culled. This represents an enormous waste of resources that were used in breeding and rearing parent chickens, fertile egg handling and hatching of eggs. Unlike mammals, sex in avian species is determined by the female and favoring female offspring would provide tremendous cost-saving to the poultry egg industry. Despite the commercial importance of sex selection to the poultry industry, there are only a few studies that have investigated the possibility of altering the sex ratio. Sex determination in avian species occurs during the first meiotic division that typically happens 2–4 h before ovulation (Olson and Fraps 1950). Among various factors co-incident during this period, a dramatic surge in circulating progesterone levels, predominantly emerging from the preovulatory follicles, is highly critical for the induction of ovulation. Using commercial Leghorn chickens, Correa et al. attempted to further elevate circulating progesterone levels during the critical window of sex determination (Correa et al. 2005). In this study, administering progesterone 2 mg (high dose) or 0.25 mg (low dose) to White Leghorn hens (Babcock B300 strain) 4 h prior to the end of the light cycle resulted in far fewer males (25 %) from hens treated with high dose of progesterone compared with the number of males from low dose progesterone or sesame oil-treated hens (61–63 %). While this study confirms the proof-of-principle that progesterone can affect sex ratio in the first egg, more research needs to be done to determine the effect of progesterone on sex ratio and its impact on egg production efficiency over a longer time period. Two recent studies adopted a similar approach of elevating circulating testosterone (Pinson et al. 2011b) or corticosterone (Pinson et al. 2011a) levels in White Leghorn hens (Hyline strain) during the predicted window of sex determination. A single dose of testosterone or corticosterone was administered to hens 5 h prior to the predicted time of ovulation and sex of the resultant offspring was determined. Interestingly, testosterone treatment resulted in a significantly higher proportion of male chicks compared to the control (Pinson et al. 2011b). Similar to testosterone, corticosterone treatment resulted in over 80 % of the chicks being male compared to only 40 % in untreated control hens (Pinson et al. 2011a). Based on the foregoing, altering sex ratio and hatching more females in the commercial poultry egg industry seems plausible in the future. It is, however, important that the overall egg production efficiency is not compromised while we attempt sex selection. Future studies should focus on using non-hormonal feed supplements that will accomplish sustained sex selection.

Transgenic Chickens

Modifying the chicken genome through transgenic technology has tremendous potential for imparting disease resistance and for expression of novel compounds in meat and eggs. Development of tools used in transgenic technology will also facilitate conservation and long-term preservation of PGC and embryonic and stem cells. Comprehensive reviews of various methods used to create transgenic chicken can be found elsewhere (Han 2009; Mozdziak and Petitte 2004; Park and Han 2012a; Petitte et al. 2004). Replication incompetent retroviral vectors including lentiviral vectors have been used for integrating transgenes into the chicken genome. Using this technique, several elegant studies have demonstrated the stable integration, germ line transmission, and expression of transgenes in chickens. Transduction of transgene constructs was typically achieved by infecting blastodermal cells or PGC derived from embryonic gonads or embryonic blood circulation with retroviral vectors (Harvey et al. 2002; Kamihira et al. 2005; Lillico et al. 2007; McGrew et al. 2004; Scott and

Lois 2006). Using lentiviral vector, a novel strategy to develop chickens that are genetically resistant to avian influenza was described (Ding et al. 2005). In this study, transgenic chickens were created overexpressing a short-hairpin RNA driven by the U6 promoter that inhibits and blocks influenza virus polymerase and prevents virus propagation. Although viral methods of creating transgenic chickens are feasible, such methods would render the transgenic chicken unsuitable for agriculture use. To overcome this disadvantage, recent studies have used DNA transposons to integrate transgenes into the chicken genome. PiggyBac, a DNA transposon isolated from the cabbage looper moth *Trichoplusia ni* (Cary et al. 1989), has been widely used for creating genetic modifications in mice (Ding et al. 2005) and in chicken embryos (Lu et al. 2009). Recently, transgenic chickens were successfully produced by microinjecting DNA constructs encoding GFP and piggyBac transposase into the sub-germinal cavity of newly laid eggs (Liu et al. 2012). Using non-virally transfected gonadal PGC with GFP and piggyBac DNA elements, transgenic chickens overexpressing GFP were created at a very high rate of transgene integration (437 transgenic chickens created out of 459 total hatched chicks; Park and Han 2012b). Similarly, piggyBac or Tol2, another transposon isolated from the medaka fish genome, was used to integrate transgene into PGC derived from embryonic blood that was then utilized to develop transgenic chickens (Macdonald et al. 2012). Taken together, non-viral methods for creating transgenic chickens offer tremendous potential for improving nutritive value of poultry meat or egg and for imparting disease resistance to chickens.

Conclusion

In conclusion, there is tremendous potential for improving the reproductive efficiency of broiler breeders and to help meet the increasing global demand for poultry meat. Understanding the role of hormones in ovarian follicular recruitment and ovulation is critical for improving reproductive efficiency. Selecting for female chicks in the egg production industry will help to conserve resources and lower costs. Some of the emerging technologies for cryopreservation of semen and PGC are likely to improve poultry production efficiency. Successful development of transgenic chickens that selectively overexpress certain microRNA and enzymes can prevent disease epidemics and disease-free flocks to allow uninterrupted food production.

References

Barash IA et al (1996) Leptin is a metabolic signal to the reproductive system. Endocrinology 137:3144–3147

Blanco JM, Long JA, Gee G, Wildt DE, Donoghue AM (2012) Comparative cryopreservation of avian spermatozoa: effects of freezing and thawing rates on turkey and sandhill crane sperm cryosurvival. Anim Reprod Sci 131:1–8

Blesbois E (2007) Current status in avian semen cryopreservation. Worlds Poult Sci Assoc 63:213–222

Blesbois E, Grasseau I, Seigneurin F (2005) Membrane fluidity and the ability of domestic bird spermatozoa to survive cryopreservation. Reproduction 129:371–378

Blesbois E et al (2008) Predictors of success of semen cryopreservation in chickens. Theriogenology 69:252–261

Bluher S, Mantzoros CS (2007) Leptin in reproduction. Curr Opin Endocrinol Diabetes Obes 14:458–464

Bruggeman V, Vanmontfort D, Renaville R, Portetelle D, Decuypere E (1997) The effect of food intake from two weeks of age to sexual maturity on plasma growth hormone, insulin-like growth factor-I, insulin-like growth factor-binding proteins, and thyroid hormones in female broiler breeder chickens. Gen Comp Endocrinol 107:212–220

Bruggeman V et al (1998a) The effect of food intake from 2 to 24 weeks of age on LHRH-I content in the median eminence and gonadotrophin levels in pituitary and plasma in female broiler breeder chickens. Gen Comp Endocrinol 112:200–209

Bruggeman V et al (1998b) Effect of long-term food restriction on pituitary sensitivity to cLHRH-I in broiler breeder females. J Reprod Fertil 114:267–276

Cary LC et al (1989) Transposon mutagenesis of baculoviruses: analysis of Trichoplusia ni transposon IFP2 insertions within the FP-locus of nuclear polyhedrosis viruses. Virology 172:156–169

Cassy S et al (2004) Leptin receptor in the chicken ovary: potential involvement in ovarian dysfunction of ad libitum-fed broiler breeder hens. Reprod Biol Endocrinol 2:72

Chen SE, McMurtry JP, Walzem RL (2006) Overfeeding-induced ovarian dysfunction in broiler breeder hens is associated with lipotoxicity. Poult Sci 85:70–81

Cook CL, Siow Y, Brenner AG, Fallat ME (2002) Relationship between serum mullerian-inhibiting substance and other reproductive hormones in untreated women with polycystic ovary syndrome and normal women. Fertil Steril 77:141–146
Correa SM, Adkins-Regan E, Johnson PA (2005) High progesterone during avian meiosis biases sex ratios toward females. Biol Lett 1:215–218
Ding S et al (2005) Efficient transposition of the piggyBac (PB) transposon in mammalian cells and mice. Cell 122:473–483
Donoghue AM, Wishart GJ (2000) Storage of poultry semen. Anim Reprod Sci 62:213–232
FAO (2009) FAO's Director-General on how to feed the world in 2050. Poupl Dev Rev 35(4):837–839
Froman DP, Feltmann AJ (2010) A new approach to sperm preservation based on bioenergetic theory. J Anim Sci 88:1314–1320
Froman DP et al (2011) Physiology and endocrinology symposium: a proteome-based model for sperm mobility phenotype. J Anim Sci 89:1330–1337
Goerzen PR, Julsrud WL, Robinson FE (1996) Duration of fertility in ad libitum and feed-restricted caged broiler breeders. Poult Sci 75:962–965
Han JY (2009) Germ cells and transgenesis in chickens. Comp Immunol Microbiol Infect Dis 32:61–80
Harvey AJ, Speksnijder G, Baugh LR, Morris JA, Ivarie R (2002) Expression of exogenous protein in the egg white of transgenic chickens. Nat Biotechnol 20:396–399
Hocking PM, Bernard R, Wilkie RS, Goddard C (1994) Plasma growth hormone and insulin-like growth factor-I (IGF-I) concentrations at the onset of lay in ad libitum and restricted broiler breeder fowl. Br Poult Sci 35:299–308
Johnson PA, Wang SY, Brooks C (1993) Characterization of a source and levels of plasma immunoreactive inhibin during the ovulatory cycle of the domestic hen. Biol Reprod 48:262–267
Johnson PA, Kent TR, Urick ME, Trevino LS, Giles JR (2009) Expression of anti-Mullerian hormone in hens selected for different ovulation rates. Reproduction 137:857–863
Kamihira M et al (2005) High-level expression of single-chain Fv-Fc fusion protein in serum and egg white of genetically manipulated chickens by using a retroviral vector. J Virol 79:10864–10874
Lillico SG et al (2007) Oviduct-specific expression of two therapeutic proteins in transgenic hens. Proc Natl Acad Sci U S A 104:1771–1776
Liu X et al (2012) Efficient production of transgenic chickens based on piggyBac. Transgenic Res 22(2):417–423
Long JA (2006) Avian semen cryopreservation: what are the biological challenges? Poult Sci 85:232–236
Lovell TM, Knight PG, Groome NP, Gladwell RT (2001) Changes in plasma inhibin A levels during sexual maturation in the female chicken and the effects of active immunization against inhibin alpha-subunit on reproductive hormone profiles and ovarian function. Biol Reprod 64:188–196
Lu Y, Lin C, Wang X (2009) PiggyBac transgenic strategies in the developing chicken spinal cord. Nucleic Acids Res 37:e141
Macdonald J et al (2012) Efficient genetic modification and germ-line transmission of primordial germ cells using piggyBac and Tol2 transposons. Proc Natl Acad Sci U S A 109:E1466–E1472
McGrew MJ et al (2004) Efficient production of germline transgenic chickens using lentiviral vectors. EMBO Rep 5:728–733
Moce E, Grasseau I, Blesbois E (2010) Cryoprotectant and freezing-process alter the ability of chicken sperm to acrosome react. Anim Reprod Sci 122:359–366
Mozdziak PE, Petitte JN (2004) Status of transgenic chicken models for developmental biology. Dev Dyn 229:414–421
Ohkubo T, Tanaka M, Nakashima K (2000) Structure and tissue distribution of chicken leptin receptor (cOb-R) mRNA. Biochim Biophys Acta 1491:303–308
Olson MW, Fraps RM (1950) Maturation changes in the hen's ovum. J Exp Zool 144:485–487
Onagbesan OM, Decuypere E, Leenstra F, Ehlhardt DA (1999) Differential effects of amount of feeding on cell proliferation and progesterone production in response to gonadotrophins and insulin-like growth factor I by ovarian granulosa cells of broiler breeder chickens selected for fatness or leanness. J Reprod Fertil 116:73–85
Paczoska-Eliasiewicz HE et al (2003) Attenuation by leptin of the effects of fasting on ovarian function in hens (Gallus domesticus). Reproduction 126: 739–751
Pan YE et al (2012) Ceramide accumulation and upregulation of proinflammatory interleukin-1beta exemplify lipotoxicity to mediate declines of reproductive efficacy of broiler hens. Domest Anim Endocrinol 42:183–194
Park TS, Han JY (2012a) Genetic modification of chicken germ cells. Ann N Y Acad Sci 1271:104–109
Park TS, Han JY (2012b) PiggyBac transposition into primordial germ cells is an efficient tool for transgenesis in chickens. Proc Natl Acad Sci USA 109:9337–9341
Pelaez J, Bongalhardo DC, Long JA (2011) Characterizing the glycocalyx of poultry spermatozoa: III. Semen cryopreservation methods alter the carbohydrate component of rooster sperm membrane glycoconjugates. Poult Sci 90:435–443
Petitte JN, Liu G, Yang Z (2004) Avian pluripotent stem cells. Mech Dev 121:1159–1168
Pinson SE, Parr CM, Wilson JL, Navara KJ (2011a) Acute corticosterone administration during meiotic segregation stimulates females to produce more male offspring. Physiol Biochem Zool 84:292–298
Pinson SE, Wilson JL, Navara KJ (2011b) Elevated testosterone during meiotic segregation stimulates laying

hens to produce more sons than daughters. Gen Comp Endocrinol 174:195–201

Renema RA, Robinson FE, Newcombe M, McKay RI (1999) Effects of body weight and feed allocation during sexual maturation in broiler breeder hens. 1. Growth and carcass characteristics. Poult Sci 78:619–628

Robinson FE, Zuidhof MJ, Renema RA (2007) Reproductive efficiency and metabolism of female broiler breeders as affected by genotype, feed allocation, and age at photostimulation. 1. Pullet growth and development. Poult Sci 86:2256–2266

Satterlee DG, Cadd GG, Fioretti WC (2002) Active immunization of broiler breeder hens with a recombinant chicken inhibin fusion protein enhances egg lay. Poult Sci 81:519–528

Scott BB, Lois C (2006) Generation of transgenic birds with replication-deficient lentiviruses. Nat Protoc 1:1406–1411

Sharp PJ, Dunn IC, Waddington D, Boswell T (2008) Chicken leptin. Gen Comp Endocrinol 158:2–4

Simon J, Rideau N, Taouis M (2009) Reply to viewpoints by PJ Sharp, IC Dunn, D Waddington and T Boswell [Chicken Leptin: General and Comparative Endocrinology, 158, 2–4 (2008)]. Gen Comp Endocrinol 161:159

Slappey SN, Davis AJ (2003) Expression pattern of messenger ribonucleic acid for the activin type II receptors and the inhibin/activin subunits during follicular development in broiler breeder hens. Poult Sci 82:338–344

Sun JM et al (2006) The relationship of body composition, feed intake, and metabolic hormones for broiler breeder females. Poult Sci 85:1173–1184

U.S. Census Bureau (2012) U.S. and World Poulation Clock. http://www.census.gov/popclock/

Vanmontfort D, Rombauts L, Decuypere E, Verhoeven G (1992) Source of immunoreactive inhibin in the chicken ovary. Biol Reprod 47:977–983

Vanmontfort D, Berghman LR, Rombauts L, Verhoeven G, Decuypere E (1995) Developmental changes in immunoreactive inhibin and FSH in plasma of chickens from hatch to sexual maturity. Br Poult Sci 36:779–790

Wojtusik J, Johnson PA (2012) Vitamin D regulates anti-Mullerian hormone expression in granulosa cells of the hen. Biol Reprod 86:91

Xie YL et al (2012) Palmitic acid in chicken granulosa cell death-lipotoxic mechanisms mediate reproductive inefficacy of broiler breeder hens. Theriogenology 78(9):1917–1928

You S, Bridgham JT, Foster DN, Johnson AL (1996) Characterization of the chicken follicle-stimulating hormone receptor (cFSH-R) complementary deoxyribonucleic acid, and expression of cFSH-R messenger ribonucleic acid in the ovary. Biol Reprod 55:1055–1062

3 Current and Future Assisted Reproductive Technologies for Fish Species

Gregory M. Weber and Cheng-Sheng Lee

Abstract

The Food and Agriculture Organization of the United Nations (FAO) estimates that in 2012 aquaculture production of fish will meet or exceed that of the capture fisheries for the first time. Thus, we have just turned the corner from a predominantly hunting gathering approach to meeting our nutritional needs from fish, to a farming approach. In 2012, 327 finfish species and five hybrids were covered by FAO aquaculture statistics, although farming of carps, tilapias, salmonids, and catfishes account for most of food-fish production from aquaculture. Although for most major species at least part of production is based on what might be considered domesticated animals, only limited production in most species is based on farming of improved lines of fish or is fully independent of wild seedstock. Consistent with the infancy of most aquaculture industries, much of the development and implementation of reproductive technologies over the past 100 years has been directed at completion of the life cycle in captivity in order to increase seed production and begin the process of domestication. The selection of species to farm and the emphasis of selective breeding must also take into account other ways to modify performance of an animal. Reproductive technologies have also been developed and implemented to affect many performance traits among fishes. Examples include technologies to control gender, alter time of sexual maturation, and induce sterilization. These technologies help take advantage of sexually dimorphic growth,

G.M. Weber (✉)
National Center for Cool and Coldwater Aquaculture, ARS/USDA, 11861 Leetown Road, Kearneysville, WV 25430, USA
e-mail: Greg.weber@ars.usda.gov

C.-S. Lee
Center for Tropical and Subtropical Aquaculture, c/o The Oceanic Institute, 41-202 Kalanianaole Hwy., Waimanalo, HI 96795, USA
e-mail: cslee@oceanicinstitute.org

G.C. Lamb and N. DiLorenzo (eds.), *Current and Future Reproductive Technologies and World Food Production*, Advances in Experimental Medicine and Biology 752, DOI 10.1007/978-1-4614-8887-3_3,

Improved lines are available for many aquatic species including common carp, Atlantic salmon (*Salmo salar*), rainbow trout (*Oncorhynchus mykiss*), tilapias (*Oreochromis* species), and channel catfish (*Ictalurus punctatus*). On the other hand, it has been estimated that only about 8.2 % of aquaculture production is based on family breeding programs (Gjedrem et al. 2012).

Unlike in terrestrial agriculture which is based on many races of a limited number of well-established species, the predominant fish species being cultured is in flux. New species are being evaluated all the time. In 2012, 327 finfish species and five hybrids were covered by FAO aquaculture statistics (FAO 2012). Currently, most species being mass cultured were originally developed due in large part to availability of wild seedstock or ease of reproduction and larval rearing. The carps, salmonids, tilapias, channel catfish, and milkfish are such examples, with most exhibiting both attributes. As improvements were made in inducing spawning and larval rearing technologies, more species, particularly marine species, could be considered for farming based on their production or profit potential. Rarely the availability of seedstock due to technical incompetence is the cause to dismiss a species, although the cost of seed production might be prohibitive. Currently, a critical debate is whether to continue exploring interspecies diversity for the best production animals, or turn to intraspecies diversity and concentrate on selective breeding of established species. Both options are made possible by recent advancements in reproductive technologies.

The selection of species to farm and the emphasis of selective breeding must also take into account other ways to modify performance of an animal. Reproductive technologies have been developed and implemented to affect many performance traits among fishes. Examples include technologies to control gender, alter time of sexual maturation, and induce sterilization. These technologies help take advantage of sexually dimorphic growth, overcome problems with growth performance and flesh quality associated with sexual maturation, and genetic containment. The large investments in gonads, often greater than 20 % body weight, and ability to escape into the wild, are problems of great concern in fish that are not faced with farming of most terrestrial species.

Another difference between the terrestrial and aquaculture industries that affects the application of reproductive technologies is that most of aquaculture is being conducted and is expanding rapidly in what are considered developing countries. In 2010 Asia accounted for 89 % of global aquaculture production by volume with China accounting for 60 % on its own (FAO 2012). In recent years, China has even begun to lead in the production of fish species not native to China such as the tilapia from Africa, and even produces channel catfish from North America. Most of aquaculture is conducted using extensive or semi-intensive approaches in Asian countries; however, as has been the case for most terrestrial animal industries, there is a trend towards increased intensification. In general, intensification requires greater control and manipulation of reproduction. As intensive aquaculture has increased, so has the development of reproductive control technologies directed towards increasing overall production efficiency.

The objective of the current chapter is to present the reproductive technologies used in fish farming and their contributions in food-fish production. We will start by presenting reproductive technologies available to the fish farmer and where they are applied in terms of production objectives. We will present an account of the expansion of culture and use of reproductive technologies in several groups of fish that are major sources of food-fish and also represent diverse and interesting applications of technologies. Finally, we will present some thoughts regarding future directions for reproductive technologies and their applications in finfish aquaculture.

Technologies for Reproductive Control

Control of reproduction takes many forms in fish farming as it does in farming terrestrial animals. Most broadly it includes control of gamete formation and release, and control of gender. Such control is sought for seedstock production, including abundance, quality, and year-round availability. Control is also sought for production efficiency and product quality, including limiting negative impacts of gamete production on growth and flesh composition, or taking advantage of sexually dimorphic growth, morphology, or behaviors. Control is also sought to improve efficiency of genetic selection and for genetic containment. As would be expected considering the diversity of life histories among fishes, the application of technologies to achieve desired outcomes varies considerably among cultivars and even among production methods for the same species. There are many technologies available to assist in controlling reproduction. It is not our objective to describe each in detail or provide instruction, but instead to provide a brief overview of what technologies are available and in use.

Assessing Maturity

A first step in controlling reproduction is assessing maturity. This starts with identification of gender, followed by monitoring reproductive maturation. Sexing of fish and determining stage of maturity involves both distinct and overlapping technologies. In some cases fish may even be sexed and staged for maturation at the same time using a single procedure such as part of a first evaluation of potential broodstock. More often, the use of the technologies are for gathering information for disparate purposes such as sexing fish for monosex grow-out, or for staging fish for proper application of induced-spawning procedures.

Sex Identification

Many fish species do not exhibit recognizable or easily recognizable sexual dimorphism, or the dimorphism does not become evident until late in the reproductive cycle. Genetic control of sex is often complex, making sex-linked DNA markers difficult to identify (see Devlin and Nagahama 2002). Furthermore, sex is determined or heavily influenced by environment in many cultured species. Finally, although most fish are gonochoristic, some cultured species such as gilthead seabream (*Sparus aurata*) are even sequential hermaphrodites. The four basic methods for gender identification are the use of external morphological features, genetic markers, biochemical markers, and examination of the gametes.

Sexually dimorphic external features are available and used for gender identification in some species. Tilapias are among the few fish that can be sexed early enough for monosex culture. Tilapia fingerlings can be distinguished by the shape of the genital papilla and location of pores at about 15 g, but this requires careful inspection and so is rarely used (Yashouv and Hefez 1959). Features that are not present or sufficiently distinct until the animals approach spawning season are not useful for sorting for monosex grow-out but can be used for sexing and sorting broodstock. Secondary sexually dimorphic traits such as the development of the hooked jaw or kype in male salmonids, or the greater width and muscularity of the head in male channel catfish, are examples of such traits. As is the case with most of the carps, sex is sometimes not easily identified until the males express sperm or milt when squeezed and females exhibit abdominal swelling from egg growth.

What is used in lieu of sexually dimorphic external features depends on internal anatomy of the fish, available knowledge of the species' physiology, availability of assays and reagents, and the basis of the need to differentiate gender. Use of gamete examination and the use of known biochemical markers for sex determination require a certain degree of gonadal development. A genetic marker is the only alternative if there is a need to know gender before such development, but few markers are available because of the complex mechanisms involved in sex determination. Salmonids are an example of fish that do not have cytogenetically distinguishable sex chromosomes. Y-chromosomal DNA probes have, however, been identified that are linked to sex in salmonids (Devlin et al. 1991, 2001; Forbes et al. 1994; Brunelli et al. 2008) including one based on the identification of the sdY gene as the master sex-determining gene in rainbow trout (Yano et al. 2012). Sex markers are of use in the establishment of monosex lines (Devlin et al. 1994a) and will be discussed later.

Sex-specific biochemical profiles may include blood or even skin mucus levels of sex steroids or the egg yolk precursor vitellogenins. The androgen 11-ketotestosterone (11KT) is male-specific in many fish species (Borg 1994) and has been identified most often as a suitable sex-specific marker in fish (Idler et al. 1971; Sangalang et al. 1978; Feist et al. 2004; Schultz et al. 2005; Kucherka et al. 2006). Steroids present in females are also present in appreciable levels in males. Vitellogenins have therefore been identified as the most suitable female-specific marker for sex in many species (Le Bail and Breton 1981; Craik and Harvey 1984; Gordon et al. 1984; Kishida et al. 1992; Matsubara and Sawano 1992; Kishida and Specker 1994; Takemura et al. 1996; Takemura and Oka 1998). However, vitellogenin production is very sensitive to estrogenic compounds, and thus, vitellogenins can be produced in males in response to environmental endocrine disruptors or even estrogens in fish feed (Sumpter 2005; Davis et al. 2009). For this reason vitellogenin levels in fish have been used as a marker for endocrine disruption (Heppell et al. 1995; Arcand-Hoy and Benson 1998; Denslow et al. 1999; Segner et al. 2003; Van Veld et al. 2005). A second problem with using biochemical markers is that most biochemical markers cannot be detected immediately and thus the fish must be tagged for identification so that they can be segregated by sex after the samples are analyzed in the laboratory. One advantage of vitellogenins and steroidogenic compounds being of interest to the endocrine disrupter community is their interest in developing on-site or dip-stick methods of identifying these metabolites (Mandich et al. 2005). Although biochemical markers are usually expressed too late to allow monosex culture in most species, they offer an advantage in sturgeon (family Acipenseridae) culture. Whereas female sturgeons are most valuable as a source of caviar, sturgeon meat is also marketable. Thus, even though it may take until the sturgeon are 1–3 years of age for the markers to be effective, depending on species, markers can be used to separate males to be grown out or immediately harvested for meat, and females that will be grown for several additional years for caviar (Feist et al. 2004; Mola et al. 2011).

Methods for examining gametes include direct examination of gametes and imaging technologies. Expression of milt is commonly used to identify and separate male broodstock animals. Depending upon species, the males might begin expressing milt weeks or even years earlier than the females mature, allowing males to be pulled from the culture population before females are ready to spawn. In common carp for example, a level of controlled spawning is achieved by separating the milt expressing males from females just before the spawning season in order to prevent the males from interacting with the females and inducing the females to ovulate in response to their courtship behaviors (Bardach et al. 1972; Jhingren and Pullin 1985). Biopsy is required to obtain gametes from immature fish. The ease of gonadal biopsy varies considerably with species, stage of reproductive maturation, and sex. Often a gonadal biopsy sample can be obtained by inserting a catheter through the genital pore to reach the gonad (Shehadeh et al. 1973; Ross 1984). For nearly mature fish with short gonadal ducts, something as simple as a hematocrit tube can inserted through the genital pore into the gonad followed by placing a finger on the end of the tube to create a vacuum to retain the gametes or gonadal tissue sample before being withdrawn (Hodson 1995). A longer flexible polyethylene catheter, often with the aide of aspiration to draw the tissue into the catheter, is commonly used in early stage fish, fish with longer ducts or ovaries, or fish with curves in the duct (Kuo et al. 1974; Lee et al. 1986a). Ovarian biopsy samples are usually easier to obtain than testicular samples. However, it is also important that the females are cystovarians, the ovarian capsule being continuous with the oviduct, so that the catheter can follow the oviduct to the gonad and suction can be created within the gonad if needed. Most teleost fishes exhibit the cystovarian condition whereas acipenserids (sturgeons) and salmonids are gymnovarians and secondary gymnovarians respectively, both releasing eggs into the coelom before reaching the oviducts, and therefore requiring an incision for biopsy (e.g., Conte et al. 1988; Chapman and Park 2005).

Different forms of gonadal imaging including laparoscopy, endoscopy, and ultrasonography have been used to visualize the gonads or even the gametes. Although laparoscopy and endoscopy procedures have been reported for cystovarian species such as cyprinids (Macri et al. 2011), these procedures have been described for use mostly in gymnovarian and secondary gymnovarian fishes, particularly salmonids (Moccia et al. 1984; Ortenburger et al. 1996) and acipenserids (Kynard and Kieffer 2002; Hurvitz et al. 2007; Falahatkar et al. 2011). Laparoscopy and endoscopy are most commonly described for use in sturgeons where an individual fish can be highly valuable. It is also worth noting laparoscopy can be used to obtain biopsy samples as described by Falahatkar et al. (2011) for great sturgeon (*Huso huso*), although obtaining a biopsy sample was only necessary to determine sex of fish with small gonads. The use of ultrasonography is the most popular means of imaging in aquaculture and its use is increasing with the reduced cost and increased sensitivity of the instruments, particularly water-resistant portable units (see Novelo and Tiersch 2012). Gender determination using ultrasound has been described for a wide diversity of species (e.g., Martin et al. 1983; Karlsen and Holm 1994; Matsubara et al. 1999; Masoudifard et al. 2011).

Monitoring Reproductive Maturation

Assessing stage of reproductive maturation is necessary for efficient management of broodstocks and breeding. Synchronization of maturation and spawning among individuals in a population varies considerably depending not only on species but also the environmental conditions in which the fish are maintained. The spawning season for some fish can last only weeks, particularly for those that spawn a single batch of eggs in a year or even a lifetime. On the other hand, the spawning season may last for months, particularly for species that spawn multiple batches of eggs in a season. Few fish spawn year round in their natural habitat but as is the case for common carp and Nile tilapia (*Oreochromis niloticus*), some species of fish will spawn year round if conditions such as temperatures are appropriate. Individual fish of the same species and population may also mature at different ages depending on sex or even within a sex. These differences in patterns of reproductive maturation and spawning place different demands on the hatchery operator. Identification of individuals that will mature in the current spawning season, or mature early versus late in the season, is important for species that must be conditioned for maturation and spawning or maintained in broodstock facilities different from that used for grow-out. Identification of animals that are ready to release gametes is critical for synchronized fertilization production. Staging can be very important in paired matings in species in which mature animals will attack immature animals in confined spaces, as with channel catfish spawned in pens or tanks. Finally, staging is critical for strip-spawning or hormonally induced spawning since inappropriately timed stripping of eggs or hormone treatment can disrupt maturation and result in egg loss from the animals (see Sullivan et al. 2003; Mylonas et al. 2010).

Monitoring reproductive maturity in males primarily consists of monitoring testis size; and milt viscosity and sperm motility as indicators of completion of spermiation (Billard et al. 1995; Cabrita et al. 2010; Mylonas et al. 2010). Stages of reproductive maturity assessed as part of hatchery management and controlling reproduction in females includes the transitions from primary to secondary oocyte growth, previtellogenic growth to vitellogenic growth, vitellogenic growth to follicle maturation; the onset of ovarian atresia, and ovulation (see Patiño and Sullivan 2002; Mylonas et al. 2010). Vitellogenic growth is marked by the uptake of liver-derived yolk protein precursors, vitellogenins. Ovarian follicle maturation consists of two stages in teleost fishes (Patiño and Sullivan 2002; Lubzens et al. 2010). The first stage is the acquisition of oocyte maturational competence (OMC), in which the follicle acquires the ability to produce and respond to the maturation-inducing steroid (MIS). The second stage is meiotic resumption, or oocyte maturation, the release of the oocyte from meiotic arrest in response to MIS produced by the somatic follicle cells. The MIS induces meiotic resumption by activating a ubiquitous maturation promoting factor in the oocyte (Nagahama 1997; Nagahama and Yamashita 2008). Although ovarian atresia can occur at any stage of gonadal development, of greatest concern to hatchery managers is usually ovarian atresia resulting from failure to initiate or complete follicle maturation following completion of oocyte growth (Sullivan et al. 2003). Assessing stage of follicle maturation is critical for predicting the timing of ovulation in females and for induction of spawning in females. Therefore, technologies have focused on identifying when follicles have completed oocyte growth or vitellogenesis, and also identifying stages of follicle maturation. The stage of maturation is often used to determine choice or timing of application of exogenous hormone therapies.

Approaches to monitoring reproductive maturation are similar to many of those used for sex identification. As with gender determination, often there are morphological changes in the animal that supply sufficient information about reproductive condition to allow the hatchery manager to properly manage the animals. Development of nuptial coloration and secondary sexual characteristics such as the kype in salmonids can be used to identify animals that will mature in the coming spawning season. Among the most commonly used indicators of reproductive condition are the swelling and feel of the abdomen with egg growth, swelling and increased redness of the genital papilla, number of eggs released or volume and viscosity of expressed milt upon application of abdominal pressure. Fish may also display behaviors indicating they or their partners are ready to spawn. The use of external morphological features and abdominal palpation is preferred and most commonly used for assessing maturation stage in most species. Comprehensive reproductive cycle profiles have been generated for many biomolecules for most major aquaculture species. Biomarkers such as changes in vitellogenin and steroid hormone profiles; primarily testosterone (T), estradiol-17β (E2), 11KT; measured in blood or mucus can be used to identify fish that have reached key stages of gonadogenesis. Sex steroids released into the water can even be measured as a means to assess the reproductive stage of a population of fish (Scott and Ellis 2007). Profiles for gonadotropins (GtHs) have been described for some species and changes in follicle stimulating hormone (FSH) and luteinizing hormone (LH) levels with approaching spawning have been described. Similarly, the MIS, primarily a progestin, has been identified in many species and its blood profile has been correlated with stages of follicle maturation. Nonetheless, most biomarker assays are either not commercially available or considered cost effective. As an example, decreases in T and E2, two steroids shared with mammals and for which commercial assay kits and assay services are available, were identified as early markers of the onset of atresia in sturgeon (Webb et al. 2001; Talbott et al. 2011). Monitoring changes in the steroids is a minimally invasive means to identify animals that are on the verge of atresia which compromises egg quality as caviar, yet monitoring the steroids is considered too expensive for industry to implement (Talbott et al. 2011).

Infrared spectroscopy has recently been applied to both biochemical analyses of plasma and follicles, and also noninvasive monitoring of reproductive maturation. Fourier infrared spectroscopy has been shown to simultaneously detect a myriad of biochemicals including steroids, vitellogenin, and lipid proteins in a follicle or plasma sample. When combined with principle component analysis, stages of vitellogenesis, follicle maturation, and atresia can be predicted with good accuracy (Lu et al. 2010, 2011). Short wavelength near infrared spectroscopy was first applied as a noninvasive means to monitoring maturation as well as gender identification in Chinook salmon (*Oncorhynchus tshawytscha*; Davis et al. 2006). More recently this approach has been used noninvasively as an abdominal scan to evaluate ovarian follicular atresia and thereby caviar quality in white sturgeon (*Acipenser transmontanus*) based on changes in intensity of lipid bands (Servid et al. 2011). Even if biomarker assays were cost effective, more immediately available information is usually preferred for selecting broodstock, spawners, and even fish for caviar harvest. To date, biomarkers have been and continue to be primarily used in species for which reproductive maturation and spawning procedures are being developed, but the assays do not seem to be maintained as part of routine hatchery use. Thus, although not usually critical in hatchery operations, they contribute significantly to the development of hatchery operations and practices used to control reproduction.

In hatchery operations where the use of morphological features and abdominal palpation are insufficient, gonadal size or even oocyte size are often used as first estimates of when fish will reach maturity, if they will mature in the coming spawning season, and if they are competent to respond to hormonal therapies to induce spawning. As with identifying gender, ultrasonography is the most popular imaging approach for monitoring reproductive maturation. Laparoscopy and endoscopy have been primarily applied to sturgeon culture where the stage of the eggs is critical for determination of caviar quality and therefore determining when to harvest (Hurvitz et al. 2007; Falahatkar et al. 2011; Webb and Doroshov 2011). Although ultrasonography is less sensitive than endoscopy or laparoscopy, it appears to be increasing in popularity particularly since invasive methods can be destructive to the developing tissues, gonads, and ducts, or have negative impacts on egg quality or spawning success due to the stress of the procedures (Webb and Doroshov 2011; Novelo and Tiersch 2012). In addition, ultrasonography may be the only imaging option in smaller fishes. Noninvasive measuring of not only gonad size but even follicle diameter by ultrasonography is possible in some species at more advanced stages of egg growth. Furthermore, the images can be easily stored as electronic records for later analysis or comparison with future measures of the same individuals. As with the use of ultrasonography for sex determination, ultrasonography has been used to monitor reproductive maturation in a wide variety of species (e.g., Mattson 1991; Shields et al. 1993; Blythe et al. 1994; Martin-Robichaud and Rommens 2001; Moghim et al. 2002; Petochi et al. 2011).

Assessment of reproductive maturation is most routinely conducted later in the spawning season to identify female fish close to ovulating. This includes selection of animals for natural spawning and also for hormone-induced breeding. Thus, simple biopsy procedures as described above involving catheterization through the genital pore and ducts continues to be the most practical means of assessing maturation stage when the use of external morphological features are not sufficiently informative. In addition to being technically simple and quick, the collection of the biopsy allows the measurement of not only follicle diameter but also macro- and microscopic examination of the oocytes for characterizing stage of follicle maturation based on cytoplasmic changes, nucleus or germinal vesicle migration and breakdown, and signs of atresia. Obtaining a sample also allows for biochemical and transcript analysis, and even performance of bioassays designed to assess competence to respond to hormone treatments. The same advantages in terms of analyses of samples also exist for samples surgically biopsied as in sturgeon. Because assessments of later stages of maturation are closely tied to induced-spawning procedures, we will discuss the use of these markers and measures in section "Induced Maturation and Spawning."

Artificial Insemination

Artificial insemination is not only a routine technology in hatchery operations for many species but it is also often the only means to obtain fertilized eggs. External fertilization combined with high fecundity in fish makes artificial insemination or strip-spawning very efficient. Timing of several events associated with spawning are critical in artificial insemination; eggs eventually and often quickly become overripe after ovulation and diminish in vitality, and sperm is activated almost immediately upon contact with culture water and activity usually diminishes within minutes. Strip-spawning was first introduced in the 1700s and procedures published in 1773 by Ludwig Jacobi in Germany (Nash 2011). Ovulated eggs and milt were simultaneously stripped from brown trout under water. In the mid-1800s hatcheries began being built throughout Europe and North America as artificial insemination methods were expanded to many freshwater species including rainbow trout in the US. Hatchery production soon led to global distribution of certain species such as dissemination of rainbow trout throughout Europe starting in 1879, and introduction of carp from Germany to California in 1872. The "dry method" was also developed about this time in Russia by Vrassky and Knoch (Nash 2011). The "dry method" involves stripping the gametes separately into dry containers, then mixing the gametes before adding water. The "dry method" or variations of the "dry method" are still primarily used today.

Successful artificial insemination is dependent upon the synchronization of the activation of the sperm and contact with the egg to allow the sperm to locate and traverse the egg micropyle. Thus, most of the technologies for artificial insemination center on activation of the sperm. Species-specific diluents that serve as sperm immobilizers, sperm activators, and sperm extenders are developed and even commercially available for some species (e.g., IVM, France; Cryogenetics AS, Norway; Alavi and Cosson 2005, 2006). In certain cases sperm is obtained from surgically removed testes instead of being obtained as expressed milt. Such a case is the use of sperm from sex-reversed rainbow trout. Genetic female rainbow trout sex-reversed with androgens to make phenotypic males (neomales) usually do not form functional ducts for sperm release (Bye and Lincoln 1986; Feist et al. 1995). The sperm stripped from surgically removed testes are usually not fully matured (Geffen and Evans 2000). In this case the milt is incubated in a high-pH solution to increase the competence of the sperm, before activation with culture water (Kobayashi et al. 2004). In some species such as catfishes, secretions from the seminal vesicle have been shown to participate in maturation of the sperm (Chowdhury and Joy 2007). In species such as the carps, a gel coating on the eggs must also be removed with dissolving solutions after insemination (e.g., Billard et al. 1995).

Fish gametes can often be chilled to extend in vitro viability. Chilled storage slows development of the gametes and also inhibits bacterial growth which can block the egg micropyle. Sperm of most species, including warm water species, can be stored for days or weeks at temperatures at or just above 0 °C with minimal impact on viability (see Bobe and Labbé 2009). Oocytes are much less amenable to chilled storage. Eggs of salmonids are perhaps the most amenable to cold storage and can be stored for 2 weeks post-ovulation or longer, and as with most species, do better between 0 and 4 °C than at the optimal temperature for the female (Withler and Morley 1968; Jensen and Alderdice 1984). Appropriate species-specific diluents approximating the composition of seminal fluids or coelomic fluids are often used for storage and dilution, and can even be critical. Cryoprotectants at these temperatures are not necessary although shown to be beneficial in some cases. Addition of antibiotics to diluents or even undiluted gametes has been recommended (Bobe and Labbé 2009). Although the metabolism is slowed at the reduced temperatures, respiration is still occurring so attention must be paid to gas exchange and available oxygen. Gametes are therefore usually stored at minimal depths and under oxygen enrichment.

Induced Maturation and Spawning

Many species of fish do not complete reproduction under production conditions for grow-out. The basis for this failure is likely missing environmental or social cues, and possibly stress (see Zohar and Mylonas 2001; Mylonas et al. 2010). In some situations it might be impossible or at least cost prohibitive to provide or mimic the requisite cues, such as the apparent need for a 5,000 mile spawning migration in European eels (*Anguilla anguilla*; Palstra and van den Thillart 2010) or for fish that spawn at great depths. In many cases providing appropriate conditions can be easily achieved. A good example is providing grass mats as substrate for egg deposition in crucian carp (*Carassius* species) and common carp culture, or milk jugs as nest sites in channel catfish culture. Without the spawning substrate the mats provide or the nesting site the jugs provide, the fish will not release gametes. An abrupt change in temperature is used to induce or facilitate spawning in some species such as gilthead seabream (Colombo et al. 1989). Many fish require the presence of the opposite sex to complete maturation or ovulation. In such cases fish previously separated by sex that are then judged to be sufficiently mature, are placed together for synchronized egg production or selective breeding. Failure to spawn in the grow-out conditions is often advantageous in that it prevents over population in production systems and allows greater control by the hatchery manager in terms of when fertilized eggs or fry are made available, which individuals are crossed for selective breeding, or even greater ability to distribute gametes among breeding partners.

Hormonal Intervention

Exogenous hormone treatments are valuable reproductive aides in fish culture. A primary use is to overcome impediments to reproductive maturation and spawning in those species that will not otherwise spawn in captivity. Even in species that spawn in captivity such as crucian and common carp, gilthead seabream, European sea bass (*Dicentrarchus labrax*) or channel catfish; the use of hormones to synchronize egg production, increase the rate of successful spawning, or increase flexibility in hatchery operations, is common (Jhingren and Pullin 1985; Barbaro et al. 1997; Moretti et al. 1999; Forniés et al. 2001; Chatakondi et al. 2011). Hormone treatment is most often used to increase volume and decrease viscosity of milt in males and induce final stages of ovarian maturation such as follicle maturation, ovulation, and oviposition in females. Only limited success has been achieved in advancing gonadal development in female fish that undergo gonadal arrest at much earlier stages in captivity.

Gonadotropin or gonadotropin releasing hormone (GnRH)-based therapies are predominantly used in fish culture. Among the greatest strides in the expansion of aquaculture was the use of homogenized carp pituitaries to spawn carps, particularly what are referred to as the Chinese carps; grass carp, silver carp, and bighead carp; and the Indian major carps; catla (*Catla catla*), rohu (*Labeo rohita*), and mrigal carp (*Cirrhinus mrigala*) that would not spawn naturally in captivity, but also the crucian and common carp (Chaudhuri and Alikunhi 1957; Jhingren and Pullin 1985). This process of hypophysation was developed in the 1930s in Brazil with fresh pituitary tissues from mature donors of the same species (Houssay 1931; von Ihering 1935, 1937). The technique was expanded to other species and other sources of GtH such as acetone dried pituitaries, pituitary extracts, and human chorionic gonadotropin (hCG) which greatly increased the accessibility of GtH preparations. Furthermore, like hCG, pituitary extracts are often calibrated for efficacy which improves the ability to predict or predetermine more precisely when the fish will ovulate.

The aquaculture industry was quick to incorporated GnRH into induced-spawning procedures shortly after its discovery in 1971 (Matsuo et al. 1971; Burgus et al. 1972). GnRH and its analogs were first used alone as injectates (Hirose and Ishida 1974; Anon. 1977; Donaldson et al. 1981), followed shortly thereafter by its incorporation into long acting cholesterol pellet implants (Weil and Crim 1983; Crim and Glebe 1984). The "Linpe" method developed in the mid-1980s combined GnRH with a dopamine antagonist (Lin and Peter 1986; Peter et al. 1988). This method is still pretty much what is used today although there have been some modifications in terms of the GnRH analogs (GnRHa) used and also innovations in design of the implant devices such as microencapsulation (Mylonas et al. 1995; Mylonas and Zohar 2000).

Carp and salmon crude pituitary homogenate extracts are still used routinely as a source of exogenous GtH. Some of the other commercially available GtH and GnRH-based products that are at least conditionally approve for induction of spawning of food-fishes in the USA or some European countries include Chorulon®

(Intervet International B. V. Boxmeer, the Netherlands), which is hCG; an injectate, Ovaprim®, and an implant, Ovaplant® (Syndel Laboratories LTD. Qualicum Beach, BC Canada), which are a combination of a salmon GnRHa and the dopamine inhibitor domperidone; Ovopel (Unic-Trade, Hungary), which is a combination of a mammalian GnRHa and the dopamine receptor antagonist metoclopramide; and GonazonTM (Intervet International B. V. Boxmeer, the Netherlands), a GnRHa. In general, the GtH products are being replaced by the use of GnRH-based products because of cost, they do not generate an immune response, they don't present a disease transmission risk, they allow more modulation by the fish's endocrine system, and the availability of long-lasting implants (see Mylonas and Zohar 2000).

As mentioned in a previous section, success in the use of hormones to induce spawning is tightly tied to correct staging of maturation. The hormone therapy and expected time of ovulation depend on the stage of maturation of the fish at the time of examination. Many criteria may be used to determine stage of maturation for induced spawning. As mentioned, external signs include swelling and color of the genitalia, and swelling and firmness of the abdomen are often all that are used. Oocyte diameter usually based on biopsy is sometime sufficient. In many species more information is needed of the biopsy sample. Examination of a biopsy sample, sometimes requiring chemical clearing of the yolk or even boiling of the oocyte, is often needed to determine if the oocyte has completed vitellogenic growth and initiated follicle maturation. Recognizing an oocyte has completed vitellogenic growth is difficult. Unless there are signs that follicle maturation has been initiated, oocyte diameter is often the only option. Unfortunately, fish often arrest just after completion of vitellogenesis because failure to spawn in captivity is usually due to a failure of the LH surge required to initiate follicle maturation, ovulation, and spawning (Zohar and Mylonas 2001). A primary concern is treatment of fish with exogenous hormones before they have completed vitellogenesis can often disrupt reproduction and render the fish unspawnable (Sullivan et al. 2003; Mylonas et al. 2010). Likewise, disruption of normal hormonal signaling at any point during gonadal development will often result in the onset of atresia. Once a fish has initiated follicle maturation, spawning options are much more reliable. Markers for the initiation of follicle maturation, specifically oocyte maturation, include changes in yolk granule and oil droplet coalescence and migration of the oocyte nucleus or germinal vesicle towards the periphery of the oocyte. The migration of the germinal vesicle is often used as the principal indicator of hormone treatment or when to check the fish for ovulation after treatment. In sturgeon for example, hormone treatment for induction of ovulation is based on what is called the oocyte polarity index which is a measure of germinal vesicle migration (Chapman and Van Eenennaam 2007a). In vitro response of biopsied follicles to hormones can also serve as an indicator of the response of the fish to in vivo hormone treatment and of egg quality (Lutes et al. 1987; Weber et al. 2000; Chapman and Van Eenennaam 2007b).

Many factors contribute to the selection of a hormonal therapy for induction of spawning (see Mylonas et al. 2010). Although primary is the stage of maturation at which the fish is arrested, considerable species-to-species variation also exists in terms of how fish respond to different hormones at the different stages of maturation. Mode of reproduction is one factor that affects response, due in part to the pattern of GtH release required for optimal egg production. Fish can be synchronous spawners, spawning once in a lifetime; single-batch group-synchronous spawners, spawning a single batch of eggs once a year; multiple-batch group-synchronous spawners, spawning multiple batches of eggs during a spawning season; or asynchronous spawners. The suppressive role of dopamine on GtH release varies considerably among species (Dufour et al. 2010). Heterologous hormones have different potencies and receptor specificity among species (Aizen et al. 2012). As an example hCG is recognized by the LH-receptor of many fish species and therefore works well in them as an ovulating agent, but is hardly recognized by GtH receptors in salmonids.

As in the use of biomarkers in establishing hatchery procedures for reproductive control, hormone therapies were and are used in the establishment of broodstocks for new species and sometimes later found to be unnecessary due to improved conditioning practices or domestication. Different hormone therapies are often required even for the same species under different conditions, often corresponding with the maturation of an industry. The striped bass (*Morone saxatilis*) is such an example. The hybrid striped bass industry is a young industry that has a selective breeding program based on captive breeders and also depends on wild adults captured along the spawning migration (Sullivan et al. 1997; Garber and Sullivan 2006). The industry started in the 1980s by using hCG to spawn wild fish that had initiated oocyte maturation as indicated by the coalescing of the oocyte oil droplet (Harrell et al. 1990). A single female can produce over two million eggs (Jennings et al. 2005) so it did not take many fish to supply a farm. Striped bass need to be strip-spawned within a 1 h window after ovulation to avoid overripening. Careful monitoring of the oocytes for stages of maturation together with careful monitoring of reproductive behaviors is required for proper strip-spawning of the females. Next, it was found that some earlier-stage wild fish with eggs devoid of signs of oocyte maturation could be induced to spawn with a combination of a slow release and faster release GnRH pellet, sometimes followed by an hCG injection (Hodson and Sullivan 1993). Domestic broodstocks were developed from reconditioning wild fish and progeny of initial spawns. Most early generation fish failed to complete maturation in captivity and would undergo atresia shortly after completing vitellogenesis. Moreover, considerable variability that existed among the females as to the size of fully grown eggs made it difficult to determine when the females were responsive or eligible for GnRH implant treatment. Many females were lost to either treatment being too early or too late. It was later determined that fish whose biopsied follicles responded to a combination of insulin-like growth factor-I (IGF-I) and the MIS in vitro, were fish that were competent to respond to the GnRH treatment in vivo (Weber et al. 2000). After several generations, the captive striped bass progress to later stages of follicle maturation before needing hormone treatment and are readily identified for GnRH implants or hCG injection.

Only limited success has been achieved in advancing gonadal development in female fish that undergo gonadal arrest at stages earlier than the completion of vitellogenesis in captivity such as at early vitellogenesis in the milkfish (Lee et al. 1986a, b) and even pre-vitellogenesis in the Japanese eel (*Anguilla japonica*; Ohta et al. 1997; Kagawa et al. 2005). In the milkfish GnRHa implants together with T implants advanced maturation and induced spawning, and in the Japanese eel salmon pituitary extract (SPE) injections followed by a combination of SPE and the MIS injections led to successful spawning. Steroids have been shown to be potent stimulators of gonadal development in males of some species. In grey mullet (*Mugil cephalus*) for example, 17α-methyltestosterone (17MT) has been shown to induce spermatogenesis and spermiation in what are considered undersized fish, in fresh or seawater, at any time of year, and a single 10 mg implant can keep a male spermiating for up to a year (Weber and Lee 1985; Lee and Weber 1986; Lee et al. 1992).

Photothermal Manipulation

Particularly in temperate species, seasonally changing photoperiod is generally considered the primary environmental factor cueing reproduction in fish, with temperature also often having an important influence. Photoperiod or photothermal conditioning is used to simulate optimal natural conditions to maximize reproductive success, synchronize spawning, extend the spawning season, induce fish to mature and spawn out-of-season, mature and spawn nonnative species, compress reproductive cycles, or even inhibit reproduction (see Bromage et al. 2001; Migaud et al. 2010; Wang et al. 2010). Light treatments are most common due in part to the low cost and ability to apply the treatment to large facilities including grow-out facilities. The use of photoperiod and temperature manipulations are used in commercial hatcheries for many species such as gilthead seabream and European sea bass to induce in-season and year-round spawning (Carrillo et al. 1989; Moretti et al. 1999). A shortened photothermal cycle is sometimes used in rainbow trout aquaculture to remature fish that had spawned at 2 years of age, to remature out-of-season at less than 3 years of age and thereby extend the availability of eggs (Bromage 1995). Continuous lighting during 6 months of the second year of growth-out of Atlantic salmon has been shown to inhibit premature gonadal maturation for the purpose of improving growth rate and flesh quality (Porter et al. 1999; Endal et al. 2000; Leclercq et al. 2011). Preliminary studies support continuous lighting at critical times of the year can also be used to reduce early maturation in other species including Atlantic cod (*Gadus morhua*) (Taranger et al. 2006; Cowan et al. 2011) and European sea bass (Beghtashi et al. 2004; Felip et al. 2008). Timing of application, location of lighting, as well as light intensities and wavelengths are all important factors in achieving desired results on reproduction without having detrimental effects on fish health and behaviors important for optimal grow-out (Vera and Migaud 2009; Vera et al. 2010; Cowan et al. 2011; Leclercq et al. 2011; Oppedal et al. 2011).

Low temperatures can also be used for slowing maturation and preventing atresia in some species. Male and female white bass (*Morone chrysops*) and striped bass, used in the production of hybrid striped bass, can be maintained at temperatures several degrees below normal spawning temperatures to extend the spawning season by several months (Sullivan et al. 2003). This practice is referred to as coldbanking. Fish that had completed vitellogenic growth before coldbanking are warmed up to spawning temperatures for induced spawning as needed.

Sterilization

Sterilization can be an important form of reproductive control to many aquaculture industries although few have found a reliable cost-effective means to achieve it. Sterilization can be used to prevent over population in grow-out facilities such as with tilapias; inhibit the negative impacts of gonadal development on growth, flesh quality, and immunodepression such as in salmonids; achieve genetic containment for environment protection due to reproduction of exotic or genetically distinct escapees such as tilapias, carps, and salmonids; and for property protection purposes such as products of selective breeding programs or genetically modified organisms (GMOs). Although many procedures to induce sterilization are possible, including androgen immersion of fry (Piferrer et al. 1994) and irradiation of eggs by cobalt-60 irradiation (Konno and Tashiro 1982; Villarreal and Thorpe 1985), only interspecies hybridization and induced polyploidy are currently in use. The impact of hybridization on fertility results primarily from problems with gonad development and chromosome pairing (Bartley et al. 2001). Other than for cyprinids, few hybrids are raised specifically for the purpose of sterility, although reduced reproductive capacity of some hybrids may contribute to improvements in performance traits such as growth (see Bartley et al. 2001).

In cyprinids, interspecific hybridization can lead to triploid sterility or functional polyploidy depending on the species crossed (Cherfas et al. 1994b; Gomelsky 2003; Liu et al. 2001, 2007). Species of carp differ in chromosome number due to an evolutionary genome duplication event. Ploidy of offspring derived from hybridization is dependent upon phylogenetic distance of the species and whether they have a duplicated genome. Most notably, distant hybridization between female red crucian carp (*Carassius auratus* red variety) and male common carp (*Cyprinus carpio* L.) has been shown to result in unreduced gametes in the second-generation animals, thought to be related to premeiotic endoreduplication, endomitosis, or fusion of germ cells (Liu 2010). This distant hybridization has been used to establish stable allotetraploid lines of the hybrid, which being fertile, can then be crossed with diploids to produce sterile all-triploid progeny (Liu et al. 2001; Luo et al. 2011).

The use of physical means of inducing polyploidy, primarily by the application of a pressure

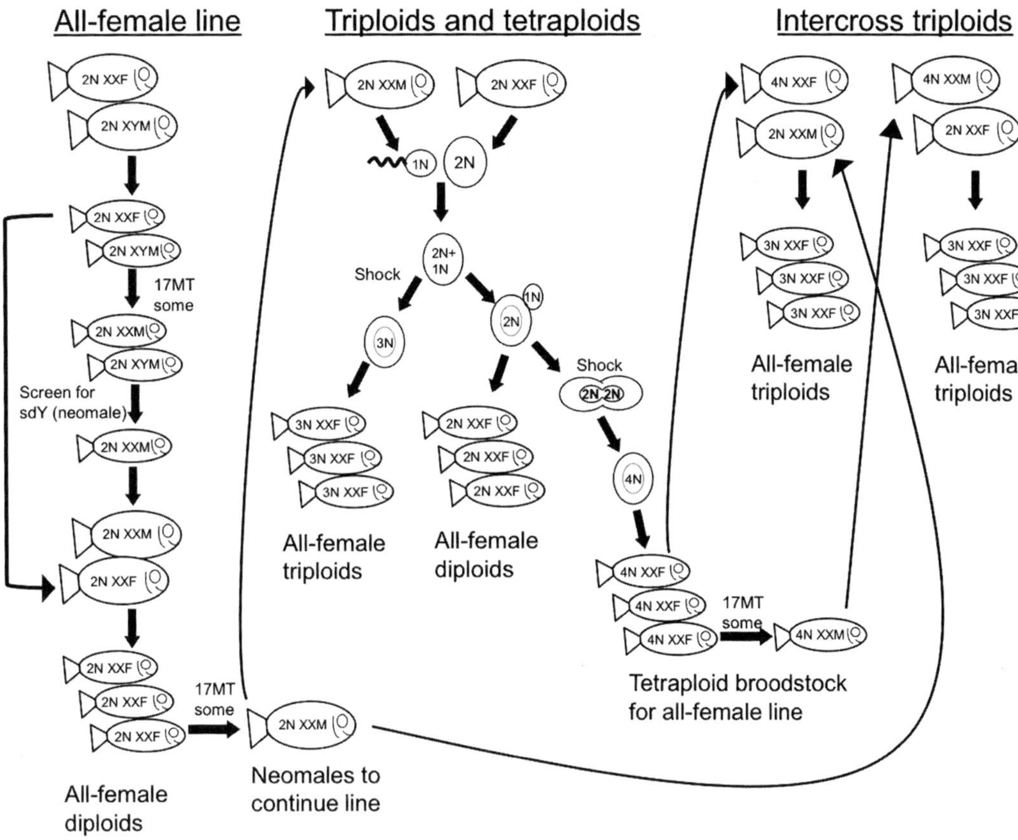

Fig. 3.2 Manipulation of gender and ploidy in rainbow trout. Shown are possible approaches for (1) creating all-female lines of diploid (2N) trout, (2) using all-female lines of trout to make all-female triploid (3N) and all-female tetraploid (4N) trout, and (3) using all-female 2N and 4N trout to make all-female 3N trout. Under (1) All-female line, 2N genetic and phenotypic females (XXF) and 2N genetic and phenotypic males (XYM) are crossed, and the offspring are fed 17MT as fry to sex reverse XXFs into genetic females but phenotypic males (XXM), also known as neomales. The neomlaes are identified by their failure to develop functional ducts or by PCR for genetic markers such as sdY. The neomales are crossed with normal XXF animals to produce all-female progeny. Some of the progeny are fed 17MT to make the next generation of neomales to continue the line. Under (2) Triploids and tetraploids, all-female 3Ns or all-female 4Ns are produced by first crossing 2N females (2N XXF) with 2N neomales (2N XXM). The zygotes are either left to develop untreated to become all-2N XXFs, shocked to cause retention of the second polar body to become all-3N XXFs, or shocked to prevent cell cleavage of the embryos following chromosome duplication to produce all-4N XXFs. Some 4N XXFs are fed 17MT as fry to make 4N-neomales (4N XXM) to continue the all-4N XXF line, or be used to make intercross triploids. Under (3) Intercross triploids, all 3N XXFs are produced by crossing either 4N XXFs with 2N XXMs or crossing 4N XXMs with 2N XXFs

or temperature shock to disrupt the expulsion of the second polar body, is more widely used among species in aquaculture than biological methods through hybridization (see Fig. 3.2). Physical induction of triploidy has been generally regarded as the most effective means for producing sterile progeny for most aquaculture industries (see Benfey 1999; Piferrer et al. 2009). Triploid induction by physical means has been used in recreational fisheries to prevent breeding of nonnative species (e.g., grass carp, Pierce 1983; Zajicek et al. 2011) or stocked fish with native species (e.g., salmonids; Loopstra and Hansen 2008, 2010) for many years and is part of routine practices in many hatcheries for restocking purposes in the US. Production of induced triploids is inexpensive, technically simple, survival through hatching is similar to that of diploids, and induction rates approaching 100 % are routinely observed (Devlin et al. 2010).

Furthermore, high-throughput approaches for ploidy verification using flow cytometry of erythrocyte nuclear DNA content, or by measurement of erythrocyte size using a *Coulter* counter, are developed for when verification of triploidy is required (Thorgaard et al. 1982; Wattendorf 1986; Devlin et al. 2010). The use of all-female triploid fish has been adopted by some segments of the rainbow trout food-fish industry primarily for maintaining flesh quality when raising fish to larger sizes, particularly for fillet markets, and to a lesser extent for increasing growth rate and efficiency when fish are raised through spawning age (see Benfey 1999; Piferrer et al. 2009). In rainbow trout, as in many fish, female triploids are raised because, owing to the difference in the timing of meiotic initiation between the sexes, ovarian growth is severely inhibited whereas testicular growth is not (Lincoln and Scott 1983; Krisfalusi and Cloud 1999).

Physical induction of tetraploidy has been developed in some species including rainbow trout by suppressing early cell cleavage following chromosome doubling (Chourrout 1982, 1984; Zhang and Onozato 2004). As with biological tetraploids, induced tetraploids can be crossed with diploids to generate sterile triploid progeny (see Fig. 3.2). There have been problems encountered with the approach which has impeded the adoption of induced tetraploidy by industry. Most species do not appear to survive as tetraploids, and even in those that do survive, survival is low and deformity rates high. Similarly, fertility of first-generation tetraploids and survival of progeny are also low due to the sperm head often being too large to penetrate the egg micropyle, and to poor egg quality (Chourrout et al. 1986; Arai 2001; Weber and Hostuttler 2012). Egg quality, however, is greater in second-generation tetraploids and sperm size is heritable allowing for selection of males that produce sufficiently small sperm (Blanc et al. 1993; Weber and Hostuttler 2012). Unpredicted levels of euploidy and aneuploidy have also been reported in progeny of induced tetraploid rainbow trout females, attributed to failure in chromosome reduction of the ova (Chourrout and Nakayama 1987). Nevertheless, unexpected levels of euploidy were not observed in an additional study in progeny of induced tetraploids, or second-generation tetraploids derived from crossing first-generation tetraploids (Weber and Hostuttler 2012). In addition, diploid/tetraploid mosaicism was reported based on erythrocyte size in progeny of crosses between second-generation tetraploids and diploids (Arai 2001). Considering the relative ease and high induction rate of the direct method of making triploids, making triploids from crosses of tetraploids with diploids has little advantage if the offspring are not 100 % sterile triploids or at least exhibit a significantly higher rate of sterility. The use of tetraploidy has been commercialized in Japan based on work at the Nagano Prefectural Fisheries Experimental Station, where female tetraploid rainbow trout are crossed with diploid neomale brown trout to produce what is referred to as a "Shinshu salmon," which are sterile (see Sakao et al. 2006; Denda 2007).

Gender Control

Technologies for controlling gender are widely applied in aquaculture for maximizing production efficiency and product quality, and for restricting reproduction. In most species one sex has a culture advantage over the other. Almost all cultured fishes display sexually dimorphic growth including the carps, salmonids, tilapias, and catfishes. In addition, in most of these species, differences in growth and feed efficiency are due in part to one sex maturing earlier than the other within the production cycle. Many objectives of gender control overlap with those of sterilization. Grow-out of a monosex population has many of the benefits of sterilization including preventing overpopulation in the grow-out facilities, avoidance of negative impacts of gonadal maturation on growth and flesh quality, and protecting the environment from establishment of exotic species assuming inability to hybridize with native species. As examples, in tilapia all-male populations are farmed to prevent overpopulation and loss of growth to reproduction, but also males grow faster than females. On the other hand, all-female rainbow trout populations are farmed because females grow slightly faster than males, males often display premature maturation resulting in reduced growth and flesh quality, and gonad development is inhibited only in triploid females. Thus, in both species monosex populations have multiple advantages.

Several approaches can be used to control sex depending on species and objective. The labile nature of sex determination in fishes and the role of sex steroids in the process are at the basis of the most widely used methods for gender control (see Devlin and Nagahama 2002). The ability of exogenous sex steroid treatments to affect phenotypic sex in most fish species is well established and used in two general approaches for endocrine control of gender, direct and indirect hormone-induced sex reversal. Direct sex reversal, either masculinization with androgens or feminization with estrogens during early development, is the simplest. Direct masculinization is widely used for monosex production of male tilapia (Phelps 2006). An example treatment is to feed larval tilapia 17MT-treated feed at 60 mg 17MT/kg feed (or 9 mg 17MT/kg body weight/day) for 28 days.

Indirect methods of hormone-induced sex reversal are used in species with heterogametic sex determination. An indirect approach to all-female production is routinely used in rainbow trout in which the homogametic sex (XX) is female and the heterogametic sex (XY) is male. Female fry are treated with androgens (e.g., 2 mg MT/kg feed 60 days starting at first feeding) to make neomales, which are then crossed with females to make all-female progeny (see Fig. 3.2). As previously mentioned, gender control is also used in combination with triploidy in rainbow to avoid gonadal growth. A similar approach can be used even if the desired sex is the heterogametic sex. Such an indirect approach to endocrine control of gender was used to develop a now commercial line or lines of "YY" supermale Nile tilapia (Mair et al. 1997; GMT®, Fishgen, LTD). The male is the heterogametic sex (XY) and female the homogametic sex (XX) in Nile tilapia. The continuation of YY supermale lines requires feminization of YY supermales as broodstock. Similar approaches are being applied to other species such as the recent development of "YY" supermales for all-male production of yellow catfish (*Pelteobagrus fulvidraco*) which is increasing rapidly in popularity in China (Liu et al. 2012). Advantages to the indirect approach are reduced hormone use since a single treated broodstock animal can usually generate thousands of offspring, and the consumer product is not exposed to exogenous hormone treatment. Unfortunately each of the procedures described above are not 100 % effective so gender control alone is not sufficient to insure reproductive containment. Sex determination in tilapia is complicated and is well established to be influenced by environment, particularly temperature, as well as autosomic influences (Scott et al. 1989; Baroiller and D'Cotta 2001). Occasional males are also observed in all-female lines of rainbow trout, thought to be the result of a recessive autosomal mutation (Quillet et al. 2002).

In some cultured species sex ratios are sufficiently responsive to environment and social interactions so that these factors can be exploited to shift sex ratio in a favorable direction. Early rearing temperature can significantly affect sex ratio even in species with heterogametic sex determination. In general, sex ratios are shifted towards males with higher temperatures and towards females with lower temperatures (see Baroiller et al. 1999; Sandra and Norma 2010). Gender in most commercially important tilapine species, for example, can be affected by temperature. *Oreochromis niloticus* (Wang and Tsai 2000) and *O. mossambicus* (Baroiller et al. 1999) which both have the XY system of sex determination, and *O. aureus* (Desprez and Mélard 1998) which has the ZW system, all increase in male proportion when exposed to elevated temperatures during early development. Channel catfish, also with a XY system, are an exception in that higher temperatures increase the proportion of females (Patiño et al. 1996). Nevertheless, temperature effects do not appear to be a major concern or an often exploited asset in tilapia or channel catfish farming.

Temperature effects on sex ratio are a concern in production of European sea bass and many flat fishes such as the Japanese flounder (hirame;

Paralichthys olivaceus) where sex determination appears more polyfactorial and weighted more heavily towards temperature determination (Tabata 1991; Pavlidis et al. 2000; Luckenbach et al. 2009). In both sea bass and Japanese flounder, females grow faster than males and males also exhibit premature maturation further slowing growth and impacting product quality (Yamamoto 1999; Saillant et al. 2001). In European sea bass sex ratio is heavily skewed towards males at warmer temperatures preferred for early rearing due to improved growth. Thus, early growth must be sacrificed to obtain more acceptable sex ratios at lower temperatures which still only approach 1:1 (Navarro-Martín et al. 2009). Standard approaches to direct but not indirect hormone-induced feminization are possible in both groups (Gorshkov et al. 2004; Yamamoto 1999), but industry prefers to avoid hormone treatment in these premium market products. Nevertheless, temperature control of sex has been integrated with indirect endocrine methods of gender control to produce all-female Japanese flounder. In all flounder of the *genus Paralichthys*, sex ratio is skewed towards males at both high and low temperatures and at best a 1:1 ratio can be obtained at intermediate temperatures (Luckenbach et al. 2009]. Similar to approaches used with rainbow trout, homogametic females (XX) are sex-reversed with 17MT and these neomales are crossed with normal females to make all-genetic female offspring. Due to the unstable nature of the XX animals in terms of phenotypic sex, fry for seedstock are reared at strict intermediate temperatures during the labile period to insure all-females, and at higher temperatures to get neomales without additional hormone treatment for continuation of the line (Yamamoto 1999).

Hybridization among certain species is used to produce monosex or skewed sex ratios. Certain crosses among species of tilapia with heterogametic male system (XY) and the heterogametic female system (WZ) result in male-skewed sex ratios that sometimes approach 100 % (Wohlfarth 1994; Hulata 2001). In addition, certain species and subspecies of carp exhibit natural androgenesis (all-paternal inheritance) and gynogenesis (all-maternal inheritance). Hybridization involving gynogenic species is taken advantage of in aquaculture to produce all-female lines with improved growth performance (Wu 1990; Cherfas et al. 1994b; Hulata 1995). The most popular cross is between the female crucian carp and male common carp (Wu 1990; Hulata 1995).

Induced gynogenesis and androgenesis are also practiced in aquaculture but primarily for research purposes (see Thorgaard 1986; Nichols 2009). These forms of chromosome manipulation, like inducing polyploidy, are particularly easy in fish with external fertilization, and there are many approaches (see Thorgaard 1986; Pandian and Koteeswaran 1998). In general, when used in aquaculture, the genetic materials of the egg or sperm are first inactivated by irradiation, usually UV-irradiation. The treated gametes are then used to activate (sperm) or are activated (egg) by the untreated gametes of the opposite sex. In gynogenesis, irradiated sperm from a different species is often used to insure the surviving embryos are gynogens. Chromosome number is restored in meiotic gynogenesis by shocking the fertilized egg to cause retention of the second polar body or restored in mitotic gynogenesis and androgenesis by preventing cell cleavage following chromosome replication. Androgenesis and gynogenesis are used in determining mode of sex determination, which is critical information for developing the more practical means of gender control used in production as already mentioned. Furthermore, gynogenesis based on maternal chromosome doubling has been used to initiate all-female lines of fish and for developing clonal lines used in aquaculture (see Arai 2001; Komen et al. 1991). The means of gender control we previously described for Japanese flounder (Yamamoto 1999) was developed with clonal lines produced through the use of gynogenesis.

Cryopreservation

Methods for cryopreservation of sperm have been developed for most major aquaculture species (see Cabrita et al. 2010). In addition, technologies such as ASMA (computer assisted sperm morphology analysis) and CASA (computer assisted sperm analysis) systems for evaluating sperm before and after cryopreservation are validated for fish and are available to increase the quality of cryopreserved sperm (Cabrita et al. 2010; Beirão et al. 2011). Technologies for commercial-scale processing of fish sperm for cryopreservation have recently been established (Hu et al. 2011). Products to aid in fish sperm cryopreservation and their application, such as species-specific extenders, dilutors, and activators, are commercially available, as well cryopreservation services for Atlantic cod, halibut (*genus Hippoglossus*), sturgeons, and most salmonids (Haffray et al. 2008; IMV Technologies, France; Cryogenetics AS, Norway). Cryopreservation of oocytes and embryos on the other hand have been more difficult owing primarily to complications from their high water content, large yolk and lipid stores, and low membrane permeability (see Zhang and Lubzen 2009; Robles et al. 2009). Cryopreservation of blastomeres has been developed as an alternative method for embryo preservation and has been achieved in several species including the rainbow trout and common carp (Nilsson and Cloud 1993; Calvi and Maisse 1998, 1999). Nonetheless, cryopreservation of Atlantic cod embryos has recently been achieved (Cryogenetics AS, Norway).

Despite the development of technologies for sperm cryopreservation in fish, cryopreservation isn't widely used in commercial production even when in vitro fertilization (strip-spawning) is routinely employed (Caffey and Tiersch 2000; Haffray et al. 2008; Hu et al. 2011). Sperm is rarely in short supply when a species is being propagated to spawn during its normal spawning season and in many fish species, particularly freshwater species, the quality of cryopreserved sperm can still be suboptimal or unreliable (Horváth et al. 2003; Cabrita et al. 2010). Nevertheless, cryopreservation of sperm is particularly valuable and more likely employed in out-of-season spawning, hybridization, all-female culture, and selective breeding programs based on individual breeding values.

Transplantation

Transplantation and even xenotransplantation of a variety of reproduction-related cells or even organs is possible in fish. Early studies have shown the possibility of making germ-line chimeras in rainbow trout through transplantation of blastomeres into embryos (Nilsson and Cloud 1992; Takeuchi et al. 2001). The success of the procedure is actually improved in some species if cryopreserved embryos are used (Yasui et al. 2011). Primordial germ cell (PGC) transplantation was recently shown to have potential to breeding programs for yellowtail (*Seriola quinqueradiata*), a difficult to breed, high-value species that is the most commonly farmed fish in Japan. Yellowtail PGCs were transplanted from single individuals into multiple recipients that then generated gametes of the donors. This demonstration was proposed as a model for using PGC transplantation as a means to accelerate selective breeding by amplifying genetic material from superior individuals (Morita et al. 2012).

Transplantation of PGCs took a very interesting twist when it was found that PGCs from one species can be transplanted into the embryo of another, proliferate, and yield viable gametes of the donor species (Takeuchi et al. 2004; Okutsu et al. 2006, 2007; Yoshizaki et al. 2012). The procedure was termed surrogate broodstocking and was first demonstrated with trout PGCs transplanted into triploid masu salmon (*Oncorhynchus masou*) embryos. Procedures have been refined including the finding that spermatogonia can be used as transplant material and that they will dedifferentiate and then become oogonia in a female host (Okutsu et al. 2006; Kise et al. 2012). In addition, it's been shown cryopreserved PGCs or blastomeres can be used for transplants (Kobayashi et al. 2007; Kawakami et al. 2012) and PGCs can be transplanted into adults with chemically induced germ cell depleted gonads (Majhi et al. 2009; Lacerda et al. 2010). Studies have shown PGCs can be transplanted across species, genus, and even family barriers (e.g., Yoshizaki et al. 2005; Saito et al. 2008, 2010). Surrogate broodstocking technologies have been suggested to be valuable

well beyond this size range. Almost all of rainbow trout farming is all-female production with high levels of all-female triploid production in Europe. The use of triploidy is also increasing for salmonids released for recreational fishing in the USA to prevent interbreeding and to limit competition with wild native populations of species such as rainbow and brook trout.

All-female lines can be derived from induced gynogenesis followed by masculinization of the gynogens into phenotypic males or neomales (Thorgaard 1992). All-female lines are also developed by taking advantage of the fact that most genetic females treated with certain dosages of androgens develop testes but fail to develop functional ducts (Johnstone et al. 1979; Feist et al. 1995). This allows the identification of likely genetic females (neomales) among a population of genetic mixed sex fish. Thus, the all-female line is derived from breeding normal females with the androgen-treated animals that do not express milt upon the application of abdominal pressure (Bye and Lincoln 1986). This approach is not 100 % effective due in large part to difficulty distinguishing neomales from normal males that have not fully matured, and possible negative effects of the androgen treatment on duct development in genetic males. A true-breeding all-female line therefore usually requires a second generation following progeny testing for sex of the first-generation families. A Y-chromosomal DNA marker for sex (OtY1) has been used to identify neomales in development of an all-female line of Chinook salmon (Devlin et al. 1994a). Recently genetic markers have been developed for maleness in rainbow trout, one designated OmyY1 (Brunelli et al. 2008) and a second for the sdY gene (Yano et al. 2012), that should allow for the identification of the true neomales and greatly simply development of all-female lines in this species. A drawback to all-female production is since most neomales do not form functional ducts, the males must be killed to obtain sperm and the sperm must be matured by exposure to a high-pH diluent (Bye and Lincoln 1986; Miura et al. 1992; Feist et al. 1995). This is inefficient since the males cannot be used more than once, sperm yield is not as high as with expressed milt, and the killed fish may still be immature.

Interest in sterilization is increasing as more rainbow trout are being raised in sea cages and because there is increasing interest in raising fish to a larger market size, particularly for a fillet market or to compete with Atlantic salmon. Steelhead rainbow trout, a naturally anadromous strain, mature at a later age and size and are sometimes farmed when a larger fish is desired, and also in sea pens. All-female triploids are the animals of choice when sterile animals are required for food-fish production. Again, ovarian but not testicular development is inhibited in triploids. As with Atlantic salmon, performance of triploids in production has been found to be similar or inferior to that of diploids (see reviews, Benfey 1999; Piferrer et al. 2009). Triploidization in commercial production is induced with either a temperature or pressure shock to induce retention of the second polar body. Induction rates approach 100 %, often over 99.9 % (Devlin et al. 2010). In addition, highly accurate and high-throughput approaches to ploidy determination by flow cytometry have been demonstrated (Devlin et al. 2010). The use of induced tetraploidy for breeding with diploids for production of triploids has been demonstrated (Chourrout et al. 1986; Myers and Hershberger 1991; Fig. 3.2). The objective of this approach is to produce 100 % triploids for genetic containment and eliminate the need for analysis such as flow cytometry to certify fish as triploids. An additional advantage proposed is improved performance by eliminating the trauma incurred by the embryo by the induction shock (Myers and Hershberger 1991). However, efficiency of producing tetraploids and tetraploid-derived triploids has been low and progeny of tetraploids exhibited unexpected levels of euploidy and mosaicism in some studies (Chourrout et al. 1986; Chourrout and Nakayama 1987; Myers and Hershberger 1991; Blanc et al. 1993; Arai 2001; Weber and Hostuttler 2012). Sex ratio is also strongly skewed towards males in second-generation tetraploids (Chourrout et al. 1986). Moreover, evidence that tetraploid-derived triploids have a performance advantage over induced

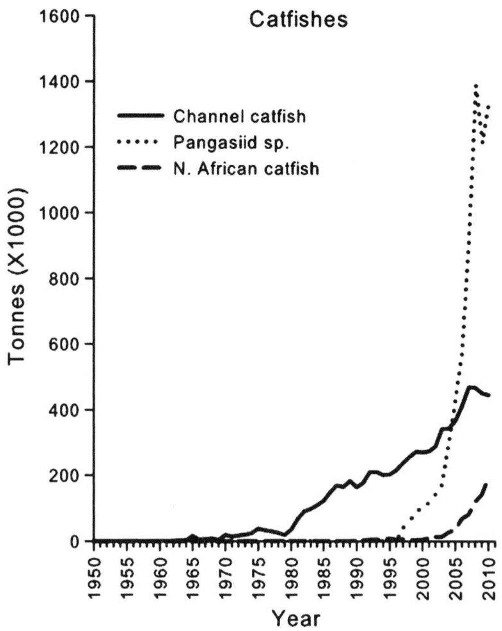

Fig. 3.6 Global aquaculture production of major farmed catfishes. Data are estimates from FAO Fisheries and Aquaculture Department, FishStat Plus software (version 2.3); data set Aquaculture Production 1950–2010 (release date: March 2012)

triploids is limited or has yet to be demonstrated for most traits (Arai 2001; Myers and Hershberger 1991; Denda 2007).

Catfish Aquaculture

The channel catfish is the primary aquaculture species in the USA, however, according to FAO (2012), in 2010 channel catfish accounted for only 13.5 % of global catfish production. Asian production of primarily *Pangasius* catfishes accounted for 73.7 %, and catfish from Africa, primarily the North African catfish (*Clarias gariepinus*; includes former *C. lazera*), accounted for 12.3 % of production (see Fig. 3.6). Channel catfish are almost exclusively grown in ponds. The channel catfish is a cavity spawner and reproduce readily in captivity in ponds only if provided with nesting sites. Hatchery production of channel catfish for stocking for recreational fishing and capture fisheries has been taken place in the USA for over 100 years, but farming catfish did not begin until the 1950s, and then rapidly increased in the 1960s and 1970s. China also began production of channel catfish in 1984, mostly in cages.

Most channel catfish fry production is conducted using broodstock ponds. The ponds are smaller than production ponds and shallow. A broodstock population is maintained over several years, replacing fish as they get older. Optimal fry production comes from fish 4 to 6 years old and 4 to 8 lb (Kelly 2004). Spawning in the broodstock ponds is controlled by the availability of nesting sites and water temperature. As pond temperatures rise to spawning temperatures, the manager will put out containers for nesting such as milk jugs. The male will take residence in the container; lure in a female for spawning, then after mating chase the female away while it guards the nest. The pond manager checks the containers every 3–4 days to collect the eggs for incubation in the hatchery. Removing the eggs also allows the nesting box to be available for another round of spawning. Seedstock are produced on site at grow-out farms or are purchased from commercial fingerling producers. As selective breeding programs have developed, purchase of fingerlings from commercial fingerling producers has increased. This has led to an increase in the use of pens within ponds that pair selected individuals for breeding, the use of aquaria spawning of channel catfish, and the use of hormone-induced spawning for more consistent and extended seed availability.

Production of the channel catfish female X blue catfish (*Ictalurus furcatus*) male hybrid has increased rapidly in recent years and may soon replace the pure channel catfish as the primary catfish farmed in the USA (see Kumar and Engle 2011. The channel catfish X blue catfish hybrid has been shown to be superior to channel catfish for a long list of traits including growth, yield, disease resistance, and low oxygen tolerance (Bosworth et al. 2004; Li et al. 2004; Dunham et al. 2008; Dunham and Masser 2012), but unfortunately the two species do not readily interbreed naturally or even after hormone injection (Dunham and Argue 2000; Dunham et al. 2000; Dunham and Masser 2012). The hybrids are primarily produced using artificial insemination (see Dunham and Masser 2012). The male catfish have small testes and do not express milt, thus requiring the surgical removal of the testis to obtain sperm. Procedures for hormone-induced spawning of channel catfish, including broodstock selection, have been developed (Phelps et al. 2012; Chatakondi et al. 2011; Dunham and Masser 2012). An additional problem with making the hybrid is the spawning seasons of the fish do not fully align. Procedures for high-throughput cryopreservation of blue catfish spermatozoa have recently been established to help alleviate shortages in blue catfish sperm due to both low testis weight and seasonality (Hu et al. 2011). Surrogate broodstocking is also being researched as a means to overcome breeding problems between the species, with attempts to transplant testicular cells of blue catfish into sterile male channel catfish (Small et al. 2011).

The production of catfishes other than ictalurids, particularly the pangasiids and clariids, has increased rapidly outside the USA. The pangasiids, *Pangasius hypophthalmus* and *P. boucourti*, also known as striped catfish and basa respectively, are raised primarily in Southeast Asia with Vietnam dominating production. The rise of the pangasiid aquaculture industry in the Mekong Delta of Vietnam has been one of the fastest rises of any animal production industry. Over the course of a decade it grew from a minor industry until in 2010 reaching over one million tonnes of production, employing 180,000 rural poor, exporting to over 100 countries, generating export income exceeding US $1.4 billion, and attaining an average production rate ranging from 200 to 400 tonnes per hectare (see De Silva and Phuong 2011; Bui et al. 2010).

Pangasiid production was based primarily on *P. boucourti* and wild seed until reliable hormone-induced spawning procedures were developed for *P. hypophthalmus* at the end of the 1990s (Cacot et al. 2002; Lazard et al. 2009). *Pangasius hypophthalmus* do not reproduce in captivity, but procedures to induce spawning using pituitary homogenates were achieved as early as 1966 (Potaros and Sitasit 1976). Unfortunately success rates were low due to difficulty identifying when females were ready for treatment since eggs of the same size can be at different stages in terms of responsiveness to hormone treatment, and because of early onset of overrippening, within 2 h of ovulation (Legendre et al. 2000). Only recently, in the 1990s, has mass production based on hormone-induced spawning been achieved (Legendre et al. 2000; Bui et al. 2010). Fish reproduce year round with a major peak from May to July, and the same individuals may be spawned multiple times (Bui et al. 2010). Fish are strip-spawned including the ability to strip milt from the males. Production is now based on captive broodstock with exchange among hatcheries and limited reintroduction from the wild (Bui et al. 2010).

The North African catfish is the major aquaculture species in sub-Saharan Africa (FAO 2012). Low levels of natural reproduction in captive fish is observed and allowed for a limited industry starting in the 1970s (De Kimpe and Micha 1974; Hogendoorn 1979). This carnivorous catfish was introduced in part as a predator to keep down overpopulation in tilapia ponds as tilapia culture was being promoted throughout the world for subsistence aquaculture (see De Graaf et al. 1996). Low levels of success with induced spawning were achieved in the late 1970s and 1980s with 11-deoxycorticosterone-acetate and with carp pituitary suspension using commercially available acetone dried and powdered carp pituitaries (De Kimpe and Micha 1974; Hogendoorn and Vismans 1980). Fish can be spawned year round and individual females can spawn multiple times, especially if maintained at higher temperatures, however males need to be sacrificed to obtain sperm (Hogendoorn and Vismans 1980; Huisman and Richter 1987; Viveiros et al. 2002). Larval rearing was also a problem early on in the industry (Hogendoorn 1980). Spawning was improved by standardizing the potency of the pituitary preparations, and the introduction of hCG and GnRH combined with dopamine inhibitors (De Leeuw et al. 1985; Huisman and Richter 1987). Nevertheless, there have been efforts to improve the use of environmental cues to induce natural spawning, particularly density, water temperature, and water depth manipulations to allow expansion of seed production in areas where a farmer might not have the resources for hormone induction (El Naggar et al. 2006).

Future Directions

Aquaculture is in flux in many ways. It is transitioning from being a distant secondary source for supplying food-fish to being the primary source. It is transitioning from an industry dependent on wild animals to one dependent on genetically improved lines. It is based primarily on semi-intensive production strategies but increasing in capabilities for high intensity. Production systems are expanding from ponds and flow-through raceways, to also include factory farms based on open ocean cage culture and fully recirculating land-based systems. Diets are transitioning from those based on fish meal and fish oil to plant-based diets. There has been a rapid expansion in the number of species identified as having farming potential. Reproductive technologies have made many of these options possible or have made the options more efficient. The future development and implementation of reproductive technologies in many ways depends on the courses aquaculture takes while at the same time will be fundamental in determining those directions.

Central to reproduction efforts are technologies for improved seed production. The application of GnRH-based technologies to allow captive breeding of new species and to improve breeding efficiency has been rapid and can be expected to continue. A limitation to current induced-spawning technologies has been the limited success in controlling stages earlier than follicle maturation. Research into the acquisition of competence to respond to GnRH could be expected to lead to new approaches to identify competence or induce competence. Similarly, the ability to induce or prevent the onset of puberty is currently limited (see Taranger et al. 2010). Technologies derived from a greater understanding of the melatonin and kisspeptin systems and how they interact with the GnRH system are likely to assist in both areas (see Filby et al. 2008; Zohar et al. 2010; Falcón et al. 2011; Cowan et al. 2012).

The interaction of the growth and reproductive axes in regulating puberty is recognized and exploited in controlling the onset of puberty in aquaculture. A greater understanding of how these pathways cross talk will likely lead to better ways of accelerating, delaying and disrupting puberty, and to maximizing growth. Recent progress on the use of lighting in the salmon industry to prevent early onset maturation (e.g., Leclercq et al. 2011); together with preliminary evidence of impact in species such as cod (e.g., Taranger et al. 2006; Cowan et al. 2011) and European sea bass (e.g., Beghtashi et al. 2004; Felip et al. 2008) supports expanded use of this technology in aquaculture. Similarly, reproductive signals, including hormones from the gonads, affect patterns of growth with many species displaying changes in somatic growth in concert with gonadal maturation (see Taranger et al. 2010; Bhatta et al. 2012). Somatic growth can be increased or decreased with the onset of maturation depending on species. Energy and nutrient partitioning can be altered including mobilization of reserves affecting which tissues grow and their composition. Furthermore, as animals are selected for improved growth it would be expected that interactions between growth and reproductive systems will be affected at multiple levels of control. In all it could be expected that what might be considered crude approaches to control these interactions such as gender control, disruption of puberty, induced sterility, will be refined to maximize the benefits of growth and reproductive axis interactions for gains in reproductive and flesh production efficiencies.

The impact of cross talk between the oocyte and follicle as mediated by intra-ovarian growth factors is increasing in appreciation (see Matzuk et al. 2002; Knight and Glister 2006; Knight et al. 2012). Actions and reproductive cycle-associated changes in transcripts have been described for the insulin/IGF family, epidermal growth factor family, and transforming growth factor-beta super family systems in fish (e.g., Ge 2005; Kamangar et al. 2006; Lankford and Weber 2010; Li and Ge 2011). Recently, bone morphogenetic proteins have been shown to down regulate FSH-receptor and up regulate

LH-receptor in the zebrafish ovary (Li et al. 2012) supporting a regulatory role in the transition from vitellogenesis to follicle maturation and spawning in fish, which is critical for successful hormone-induced spawning. The possibility that growth factors might participate in the regulation or perhaps evolutionary development of different reproductive strategies is suggested by the disparate actions of IGFs among closely related members of the genus Morone, which are hybridized to produce the farmed hybrid striped bass. In the white bass, IGFs where able to induce OMC but not the resumption of meiosis, whereas the opposite was true for follicles of the striped bass (Weber and Sullivan 2000, 2005). These growth factor actions might provide ways for hatchery managers to isolate regulation of specific components of the follicle maturation and spawning processes regulated together by GtHs. As an example, perhaps growth factor treatments can be used to induce OMC and thereby synchronize follicles before GtH is used to induce production of the MIS to induce oocyte maturation, ovulation, and spawning. Growth factors may also serve as markers for reproductive stage such as the action of IGF-I having been incorporated into a "spawning competency" test for striped bass (Weber et al. 2000).

Currently, low fertility and embryo viability is a significant problem for farmed fishes. Despite extensive research effort, it is still not clear what makes a good egg or embryo (see Kjørsvik et al. 1990; Brooks et al. 1997; Lubzens et al. 2010). Recent advances in functional genomics and proteomics approaches are making addressing these questions more feasible (see Cerdà et al. 2008). Already transcripts associated with egg quality have been identified. Egg levels of maternally derived transcripts for IGF-I, IGF-II, and IGF-receptor-1b, for example, have been correlated with embryonic survival in rainbow trout (Aegerter et al. 2004). Once variables for egg quality are identified, technologies to favorably manipulate the variables will be possible. Correspondingly, new or improved technologies to separate good gametes or embryos from bad are required. Mechanical egg pickers are very useful but need to be more efficient, and separation technologies based on factors such as differences in density or buoyancy might be improved.

More reliable technologies are sought for genetic containment for multiple reasons. Primary among them is avoidance of impact of escapees. As aquaculture increases in volume, so do the risks of escaped animals. Increasingly, aquaculture is practiced in reach of natural waterways, with exotic species, and with populations that are genetically distinct from native conspecifics. In addition, property protection is becoming more of an issue as superior lines of fish are being developed through traditional selection programs and even transgenics. Triploid induction of monosex populations is currently considered the best approach for genetic containment but both gender control and triploid induction are not always 100 % successful, and screening to confirm triploidy hampers production efforts (see Piferrer et al. 2009). Use of distant hybridization to develop tetraploid lines of carp that can then be crossed with diploid animals to generate triploids is promising for those limited number of species for which it is possible (see Luo et al. 2011). The "Shinshu salmon" developed by the Nagano Prefectural Fisheries Experimental Station, Japan, is the first commercialized production of an animal derived from induced tetraploidy of parental stocks (Denda 2007). Using induced tetraploidy to generate triploids from mating with diploids has potential for additional species if avoiding unexpected levels of euploidy and mosaicism are surmountable problems (see Chourrout et al. 1986; Chourrout and Nakayama 1987; Arai 2001; Weber and Hostuttler 2012). Also, failure to survive to maturity and low fertility in tetraploids has been a persistent problem for most species. Research has shown tetraploids of species that might be inviable, such as tetraploid masu salmon, produce tetraploid PGCs capable of migration to the germinal ridge (Sakao et al. 2009). This work was conducted as part of an ongoing effort to determine if surrogate broodstocking could be used to produce viable tetraploid gametes. Acceptance of transgenics opens the door to a wide range of possibilities to induce sterility that may even be reversible if replaceable hormones such as GnRH or GtH are targeted for knock out or disruption (see review, Wong and Van Eenennaam 2008).

Industry scale is likely to impact the implementation and further improvement of reproductive technologies currently available at the laboratory scale. Examples are cryopreservation and maturation monitoring technologies. Both technologies would benefit all-female rainbow trout production and hybrid channel catfish x blue catfish production but are not widely used. Ultrasound should be able to be used to identify fully grown testes, in males or neomales so immature fish are not sacrificed needlessly, and then all available milt can be cryopreserved, increasing the efficiency and reliability of milt production. With refinement it may be possible to assess the degree of hydration of the milt. In preliminary studies we found ultrasonography can be used to identify female rainbow trout that will ovulate in the coming week based on reduced opacity of the follicle images. The use of infrared spectrometry combined with computer analysis to draw interpretations about reproductive stage and egg quality based on analysis of ovarian biopsy samples, plasma, or even by noninvasive scans (Lu et al. 2010, 2011; Servid et al. 2011) supports the technology has potential to more widely address issues of egg quality, perhaps identifying females with good versus bad batches of eggs. If such technologies are adopted, increased automation may follow.

Sociopolitical factors play heavily into development and implementation of future reproductive technologies. Actual and perceived effects of aquaculture, including reproductive technologies, on the environment and human health are central to the development and implementation of many reproductive technologies. The costs and unpredictable nature of the approval processes for legalizing different uses of hormones in agriculture lead to great uncertainty in terms of the development and implementation of reproductive technologies. In the USA for example, the use of hCG to spawn fish is the only fully F.D.A. approved use of a hormone in aquaculture. The few other hormones available to farmers are utilized as part of I.N.A.D.s. Many uses are perceived as innocuous such as the use of GnRHa and dopamine antagonists in spawning fish. The most controversial is the use of sex steroids, particularly synthetic hormones such as 17MT for sex reversal of fry. The advantages of using the steroid in tilapia culture are clear. Male tilapia grow significantly larger than female tilapia and uncontrolled reproduction in production systems leads to stunting. Studies conducted on clearance of the steroid and its metabolites provide strong evidence supporting the hormones are rapidly cleared from the fishes' tissues (Johnstone et al. 1983; Curtis et al. 1991; Cravedi et al. 1993a, b; Chu et al. 2006). Although complete clearance of all metabolites can never be guaranteed, it seems unlikely they exist in the flesh at levels that have near the biological activity of the natural sex steroids present. In gag grouper for example, T and E2 were measured at 50 and 150 ng/kg muscle weight respectively, during the breeding season (Heppell and Sullivan 2000). On the other hand, studies have shown steroid contamination of soils when tilapia are treated in ponds (Ong et al. 2012). Reducing the risk of environmental contamination by exogenous steroids is achievable by multiple technical means such as using rubber-lined ponds or enclosed recirculation systems from which the hormones can be removed.

Technologies can even be controversial when they appear to mitigate the very problem they are perceived to cause. A good example is the use of hormones for indirect sex reversal in which a monosex population produces less sex steroids than a population of the unaltered mixed sex animals. What must be acknowledged is fish, like all sexually mature or maturing animals, produce sex steroids and those natural sex steroids are not only present in the flesh but they are also present in the mucus, and are released into the production environment The steroids are produced at sufficient levels to permit monitoring changes in reproductive stage by measuring levels of the steroids in the muscle, mucus, or tank water (e.g., Heppell and Sullivan 2000; Schultz et al. 2005;

Calvi SL, Maisse G (1998) Cryopreservation of rainbow trout (*Oncorhynchus mykiss*) blastomeres: influence of embryo stage on postthaw survival rate. Cryobiology 36(4):255–262

Calvi SL, Maisse G (1999) Cryopreservation of carp (*Cyprinus carpio*) blastomeres. Aquat Living Resour 12(1):71–74

Carrillo M, Bromage N, Zanuy S, Serrano R, Prat F (1989) The effect of modifications in photoperiod on spawning time, ovarian development and egg quality in the sea bass *Dicentrarchus labrax* L. Aquaculture 81:351–365

Cerdà J, Bobe J, Babin PJ, Admon A, Lubzens E (2008) Functional genomics and proteomic approaches for the study of gamete formation and viability in farmed finfish. Rev Fish Sci 16(suppl 2):127–135

Chakraborty SB, Banerjee S, Chatterjee S (2011) Increased androgen receptor expression in muscle tissue contributing to growth increase in androgen-treated Nile tilapia. Aquac Int 19:1119–1137

Chapman FA, Park C (2005) Comparison of sutures used for wound closure in sturgeon following a gonad biopsy. N Am J Aquac 67(2):98–101

Chapman FA, Van Eenennaam JP (2007a) Sturgeon aquaculture: specialized techniques: determining the stage of sexual maturity in female sturgeon for artificial spawning: the egg polarization index or pi. University of Florida, Institute of Food and Agricultural Sciences, Florida Cooperative Extension Service, Fisheries and Aquatic Sciences Department, Florida. Report no.: FA153

Chapman FA, Van Eenennaam JP (2007b) Sturgeon aquaculture: specialized techniques: determining the stage of sexual maturity in female sturgeon for artificial spawning: the egg maturation assay. University of Florida, Institute of Food and Agricultural Sciences, Florida Cooperative Extension Service, Fisheries and Aquatic Sciences Department, Florida. Report no.: FA154

Chatakondi NG, Yant DR, Kristanto A, Umali-Maceina GM, Dunham RA (2011) The effect of luteinizing hormone releasing hormone analog regime and stage of oocyte maturity for induced ovulation of channel catfish, Ictalurus punctatus. J World Aquac Soc 42:845–853

Chaudhuri H, Alikunhi KH (1957) Observations on the spawning in Indian carps by hormone injection. Curr Sci 26:381–382

Chen S, Wang J, Liu SJ, Qin QB, Xiao J, Duan W et al (2009) Biological characteristics of an improved triploid crucian carp. Sci China Life Sci 52(8):733–738

Chen TT, Vrolijk NH, Lu JK, Lin CM, Reimschuessel R, Dunham RA (1996) Transgenic fish and its application in basic and applied research. Biotechnol Annu Rev 2(c):205–236

Cherfas NB, Gomelsky B, Ben-Dom N, Joseph D, Cohen S, Israel I et al (1996) Assessment of all-female common carp progenies for fish culture. Isr J Aquac Bamidgeh 48:149–157

Cherfas NB, Gomelsky B, Ben-Dom N, Peretz Y, Hulata G (1994a) Assessment of triploid common carp (*Cyprinus carpio* L.) for culture. Aquaculture 127:11–18

Cherfas NB, Gomelsky BI, Emelyanova OV, Recoubratsky AV (1994b) Induced diploid gynogenesis and polyploidy in crucian carp, *Carassius auratus gibelio* (Bloch), × common carp, *Cyprinus carpio* L., hybrids. Aquac Fish Manag 25:943–954

Chistiakov DA, Voronova NV (2009) Genetic evolution and diversity of common carp *Cyprinus carpio* L. Cent Eur J Biol 4(3):304–312

Chourrout D (1982) Tetraploidy induced by heat shocks in the rainbow trout (*Salmo gairdneri* R.). Reprod Nutr Dev 22(3):569–574

Chourrout D (1984) Pressure-induced retention of second polar body and suppression of first cleavage in rainbow trout: production of all-triploids, all-tetraploids, and heterozygous and homozygous diploid gynogenetics. Aquaculture 36(1–2):111–126

Chourrout D, Chevassus B, Krieg F, Happe A, Burger G, Renard P (1986) Production of second generation triploid and tetraploid rainbow trout by mating tetraploid males and diploid females—potential of tetraploid fish. Theor Appl Genet 72:193–206

Chourrout D, Nakayama I (1987) Chromosome studies of progenies of tetraploid female rainbow trout. Theor Appl Genet 74:687–692

Chowdhury I, Joy KP (2007) Seminal vesicle and its role in the reproduction of teleosts. Fish Physiol Biochem 33(4):383–398

Christie MR, Marine ML, French RA, Blouin MS (2012) Genetic adaptation to captivity can occur in a single generation. Proc Natl Acad Sci U S A 109(1):238–242

Chu PS, Lopez M, Serfling S, Gieseker C, Reimschuessel R (2006) Determination of 17α-methyltestosterone in muscle tissues of tilapia, rainbow trout, and salmon using liquid chromatography-tandem mass spectrometry. J Agric Food Chem 54(9):3193–3198

Cleveland BM, Weber GM (2011) Effects of sex steroids on indices of protein turnover in rainbow trout (*Oncorhynchus mykiss*) white muscle. Gen Comp Endocrinol 174(2):132–142

Cnaani A, Levavi-Sivan B (2009) Sexual development in fish, practical applications for aquaculture. Sex Dev 3(2–3):164–175

Colombo L, Francescon A, Barbaro A, Belvedere P, Melotti P (1989) Induction of spawning in the gilthead seabream, *Sparus aurata* L., by elevation of water temperature and salinity and by hCG or LH-RH analogue treatments. Riv Ital Acquacol 24:187–196

Conte FS, Doroshov SI, Lutes PB, Strange EM (1988) Hatchery manual for the white sturgeon Acipenser Transmontanus Richardson with application to other North American Acipenseridae. University of California, Publications Division of Agriculture and Natural Resources, Oakland, CA

Cotter D, O'Donovan V, Drumm A, Roche N, Ling EN, Wilkins NP (2002) Comparison of freshwater and marine performances of all-female diploid and triploid Atlantic salmon (*Salmo salar* L.). Aquac Res 33(1):43–53

Cowan M, Davie A, Migaud H (2011) The effect of combining shading and continuous lighting on the suppression of sexual maturation in outdoor-reared Atlantic cod, *Gadus morhua*. Aquaculture 320(1–2):113–122

Cowan M, Davie A, Migaud H (2012) Photoperiod effects on the expression of kisspeptin and gonadotropin genes in Atlantic cod, *Gadus morhua*, during first maturation. Comp Biochem Physiol Mol Integr Physiol 163:82–94

Coward K, Bromage NR (2000) Reproductive physiology of female tilapia broodstock. Rev Fish Biol Fish 10(1):1–25

Cowx IG (1997) Introduction of fish species into European fresh waters: economic successes or ecological disasters? Bull France Peche et Piscicult 344–345:57–77

Craik JCA, Harvey SM (1984) A biochemical method for distinguishing between the sexes of fishes by the presence of yolk protein in the blood. J Fish Biol 25: 293–303

Cravedi JP, Delous G, Debrauwer L, Prome D (1993a) Biotransformation and branchial excretion of 17α-methyltestosterone in trout. Drug Metab Dispos 21(2):377–385

Cravedi JP, Delous G, Debrauwer L, Rao D, Prome D (1993b) Liquid chromatographic separation and gas chromatographic-mass spectrometric determination of 17α-methyltestosterone residues extracted from rainbow trout tissues. Anal Chim Acta 275(1–2):89–94

Crim LW, Glebe BD (1984) Advancement and synchrony of ovulation in Atlantic salmon with pelleted LHRH analog. Aquaculture 43(1–3):47–56

Curtis LR, Diren FT, Hurley MD, Seim WK, Tubb RA (1991) Disposition and elimination of 17α-methyltestosterone in Nile tilapia (*Oreochromis niloticus*). Aquaculture 99(1–2):193–201

Davis CA, Cavinato AG, Hoffnagle T (2006) Non-invasive determination of gender and maturity status in Chinook salmon by short wavelength near infrared spectroscopy. East Oreg Sci J 19:30–33

Davis LK, Fox BK, Lim C, Hiramatsu N, Sullivan CV, Hirano T et al (2009) Induction of vitellogenin production in male tilapia (*Oreochromis mossambicus*) by commercial fish diets. Comp Biochem Physiol A Mol Integr Physiol 154(2):249–254

De Graaf G, Galemoni F, Banzoussi B (1996) Recruitment control of Nile tilapia, *Oreochromis niloticus*, by the African catfish, *Clarias gariepinus* (Burchell 1822), and the African snakehead, *Ophiocephalus obscuris* I: a biological analysis. Aquaculture 146(1–2):85–100

De Kimpe P, Micha J-C (1974) First guidelines for the culture of *Clarias lazera* in Central Africa. Aquaculture 4:227–248

De Leeuw R, Goos HJT, Richter CJJ, Eding EH (1985) Pimozide—LHRHa induced breeding of the African catfish, *Clarias gariepinus* (Burchell). Aquaculture 44:295–302

De Silva SS, Phuong NT (2011) Striped catfish farming in the Mekong Delta, Vietnam: a tumultuous path to a global success. Rev Aquac 3(2):45–73

Denda I (2007) Development of new farmed fish "Shinshu Salmon" by chromosome manipulation. Biosci Bioind 65(12):596–599, Japanese

Denslow ND, Chow MC, Kroll KJ, Green L (1999) Vitellogenin as a biomarker of exposure for estrogen or estrogen mimics. Ecotoxicology 8(5):385–398

Desprez D, Mélard C (1998) Effect of ambient water temperature on sex determinism in the blue tilapia *Oreochromis aureus*. Aquaculture 162(1–2):79–84

Devlin RH (1997) Transgenic salmonids. In: Houdebine LM (ed) Transgenic animals: generation and use. Harwood Academic, Amsterdam, pp 105–117

Devlin RH, Biagi CA, Smailus DE (2001) Genetic mapping of Y-chromosomal DNA markers in Pacific salmon. Genetica 111(1–3):43–58

Devlin RH, McNeil BK, Groves TDD, Donaldson EM (1991) Isolation of a Y-chromosomal DNA probe capable of determining genetic sex in chinook salmon (*Oncorhynchus tshawytscha*). Can J Fish Aquat Sci 48:1606–1612

Devlin RH, McNeil BK, Solar II, Donaldson EM (1994a) A rapid PCR-based test for Y-chromosomal DNA allows simple production of all-female strains of chinook salmon. Aquaculture 128(3–4):211–220

Devlin RH, Nagahama Y (2002) Sex determination and sex differentiation in fish: an overview of genetic, physiological, and environmental influences. Aquaculture 208(3–4):191–364

Devlin RH, Sakhrani D, Biagi CA, Eom KW (2010) Occurrence of incomplete paternal-chromosome retention in GH-transgenic coho salmon being assessed for reproductive containment by pressure-shock-induced triploidy. Aquaculture 304(1–4): 66–78

Devlin RH, Yesaki TY, Biagi CA, Donaldson EM, Swanson P, Chan W-K (1994b) Extraordinary salmon growth. Nature 371(6494):209–210

Devlin RH, Yesaki TY, Donaldson EM, Du SJ, Hew CL (1995) Production of germline transgenic Pacific salmonids with dramatically increased growth performance. Can J Fish Aquat Sci 52(7):1376–1384

Dey MM, Rab MA, Paraguas FJ, Bhatta R, Alam MF, Koeshendrajana S et al (2005) Status and economics of freshwater aquaculture in selected countries of Asia. Aquac Econ Manag 9(1–2):11–37

Donaldson EM, Hunter GA, Dye HM (1981) Induced ovulation in coho salmon (Oncorhynchus kisutch). II. Preliminary study of the use of LH-RH and two high potency LH-RH analogues. Aquaculture 26:129–141

Duarte M, Marbá N, Holmer M (2007) Rapid domestication of marine species. Science 316:382–383

Dufour S, Sebert ME, Weltzien FA, Rousseau K, Pasqualini C (2010) Neuroendocrine control by dopamine of teleost reproduction. J Fish Biol 76(1):129–160

Dunham RA (2009) Transgenic fish resistant to infectious diseases, their risk and prevention of escape into the environment and future candidate genes for disease transgene manipulation. Comp Immunol Microbiol Infect Dis 32(2):139–161

Dunham RA, Argue BJ (2000) Reproduction among channel catfish, blue catfish, and their F1 and F2 hybrids. Trans Am Fish Soc 129(1):222–231

Dunham RA, Chitmanat C, Nichols A, Argue B, Powers DA, Chen TT (1999) Predator avoidance of transgenic channel catfish containing salmonid growth hormone genes. Mar Biotechnol 1(6):545–551

Dunham RA, Lambert DM, Argue BJ, Ligeon C, Yant DR, Liu Z (2000) Comparison of manual stripping and pen spawning for production of channel catfish × blue catfish hybrids and aquarium spawning of channel catfish. N Am J Aquac 62(4):260–265

Dunham RA, Masser M (2012) Production of hybrid catfish. Southern Regional Aquaculture Center, Stoneville. SRAC publication no.: 190

Dunham RA, Umali GM, Beam R, Kristanto AH, Trask M (2008) Comparison of production traits of NWAC103 channel catfish, NWAC103 channel catfish X blue catfish hybrids, Kansas Select 21 channel catfish, and blue catfish grown at commercial densities and exposed to natural bacterial epizootics. N Am J Aquac 70(1):98–106

Eknath AE, Doyle RW (1990) Effective population size and rate of inbreeding in aquaculture of Indian major carps. Aquaculture 85:293–305

El Naggar GO, John G, Rezk MA, Elwan W, Yehia M (2006) Effect of varying density and water level on the spawning response of African catfish *Clarias gariepinus*: implications for seed production. Aquaculture 261(3):904–907

Endal HP, Taranger GL, Stefansson SO, Hansen T (2000) Effects of continuous additional light on growth and sexual maturity in Atlantic salmon, *Salmo salar*, reared in sea cages. Aquaculture 191(4):337–349

Espinosa E, Josa A, Gil L, Malo C, Mitjana O (2012) Comparing sex steroid levels during the annual cycles of rainbow trout (*Oncorhynchus mykiss*) diploid female (XX) and triploid female (XXX) genotypic sex. Reprod Domest Anim. doi:10.1111/j.1439-0531.2012.02117.x

Falahatkar B, Tolouei Gilani MH, Falahatkar S, Abbasalizadeh A (2011) Laparoscopy, a minimally-invasive technique for sex identification in cultured great sturgeon *Huso huso*. Aquaculture 321(3–4): 273–279

Falcón J, Besseau L, Magnanou E, Herrero MJ, Nagai M, Boeuf G (2011) Melatonin, the time keeper: biosynthesis and effects in fish. Cybium 35(1):3–18

FAO (2012) The state of world fisheries and aquaculture 2012. Fisheries and Aquaculture Department, Rome. ISBN 978-92-5-107225-7

Feist G, Van Eenennaam JP, Doroshov SI, Schreck CB, Schneider RP, Fitzpatrick MS (2004) Early identification of sex in cultured white sturgeon, *Acipenser transmontanus*, using plasma steroid levels. Aquaculture 232(1–4):581–590

Feist G, Yeoh CG, Fitzpatrick MS, Schreck CB (1995) The production of functional sex-reversed male rainbow trout with 17alpha-methyltestosterone and 11beta-hydroxyandrostenedione. Aquaculture 131(1–2): 145–152

Felip A, Zanuy S, Muriach B, Cerdá-Reverter JM, Carrillo M (2008) Reduction of sexual maturation in male *Dicentrarchus labrax* by continuous light both before and during gametogenesis. Aquaculture 275(1–4):347–355

Filby AL, Van Aerle R, Duitman J, Tyler CR (2008) The kisspeptin/gonadotropin-releasing hormone pathway and molecular signaling of puberty in fish. Biol Reprod 78(2):278–289

Forbes SH, Knudsen KL, North TW, Allendorf FW (1994) One of two growth hormone genes in coho salmon is sex-linked. Proc Natl Acad Sci U S A 91(5): 1628–1631

Forniés MA, Mañanós E, Carrillo M, Rocha A, Laureau S, Mylonas CC et al (2001) Spawning induction of individual European sea bass females (*Dicentrarchus labrax*) using different GnRHa-delivery systems. Aquaculture 202(3–4):221–234

Friars GW, McMillan I, Quinton VM, O'Flynn FM, McGeachy SA, Benfey TJ (2001) Family differences in relative growth of diploid and triploid Atlantic salmon (*Salmo salar* L.). Aquaculture 192(1):23–29

Garber AF, Sullivan CV (2006) Selective breeding for the hybrid striped bass (*Morone chrysops*, Rafinesque x *M. saxatilis*, Walbaum) industry: status and perspectives. Aquac Res 37(4):319–338

Ge W (2005) Intrafollicular paracrine communication in the zebrafish ovary: the state of the art of an emerging model for the study of vertebrate folliculogenesis. Mol Cell Endocrinol 237(1–2):1–10

Geffen AJ, Evans JP (2000) Sperm traits and fertilization success of male and sex-reversed female rainbow trout (*Oncorhynchus mykiss*). Aquaculture 182(1–2):61–72

Gjedrem T, Robinson N, Rye M (2012) The importance of selective breeding in aquaculture to meet future demands for animal protein: a review. Aquaculture 350–353:117–129

Glover KA, Quintela M, Wennevik V, Besnier F, Sørvik AGE, Skaala Ø (2012) Three decades of farmed escapees in the wild: a spatio-temporal analysis of Atlantic salmon population genetic structure throughout Norway. PLoS ONE 7(8):e43129

Gomelsky B (2003) Chromosome set manipulation and sex control in common carp: a review. Aquat Living Resour 16(5):408–415

Gordon MR, Owen TG, Ternan TA, Hildebrand LD (1984) Measurement of a sex-specific protein in skin mucus of premature coho salmon (*Oncorhynchus kisutch*). Aquaculture 43(1–3):333–339

Gorshkov S, Gorshkova G, Colorni B, Gordin H (2004) Effects of natural estradiol-17β and synthetic 17-ethynylestradiol on direct feminization of European sea bass *Dicentrarchus labrax*. J World Aquac Soc 35(2):167–177

Haffray P, Labbe C, Technologies IMV, Maisse G (2008) Fish sperm cryopreservation in France: from laboratory studies to application in selective breeding programs. Cybium 32(suppl 2):S127–S129

Hallerman EM, McLean E, Fleming IA (2007) Effects of growth hormone transgenes on the behavior and welfare of aquacultured fishes: a review identifying research needs. Appl Anim Behav Sci 104(3–4):265–294

Han Y, Ma Q, Lu J, Xue Y, Xue C (2012) Optimisation for subcritical fluid extraction of 17-methyltestosterone with 1,1,1,2-tetrafluoroethane for HPLC analysis. Food Chem 135:2988–2993

Harrell RM, Kerby JH, Minton RV (1990) Culture and propagation of striped bass and its hybrids. American Fisheries Society, Bethesda, MD

Heggberget TG, Johnsen BO, Hindar K, Jonsson B, Hansen LP, Hvidsten NA et al (1993) Interactions between wild and cultured Atlantic salmon: a review of the Norwegian experience. Fish Res 18(1–2):123–146

Heppell SA, Denslow ND, Folmar LC, Sullivan CV (1995) Universal assay of vitellogenin as a biomarker for environmental estrogens. Environ Health Perspect 103(suppl 7):S9–S15

Heppell SA, Sullivan CV (2000) Identification of gender and reproductive maturity in the absence of gonads: muscle tissue levels of sex steroids and vitellogenin in gag (*Mycteroperca microlepis*). Can J Fish Aquat Sci 57(1):148–159

Hinits Y, Moav B (1999) Growth performance studies in transgenic *Cyprinus carpio*. Aquaculture 173(1–4): 285–296

Hirose K, Ishida R (1974) Induction of ovulation in the ayu, *Plecoglossus altivelis*, with LH-releasing hormone (LH-RH). Bull Jpn Soc Sci Fish 40:1235–1240

Hodson RG (1995) Farming a new fish: hybrid striped bass. North Carolina Sea Grant Publication, Raleigh, NC, p 41

Hodson R, Sullivan CV (1993) Induced maturation and spawning of domestic and wild striped bass, *Morone saxatilis* (Walbaum), broodstock with implanted GnRH analogue and injected hCG. Aquac Fish Manag 24:389–398

Hogendoorn H (1979) Controlled propagation of the African catfish, *Clarius lazera*, (C&V). I. Reproductive biology and field experiments. Aquaculture 17(4):323–333

Hogendoorn H, Vismans MM (1980) Controlled propagation of the African catfish *Clarias lazera* (C&V). II. Artificial reproduction. Aquaculture 21(1):39–53

Hogendoorn H (1980) Controlled propagation of the African catfish, *Clarias lazera* (C&V). III. Feeding and growth of fry (Artemia). Aquaculture 21(3):233–241

Horváth Á, Miskolczi E, Urbányi B (2003) Cryopreservation of common carp sperm. Aquat Living Resour 16(5):457–460

Houssay BA (1931) Action sexuelle de l'hypophyse sur les poissons et les reptiles. Soc Biol Paris 106:377–378, French

Hu E, Yang H, Tiersch TR (2011) High-throughput cryopreservation of spermatozoa of blue catfish (*Ictalurus furcatus*): establishment of an approach for commercial-scale processing. Cryobiology 62(1):74–82

Huisman EA, Richter CJJ (1987) Reproduction, growth, health control and aquacultural potential of the African catfish, *Clarias gariepinus* (Burchell 1822). Aquaculture 63(1–4):1–14

Hulata G (1995) A review of genetic improvement of the common carp (*Cyprinus carpio* L.) and other cyprinids by crossbreeding, hybridization and selection. Aquaculture 129(1–4):143–155

Hulata G (2001) Genetic manipulations in aquaculture: a review of stock improvement by classical and modern technologies. Genetica 111(1–3):155–173

Hurvitz A, Jackson K, Degani G, Levavi-Sivan B (2007) Use of endoscopy for gender and ovarian stage determinations in Russian sturgeon (*Acipenser gueldenstaedtii*) grown in aquaculture. Aquaculture 270(1–4):158–166

Idler DR, Horne DA, Sangalang GB (1971) Identification and quantification of the major androgens in testicular and peripheral plasma of Atlantic salmon (*Salmo salar*) during sexual maturation. Gen Comp Endocrinol 16(2):257–267

Jennings CA, Will TA, Reinert TR (2005) Efficacy of a high- and low-frequency ultrasonic probe for measuring ovary volume and estimating fecundity of striped bass *Morone saxatilis* in the Savannah River Estuary. Fish Res 76(3):445–453

Jensen JOT, Alderdice DF (1984) Effect of temperature on short-term storage of eggs and sperm of chum salmon (*Oncorhynchus keta*). Aquaculture 37(3): 251–265

Jhingren VG, Pullin ISV (1985) A hatchery manual for the common, Chinese and Indian major carps. Asian Development Bank, Manila, p 191. Co-published by the International Center for Living Aquatic Resources Management

Johnstone R, Macintosh DJ, Wright RS (1983) Elimination of orally administered 17α-methyltestosterone by *Oreochromis mossambicus* (tilapia) and *Salmo gairdneri* (rainbow trout) juveniles. Aquaculture 35(C): 249–257

Johnstone R, Simpson TH, Youngson AF, Whitehead C (1979) Sex reversal in salmonid culture. Part II. The progeny of sex-reversed rainbow trout. Aquaculture 18:13–19

Kagawa H, Tanaka H, Ohta H, Unuma T, Nomura K (2005) The first success of glass eel production in the world: basic biology on fish reproduction advances new applied technology in aquaculture. Fish Physiol Biochem 31:193–199

Kamangar BB, Gabillard JC, Bobe J (2006) Insulin-like growth factor-binding protein (IGFBP)-1, -2, -3, -4, -5, and -6 and IGFBP-related protein 1 during rainbow trout postvitellogenesis and oocyte maturation: molecular characterization, expression profiles, and hormonal regulation. Endocrinology 14(5):2399–2410

Karlsen O, Holm JC (1994) Ultrasonography, a noninvasive method for sex determination in cod (*Gadus morhua*). J Fish Biol 44:965–971

Kause A, Ritola O, Paananen T, Mäntysaari E, Eskelinen U (2003) Selection against early maturity in large rainbow trout *Oncorhynchus mykiss*: the quantitative genetics of sexual dimorphism and genotype-by-environment interactions. Aquaculture 228(1–4):53–68

Kawakami Y, Saito T, Fujimoto T, Goto-Kazeto R, Takahashi E, Adachi S et al (2012) Technical note: viability and motility of vitrified/thawed primordial germ cell isolated from common carp (*Cyprinus carpio*) somite embryos. J Anim Sci 90:495–500

Kelly AM (2004) Channel catfish broodfish management. Southern Regional Aquaculture Center, Stoneville. SRAC publication no.: 1802

Kise K, Yoshikawa H, Sato M, Tashiro M, Yazawa R, Nagasaka Y et al (2012) Flow-cytometric isolation and enrichment of teleost type a spermatogonia based on light-scattering properties. Biol Reprod 86(4):107

Kishida M, Anderson TR, Specker JL (1992) Induction by β-estradiol of vitellogenin in striped bass (*Morone saxatilis*): characterization and quantification in plasma and mucus. Gen Comp Endocrinol 8(1):9–39

Kishida M, Specker JL (1994) Vitellogenin in the surface mucus of tilapia (*Oreochromis mossambicus*): possibility for uptake by the free-swimming embryos. J Exp Zool 268(4):259–268

Kjørsvik E, Mangor-Jensen A, Holmefjord I (1990) Egg quality in fishes. Adv Mar Biol 26(c):71–113

Knight PG, Glister C (2006) TGF-β superfamily members and ovarian follicle development. Reproduction 132(2):191–206

Knight PG, Satchell L, Glister C (2012) Intra-ovarian roles of activins and inhibins. Mol Cell Endocrinol 359(1–2):53–65

Kobayashi T, Fushiki S, Sakai N, Hara A, Amano M, Aida K et al (1998) Oogenesis and changes in the levels of reproductive hormones in triploid female rainbow trout. Fish Sci 64(2):206–215

Kobayashi T, Fushiki S, Ueno K (2004) Improvement of sperm motility of sex-reversed male rainbow trout, *Oncorhynchus mykiss*, by incubation in high-pH artificial seminal plasma. Environ Biol Fishes 69(1–4): 419–425

Kobayashi T, Takeuchi Y, Takeuchi T, Yoshizaki G (2007) Generation of viable fish from cryopreserved primordial germ cells. Mol Reprod Dev 74(2):207–213

Kocour M, Linhart O, Gela D (2003) Results of comparative growing test of all-female and bisexual population in two-year-old common carp (*Cyprinus carpio* L.). Aquac Int 11(4):369–378

Komen J, Bongers ABJ, Richter CJJ, van Muiswinkel WB, Huisman EA (1991) Gynogenesis in common carp (*Cyprinus carpio* L.). II. The production of homozygous gynogenetic clones and F1 hybrids. Aquaculture 92(c):127–142

Konno K, Tashiro F (1982) The sterility of rainbow trout (*Salmo gairdneri*) irradiated with cobalt-60 gamma rays. J Tokyo Univ Fish 68:75–80

Krisfalusi M, Cloud JG (1999) Gonadal sex reversal in triploid rainbow trout (*Oncorhynchus mykiss*). J Exp Zool 284(4):466–472

Kucherka WD, Thomas P, Khan IA (2006) Sex differences in circulating steroid hormone levels in the red drum, *Sciaenops ocellatus* L. Aquac Res 37(14): 1464–1472

Kumar G, Engle C (2011) The effect of hybrid catfish fingerling prices on the relative profitability of hybrid channel Catfish. J World Aquac Soc 42:469–483

Kuo CM (1985) A review of induced breeding of milkfish. In: Lee CS, Liao IC (eds) Reproduction and culture of milkfish. Oceanic Institute, Hawaii, pp 57–77

Kuo CM, Nash CE, Shehadeh ZH (1974) A procedural guide to induce spawning in grey mullet (*Mugil cephalus* L.). Aquaculture 3(1):1–14

Kuwaye TT, Okimoto DK, Shimoda SK, Howerton RD, Lin HR, Pang PKT et al (1993) Effect of 17α-methyltestosterone on the growth of the euryhaline tilapia, *Oreochromis mossambicus*, in fresh water and in sea water. Aquaculture 113(1–2):137–152

Kynard B, Kieffer M (2002) Use of a borescope to determine the sex and egg maturity stage of sturgeons and the effect of borescope use on reproductive structures. J Appl Ichthyol 18(4–6):505–508

Lacerda SMSN, Batlouni SR, Costa GMJ, Segatelli TM, Quirino BR, Queiroz BM et al (2010) A new and fast technique to generate offspring after germ cells transplantation in adult fish: the Nile tilapia (*Oreochromis niloticus*) model. PLoS ONE 5(5):e10740

Lankford SE, Weber GM (2010) Temporal mRNA expression of transforming growth factor-beta superfamily members and inhibitors in the developing rainbow trout ovary. Gen Comp Endocrinol 166(2):250–258

Lazard J, Cacot P, Slembrouck J, Legendre M (2009) Fish farming of Pangasiids. Cah Agric 18(2): 164–173

Le Bail PY, Breton B (1981) Rapid determination of the sex of puberal salmonid fish by a technique of immunoagglutination. Aquaculture 22(c):367–375

Leclercq E, Taylor JF, Sprague M, Migaud H (2011) The potential of alternative lighting-systems to suppress pre-harvest sexual maturation of 1+ Atlantic salmon (*Salmo salar*) post-smolts reared in commercial sea-cages. Aquac Eng 44(2):35–47

Lee CS, Tamaru CS, Banno JE, Kelley CD (1986a) Influence of chronic administration of LHRH-analogue and/or 17α-methyltestosterone on maturation in milkfish, *Chanos chanos*. Aquaculture 59(2):147–159

Lee CS, Tamaru CS, Banno JE, Kelley CD, Bocek A, Wyban JA (1986b) Induced maturation and spawning of milkfish, *Chanos chanos* Forssakal, by hormone implantation. Aquaculture 52(3):199–205

Lee CS, Tamaru CS, Kelley CD, Miyamoto GT, Moriwake AM (1992) The minimum effective dosage of 17α-methyltestosterone for induction of testicular maturation in the striped mullet, *Mugil cephalus* L. Aquaculture 104(1–2):183–191

Lee CS, Weber GM (1986) Effects of salinity and photoperiod on 17α-methyltestosterone-induced spermatogenesis in the grey mullet, *Mugil cephalus* L. Aquaculture 56(1):53–62

Legendre M, Slembrouck J, Subagja J, Hari Kristanto A (2000) Ovulation rate, latency period and ova viability after GnRH-or hCG-induced breeding in the Asian catfish *Pangasius hypophthalmus* (*Siluriformes, Pangasiidae*). Aquat Living Resour 13(3):145–151

Li CW, Ge W (2011) Spatiotemporal expression of bone morphogenetic protein family ligands and receptors in the zebrafish ovary: a potential paracrine-signaling mechanism for oocyte-follicle cell communication. Biol Reprod 85(5):977–986

Li CW, Zhou R, Ge W (2012) Differential regulation of gonadotropin receptors by bone morphogenetic proteins in the zebrafish ovary. Gen Comp Endocrinol 176(3):420–425

Li MH, Robinson EH, Manning BB, Yant DR, Chatakondi NG, Bosworth BG et al (2004) Comparison of the channel catfish, *Ictalurus punctatus* (NWAC103 Strain) and the channel X blue catfish, *I. punctatus* X *I. furcatus*, F1 hybrid for growth, feed efficiency, processing yield, and body composition. J Appl Aquac 15(3–4):63–71

Lin HR, Peter RE (1986) Induction of gonadotropin secretion and ovulation in teleosts using LHRH analogous and catecholaminergic drugs: a review. In: Maclean JL, Dizon LB, Hosillos LV (eds) The first Asian Fisheries Forum. Asian Fisheries Society, Manila, pp 667–670

Lincoln RF, Scott AP (1983) Production of all-female triploid rainbow trout. Aquaculture 30(1–4):375–380

Liu H, Guan B, Xu J, Hou C, Tian H, Chen H (2012) Genetic manipulation of sex ratio for the large-scale breeding of YY super-male and XY all-male yellow catfish (*Pelteobagrus fulvidraco* (Richardson)). Mar Biotechnol. doi:10.1007/s10126-012-9487-7

Liu SJ (2010) Distant hybridization leads to different ploidy fishes. Sci China Life Sci 53:416–425

Liu SJ, Liu Y, Zhou GJ, Zhang X, Luo C, Feng H et al (2001) The formation of tetraploid stocks of red crucian carp × common carp hybrids as an effect of interspecific hybridization. Aquaculture 192(2–4):171–186

Liu SJ, Qin QB, Xiao J, Lu W, Shen J, Li W et al (2007) The formation of the polyploid hybrids from different subfamily fish crossing and its evolutionary significance. Genetics 176:1023–1034

Loopstra DP, Hansen PA (2008) Induction of triploidy in rainbow trout (*Oncorhynchus mykiss*) using hydrostatic pressure. Alaska Department of Fish and Game, Kodiak. Fishery data series no.: 8–22

Loopstra DP, Hansen PA (2010) Induction of triploidy in arctic grayling (*Thymallus arcticus*) using hydrostatic pressure. Alaska Department of Fish and Game, Kodiak. Fishery data series no.: 10–55

Lu X, Webb MAH, Talbott MJ, Van Eenennaam JP, Doroshov SI, Rasco BA (2011) A study of biochemical parameters associated with ovarian atresia and quality of caviar in farmed white sturgeon (*Acipenser transmontanus*) by Fourier Transform Infrared (FT-IR) Spectroscopy. Aquaculture 315(3–4):298–305

Lu X, Webb M, Talbott M, Van Eenennaam J, Palumbo A, Linares-Casenave J, Doroshov S et al (2010) Distinguishing ovarian maturity of farmed white sturgeon (*Acipenser transmontanus*) by Fourier transform infrared spectroscopy: a potential tool for caviar production management. J Agric Food Chem 58(7):4056–4064

Lubzens E, Young G, Bobe J, Cerdà J (2010) Oogenesis in teleosts: how fish eggs are formed. Gen Comp Endocrinol 165(3):367–389

Luckenbach JA, Borski RJ, Daniels HV, Godwin J (2009) Sex determination in flatfishes: mechanisms and environmental influences. Semin Cell Dev Biol 20(3): 256–263

Luo K, Xiao J, Liu S, Wang J, He W, Hu J et al (2011) Massive production of all-female diploids and triploids in the crucian carp. Int J Biol Sci 7(4):487–495

Lutes PB, Doroshov SI, Chapman F, Harrah J, Fitzgerald R, Fitzpatrick M (1987) Morpho-physiological predictors of ovulatory success in white sturgeon, *Acipenser transmontanus* Richardson. Aquaculture 66(1):43–52

Macri F, Rapisarda G, Marino G, De Majo M, Aiudi G (2011) Use of laparoscopy for the evaluation of the reproductive status of tench (*Tinca tinca*). Reprod Domest Anim 46(1):130–133

Mair GC, Abucay US, Skibinski DOF, Abella TA, Beardmore JA (1997) Genetic manipulation of sex ratio for the large-scale production of all-male tilapia, *Oreochromis niloticus*. Can J Fish Aquat Sci 54(2): 396–404

Majhi SK, Hattori RS, Yokota M, Watanabe S, Strüssmann CA (2009) Germ cell transplantation using sexually competent fish: an approach for rapid propagation of endangered and valuable germlines. PLoS ONE 4(7):e6132

Mandich A, Benfenati E, Cronin MTD, Goksøyr A, Grøsvik BE, Kloas W et al (2005) Environmental agent susceptibility assessment using existing and novel biomarkers as rapid noninvasive testing methods. Ann N Y Acad Sci 1040:381–386

Martin RW, Myers J, Sower SA, Phillips DJ, McAuley C (1983) Ultrasonic imaging, a potential tool for sex determination of live fish. N Am J Fish Manag 3:258–264

Martin-Robichaud DJ, Rommens M (2001) Assessment of sex and evaluation of ovarian maturation of fish using ultrasonography. Aquac Res 32:113–120

Martínez R, Juncal J, Zaldívar C, Arenal A, Guillén I, Morera V et al (2000) Growth efficiency in transgenic tilapia (*Oreochromis sp.*) carrying a single copy of an homologous cDNA growth hormone. Biochem Biophys Res Commun 267(1):466–472

Masoudifard M, Vajhi AR, Moghim M, Nazari RM, Naghavi AR, Sohrabnejad M (2011) High validity sex determination of three years old cultured beluga sturgeon (*Huso huso*) using ultrasonography. J Appl Ichthyol 27:643–664

Matsubara T, Sawano K (1992) Sex determination of Pacific halibut (*Hippoglossus stenolepsis* Schmidt) by immuno-dot-blotting technique using antiserum against vitellogenin. Bull Hokkaido Natl Fish Res Inst 56:17–26

Matsubara T, Watanabe K, Yamanome T, Kayaba T (1999) Application of ultrasonography to non-invasive sexing based on the sexual dimorphism in gonads of immature barfin flounder *Verasper moseri*. Fish Sci 65:244–247

Matsuo H, Baba Y, Nair RM, Arimura A, Schally AV (1971) Structure of the porcine LH- and FSH-releasing hormone. I. The proposed amino acid sequence. Biochem Biophys Res Commun 43:1334–1339

Mattson NS (1991) A new method to determine sex and gonad size in live fishes by using ultrasonography. J Fish Biol 39:673–677

Matzuk MM, Burns KH, Viveiros MM, Eppig JJ (2002) Intercellular communication in the mammalian ovary: oocytes carry the conversation. Science 296(5576): 2178–2180

McClure CA, Hammell KL, Moore M, Dohoo IR, Burnley H (2007) Risk factors for early sexual maturation in Atlantic salmon in seawater farms in New Brunswick and Nova Scotia, Canada. Aquaculture 272(1–4):370–379

McGeachy SA, O'Flynn FM, Benfey TJ, Friars GW (1996) Seawater performance of triploid Atlantic salmon in New Brunswick aquaculture. Bull Aquat Assoc Can 137:24–28

Medeiros EF, Phelps MP, Fuentes FD, Bradley TM (2009) Overexpression of follistatin in trout stimulates increased muscling. Am J Physiol Regul Integr Comp Physiol 297(1):R235–R242

Migaud H, Davie A, Taylor JF (2010) Current knowledge on the photoneuroendocrine regulation of reproduction in temperate fish species. J Fish Biol 76(1):27–68

Miura T, Yamauchi K, Takahashi H, Nagahama Y (1992) The role of hormones in the acquisition of sperm motility in salmonid fish. J Exp Zool 261(3):359–363

Moccia RD, Wilkie EJ, Munkittrick KR, Thompson WD (1984) The use of fine needle fibre endoscopy in fish for in vivo examination of visceral organs, with special reference to ovarian evaluation. Aquaculture 56:139–149

Moghim M, Vajhi AR, Veshkini A, Masoudifard M (2002) Determination of sex and maturity in *Acipenser stellatus* by using ultrasonography. J Appl Ichthyol 18:325–328

Mola AE, Hovannisyan HG, Nazari RM, Ovissipour M (2011) Early sex identification in cultured beluga (*Huso huso*) using plasma steroid hormones. Afr J Biotechnol 10(10):1959–1965

Moretti A, Pedini Fernandez-Criado M, Cittolin G, Guidastri R (1999) Manual on hatchery production of seabass and gilthead seabream, vol 1. Rome, FAO, p 194

Morita T, Kumakura N, Morishima K, Mitsuboshi T, Ishida M, Hara T et al (2012) Production of donor-derived offspring by allogeneic transplantation of spermatogonia in the yellowtail (*Seriola quinqueradiata*). Biol Reprod 86(6):176

Myers JM, Hershberger WK (1991) Early growth and survival of heat-shocked and tetraploid-derived triploid rainbow trout (*Oncorhynchus mykiss*). Aquaculture 96(2):97–107

Mylonas CC, Fostier A, Zanuy S (2010) Broodstock management and hormonal manipulations of fish reproduction. Gen Comp Endocrinol 165:516–534

Mylonas CC, Tabata Y, Langer R, Zohar Y (1995) Preparation and evaluation of polyanhydride microspheres containing gonadotropin-releasing hormone (GnRH), for inducing ovulation and spermiation in fish. J Control Release 35(1):23–34

Mylonas CC, Zohar Y (2000) Use of GnRHa-delivery systems for the control of reproduction in fish. Rev Fish Biol Fish 10(4):463–491

Nagahama Y (1997) 17α,20β-dihydroxy-4-pregnen-3-one, a maturation-inducing hormone in fish oocytes: mechanisms of synthesis and action. Steroids 62:190–196

Nagahama Y, Yamashita M (2008) Regulation of oocyte maturation in fish. Dev Growth Differ 50(suppl 1): S195–S219

Nash CE (2011) The history of aquaculture. Wiley-Blackwell, Iowa

Navarro-Martín L, Blázquez M, Viñas J, Joly S, Piferrer F (2009) Balancing the effects of rearing at low temperature during early development on sex ratios, growth and maturation in the European sea bass (*Dicentrarchus labrax*): limitations and opportunities for the production of highly female-biased stocks. Aquaculture 296(3–4):347–358

Nazar DS, Persson G, Olin T, Waters S, von der Decken A (1991) Sarcoplasmic and myofibrillar proteins in white truck muscle of salmon (*Salmo salar*) after estradiol treatment. Comp Biochem Physiol B 98(b):109–114

Nichols KM (2009) Clonal lines and chromosome manipulation for aquaculture research and production. In: Overturf K (ed) Molecular research in aquaculture. Wiley-Blackwell, Hoboken, NJ, pp 195–216

Nilsson EE, Cloud JG (1992) Rainbow trout chimeras produced by injection of blastomeres into recipient blastulae. Proc Natl Acad Sci USA 89:9425–9428

Nilsson E, Cloud JG (1993) Cryopreservation of rainbow trout (*Oncorhynchus mykiss*) blastomeres. Aquat Living Resour 6:77–80

Noh CH, Kim DS (2012) Growth response to a GH-autotransgenesis in common carp *Cyprinus carpio*. Fish Aquat Sci 15:37–41

Novelo ND, Tiersch TR (2012) A review of the use of ultrasonography in fish reproduction. N Am J Aquac 74(2):169–181

Ohta H, Kagawa H, Tanaka H, Okuzawa K, Iinuma N, Hirose K (1997) Artificial induction of maturation and fertilization in the Japanese eel, *Anguilla japonica*. Fish Physiol Biochem 17(1–6):163–169

Okutsu T, Shikina S, Kanno M, Takeuchi Y, Yoshizaki G (2007) Production of trout offspring from triploid salmon parents. Science 317(5844):1517

Okutsu T, Suzuki K, Takeuchi Y, Takeuchi T, Yoshizaki G (2006) Testicular germ cells can colonize sexually undifferentiated embryonic gonad and produce functional eggs in fish. Proc Natl Acad Sci U S A 103(8):2725–2729

Ong SK, Chotisukarn P, Limpiyakorn T (2012) Sorption of 17α-methyltestosterone onto soils and sediment. Water Air Soil Pollut 223:3869–3875

Oppedal F, Dempster T, Stien LH (2011) Environmental drivers of Atlantic salmon behaviour in sea-cages: a review. Aquaculture 311(1–4):1–18

Oppedal F, Taranger GL, Hansen T (2003) Growth performance and sexual maturation in diploid and triploid Atlantic salmon (*Salmo salar* L.) in seawater tanks exposed to continuous light or simulated natural photoperiod. Aquaculture 215(1–4):145–162

Ortenburger AI, Jansen ME, Whyte SK (1996) Nonsurgical videolaparoscopy for determination of reproductive status of the Arctic charr. Can Vet J 37:96–100

Palstra AP, van den Thillart GEEJM (2010) Swimming physiology of European silver eels (*Anguilla anguilla* L.): energetic costs and effects on sexual maturation and reproduction. Fish Physiol Biochem 36(3): 297–322

Pandian TJ, Koteeswaran R (1998) Ploidy induction and sex control in fish. Hydrobiology 384(1–3):167–243

Patiño R, Davis KB, Schoore JE, Uguz C, Strüssmann CA, Parker NC et al (1996) Sex differentiation of channel catfish gonads: normal development and effects of temperature. J Exp Zool 276(3):209–218

Patiño R, Sullivan CV (2002) Ovarian follicle growth, maturation, and ovulation in teleost fish. Fish Physiol Biochem 26:57–70

Pavlidis M, Koumoundouros G, Sterioti A, Somarakis S, Divanach P, Kentouri M (2000) Evidence of temperature-dependent sex determination in the European sea bass (*Dicentrarchus labrax* L.). J Exp Zool 287(3):225–232

Peter RE, Lin H-R, Van der Kraak GV (1988) Induced ovulation and spawning of cultured freshwater fish in China: advances in application of GnRH analogous and dopamine antagonists. Aquaculture 74:1–10

Petochi BT, Di Marco P, Donadelli V, Longobardi A, Corsalini I, Bertotto D et al (2011) Sex and reproductive stage identification of sturgeon hybrids (Acipenser naccarii×Acipenser baerii) using different tools: ultrasounds, histology and sex steroids. J Appl Ichthyol 27(2):637–642

Phelps RP (2006) Hormone manipulation of sex. In: Webster CW, Lim CE (eds) Tilapias: culture, nutrition and feeding. Hawthorn Press, New York, pp 211–252

Phelps RP, Hastey R, Broach J, Pendetar A, Linley L, Papanikos N et al (2012) Broodstock selection criteria for induced spawning of channel catfish for the production of channel × blue catfish hybrid fry and the Influence of temperature. N Am J Aquac 74:180–186

Pierce BA (1983) Grass carp status in the United States: a review. Environ Manag 7(2):151–160

Piferrer F, Beaumont A, Falguière JC, Flajšhans M, Haffray P, Colombo L (2009) Polyploid fish and shellfish: production, biology and applications to aquaculture for performance improvement and genetic containment. Aquaculture 293:125–156

Piferrer F, Carrillo M, Zanuy S, Solar II, Donaldson EM (1994) Induction of sterility in coho salmon (*Oncorhynchus kisutch*) by androgen immersion before first feeding. Aquaculture 119(4):409–423

Porter MJR, Duncan NJ, Mitchell D, Bromagea NR (1999) The use of cage lighting to reduce plasma melatonin in Atlantic salmon (*Salmo salar*) and its effects on the inhibition of grilsing. Aquaculture 176(3–4): 237–244

Potaros M, Sitasit P (1976) Induced spawning of *Pangasius sutchi* (Fowler). Freshwater Fisheries Division, Department of Fisheries, Bangkok. Technical paper no. 15

Powers DA, Hereford L, Cole T, Chen TT, Lin CM, Kight K et al (1992) Electroporation: a method for transferring genes into the gametes of zebrafish (*Brachydanio rerio*), channel catfish (*Ictalurus punctatus*), and common carp (*Cyprinus carpio*). Mol Mar Biol Biotechnol 1(4–5):301–308

Quillet E, Aubard G, Quéau I (2002) Mutation in a sex-determining gene in rainbow trout: detection and genetic analysis. J Hered 93(2):91–99

Rabanal H (1988) History of aquaculture. ASEAN/UNDP/FAO Regional Small-Scale Coastal Fisheries Development Project, Manila. Technical report no. ASEAN/SF/88/Tech.7

Rahman MA, Maclean N (1999) Growth performance of transgenic tilapia containing an exogenous piscine growth hormone gene. Aquaculture 173(1–4): 333–346

Riley LG, Hirano T, Grau EG (2002a) Disparate effects of gonadal steroid hormones on plasma and liver mRNA levels of insulin-like growth factor-I and vitellogenin in the tilapia, *Oreochromis mossambicus*. Fish Phys Biochem 26:223–230

Riley LG, Richman NH III, Hirano T, Grau EG (2002b) Activation of the growth hormone/insulin-like growth factor axis by treatment with 17α-methyltestosterone and seawater rearing in the tilapia, *Oreochromis mossambicus*. Gen Comp Endocrinol 127(3):285–292

Robles V, Cabrita E, Acker AJP, Herráez P (2009) Embryo cryopreservation: what we know until now. In: Cabrita E, Robles V, Herráez P (eds) Methods in reproductive aquaculture: marine and freshwater species. CRC Press, Boca Raton, pp 265–294

Rodgers BD, Garikipati DK (2008) Clinical, agricultural, and evolutionary biology of myostatin: a comparative review. Endocr Rev 29(5):513–534

Ross RM (1984) Catheterization: a non-harmful method of sex identification for sexually monomorphic fishes. Prog Fish Cult 46(2):151–152

Routray P, Choudhary AK, Dash SN, Verma DK, Dash C, Swain P et al (2006) Cryopreservation of dead fish spermatozoa several hours after death of Indian major carp, Labeo rohita and its successful utilization in fish production. Aquaculture 261:1204–1211

Routray P, Verma DK, Sarkar SK, Sarangi N (2007) Recent advances in carp seed production and milt cryopreservation. Fish Physiol Biochem 33(4): 413–427

Sacobie CFD, Glebe BD, Barbeau MA, Lall SP, Benfey TJ (2012) Effect of strain and ploidy on growth performance of Atlantic salmon, *Salmo salar*, following seawater transfer. Aquaculture 334–337:58–64

Sadler J, Pankhurst PM, King HR (2001) High prevalence of skeletal deformity and reduced gill surface area in triploid Atlantic salmon (*Salmo salar* L.). Aquaculture 198(3–4):369–386

Saillant E, Fostier A, Menu B, Haffray P, Chatain B (2001) Sexual growth dimorphism in sea bass Dicentrarchus labrax. Aquaculture 202(3–4):371–387

Saito T, Goto-Kazeto R, Arai K, Yamaha E (2008) Xenogenesis in teleost fish through generation of germ-line chimeras by single primordial germ cell transplantation. Biol Reprod 78(1):159–166

Saito T, Goto-Kazeto R, Fujimoto T, Kawakami Y, Arai K, Yamaha E (2010) Inter-species transplantation and migration of primordial germ cells in cyprinid fish. Int J Dev Biol 54(10):1479–1484
Sakao S, Fujimoto T, Kimura S, Yamaha E, Arai K (2006) Drastic mortality in tetraploid induction results from the elevation of ploidy in masu salmon *Oncorhynchus masou*. Aquaculture 252(2–4):147–160
Sakao S, Fujimoto T, Kobayashi T, Yoshizaki G, Yamah E, Arai K (2009) Artificially induced tetraploid masu salmon have the ability to form primordial germ cells. Fish Sci 75:993–1000
Sandra GE, Norma MM (2010) Sexual determination and differentiation in teleost fish. Rev Fish Biol Fish 20(1):101–121
Sangalang GB, Freeman HC, Flemming RB (1978) A simple technique for determining the sex of fish by radioimmunoassay using 11-ketotestosterone antiserum. Gen Comp Endocrinol 36(2):187–193
Santandreu IA, Diaz NF (1994) Effect of 17alpha-methyltestosterone on growth and nitrogen excretion in masu salmon (*Oncorhynchus masau* Brevoort). Aquaculture 124:321–333
Schultz DR, Perez N, Tan CK, Mendez AJ, Capo TR, Snodgrass D et al (2005) Concurrent levels of 11-ketotestosterone in fish surface mucus, muscle tissue and blood. J Appl Ichthyol 21(5):394–398
Scott AP, Ellis T (2007) Measurement of fish steroids in water—a review. Gen Comp Endocrinol 153(1–3):392–400
Scott AG, Penman DJ, Beardmore JA, Skibinski DOF (1989) The 'YY' supermale in *Oreochromis niloticus* (L.) and its potential in aquaculture. Aquaculture 78(3–4):237–251
Segner H, Caroll K, Fenske M, Janssen CR, Maack G, Pascoe D et al (2003) Identification of endocrine-disrupting effects in aquatic vertebrates and invertebrates: report from the European IDEA project. Ecotoxicol Environ Saf 54(3):302–314
Servid SA, Talbott MJ, Van Eenennaam JP, Doroshov SI, Struffenegger P, Webb MAH et al (2011) Rapid noninvasive characterization of ovarian follicular atresia in cultured white sturgeon (*Acipenser transmontanus*) by near infrared spectroscopy. Aquaculture 315(3–4):290–297
Sheela SG, Pandian TJ, Mathavan S (1999) Electroporatic transfer, stable integration, expression and transmission of pZpβypGH and pZpβrtGH in Indian catfish, *Heteropneustes fossilis* (Bloch). Aquac Res 30(4):233–248
Shehadeh ZH, Kuo C, Milisen KK (1973) Validation of an in vivo method for monitoring ovarian development in the grey mullet (*Mugil cephalus* L.). J Fish Biol 5:489–496
Shields RJ, Davenport J, Young C, Smith PL (1993) Oocyte maturation and ovulation in the Atlantic halibut, *Hippoglossus hippoglossus* (L.), examined using ultrasonography. Aquac Fish Manag 24:181–186
Small B, Quiniou S, Warren J, Ott L, Khoo L (2011) Testicular germ cell transplantation: can it be used for hybrid catfish fry production? Aquaculture America, 28 Feb–3 Mar 2011, New Orleans, LA, p 436
Smith CJ, Haley SR (1987) Evidence of steroidogenesis in postovulatory follicles of the tilapia, *Oreochromis mossambicus*. Cell Tissue Res 247(3):675–687
Song C, Liu S, Xiao J, He W, Zhou Y, Qin Q et al (2012) Polyploid organisms (review). Sci China Life Sci 55:301–311
Sparks RT, Shepherd BS, Ron B, Richman NH III, Riley LG, Iwama GK et al (2003) Effects of environmental salinity and 17α-methyltestosterone on growth and oxygen consumption in the tilapia, *Oreochromis mossambicus*. Comp Biochem Physiol B Biochem Mol Biol 136(4):657–665
Sullivan CV, Berlinsky DL, Hodson RG (1997) Striped bass and other morone culture. Dev Aquac Fish Sci 30:11–73
Sullivan CV, Hiramatsu N, Kennedy AM, Clark RW, Weber GM, Matsubara T et al (2003) Induced maturation and spawning: opportunities and applications for research on oogenesis. Fish Physiol Biochem 28:481–486
Sumpter JP (2005) Endocrine disrupters in the aquatic environment: an overview. Acta Hydrochim Hydrobiol 33(1):9–16
Tabata K (1991) Induction of gynogenetic diploid males and presumption of sex determination mechanisms in the hirame *Paralichthys olivaceus*. Nippon Suisan Gakkaishi 57:845–850
Takemura A, Kanemats M, Oka M (1996) Early sex distinction in greater amberjack *Seriola dumerili* using skin mucus. Nippon Suisan Gakkaishi 62(1):62–67, Japanese
Takemura A, Oka M (1998) Immunochemical sexing of living yellowfin tuna, Thunnus albacares (Bonnaterre), using a vitellogenin-like protein. Aquac Res 29(4):245–249
Takeuchi Y, Yoshizaki G, Kobayashi T, Takeuchi T (2002) Mass isolation of primordial germ cells from transgenic rainbow trout carrying the green fluorescent protein gene driven by the vasa gene promoter. Biol Reprod 67(4):1087–1092
Takeuchi Y, Yoshizaki G, Takeuchi T (2001) Production of germ-line chimeras in rainbow trout by blastomere transplantation. Mol Reprod Dev 59(4):380–389
Takeuchi Y, Yoshizaki G, Takeuchi T (2004) Surrogate broodstock produces salmonids. Nature 430(7000):629–630
Talbott MJ, Van Eenennaam JP, Linares-Casenave J, Doroshov SI, Guy CS, Struffenegger P et al (2011) Investigating the use of plasma testosterone and estradiol-17β to detect ovarian follicular atresia in farmed white sturgeon, *Acipenser transmontanus*. Aquaculture 315(3–4):283–289
Taranger GL, Aardal L, Hansen T, Kjesbu OS (2006) Continuous light delays sexual maturation and increases growth of Atlantic cod (*Gadus morhua* L.) in sea cages. ICES J Mar Sci 63(2):365–375
Taranger GL, Carrillo M, Schulz RW, Fontaine P, Zanuy S, Felip A et al (2010) Control of puberty in farmed fish. Gen Comp Endocrinol 165(3):483–515

Taylor JF, Preston AC, Guy D, Migaud H (2011) Ploidy effects on hatchery survival, deformities, and performance in Atlantic salmon (*Salmo salar*). Aquaculture 315(1–2):61–68

Thorgaard GH (1986) Ploidy manipulation and performance. Aquaculture 57(1–4):57–64

Thorgaard GH (1992) Application of genetic technologies to rainbow trout. Aquaculture 100(1–3):85–97

Thorgaard GH, Rabinovitch PS, Shen MW, Gall GAE, Propp J, Utter FM (1982) Triploid rainbow trout identified by flow cytometry. Aquaculture 29(3–4):305–309

Trewavas E (1983) Tilapiine fishes of the genera *Sarotherodon Oreochromis* and *Danakilia*. British Museum (Natural History), London

Van Veld PA, Rutan BJ, Sullivan CA, Johnston LD, Rice CD, Fisher DF et al (2005) A universal assay for vitellogenin in fish mucus and plasma. Environ Toxicol Chem 24(12):3048–3052

Vera LM, Davie A, Taylor JF, Migaud H (2010) Differential light intensity and spectral sensitivities of Atlantic salmon, European sea bass and Atlantic cod pineal glands ex vivo. Gen Comp Endocrinol 165(1):25–33

Vera LM, Migaud H (2009) Continuous high light intensity can induce retinal degeneration in Atlantic salmon, Atlantic cod and European sea bass. Aquaculture 296(1–2):150–158

Vilizzi L (2012) The common carp, *Cyprinus carpio*, in the Mediterranean region: origin, distribution, economic benefits, impacts and management. Fish Manag Ecol 19(2):93–110

Villarreal CA, Thorpe JE (1985) Gonadal growth and bimodality of length frequency distribution in juvenile Atlantic salmon (*Salmo salar*). Aquaculture 45(1–4):265–288

Viveiros ATM, Fessehaye Y, Ter Veld M, Schulz RW, Komen J (2002) Hand-stripping of semen and semen quality after maturational hormone treatments, in African catfish *Clarias gariepinus*. Aquaculture 213(1–4):373–386

von Ihering R (1935) Die wirkung von hypophyseninjektion auf den laichakt von fischen. Zool Anz 111:273–279, German

von Ihering R (1937) A method for inducing fish to spawn. Prog Fish Cult 34:15–16

Wang LH, Tsai CL (2000) Effects of temperature on the deformity and sex differentiation of tilapia, *Oreochromis mossambicus*. J Exp Zool 286(5): 534–537

Wang N, Teletchea F, Kestemont P, Milla S, Fontaine P (2010) Photothermal control of the reproductive cycle in temperate fishes. Rev Aquac 2(4):209–222

Watanabe WO, Smith SJ, Wicklund RI, Olla BL (1992) Hatchery production of Florida red tilapia seed in brackishwater tanks under natural-mouthbrooding and clutch-removal methods. Aquaculture 102(1–2):77–88

Wattendorf RJ (1986) Rapid identification of triploid grass carp with a Coulter counter and Channalizer. Prog Fish Cult 48:125–132

Webb MAH, Doroshov SI (2011) Importance of environmental endocrinology in fisheries management and aquaculture of sturgeons. Gen Comp Endocrinol 170(2):313–321

Webb MAH, Van Eenennaam JP, Feist GW, Linares-Casenave J, Fitzpatrick MS, Schreck CB et al (2001) Effects of thermal regime on ovarian maturation and plasma sex steroids in farmed white sturgeon, *Acipenser transmontanus*. Aquaculture 201:137–151

Weber GM, Hostuttler MA (2012) Factors affecting the first cleavage interval and effects of parental generation on tetraploid production in rainbow trout (*Oncorhynchus mykiss*). Aquaculture 344–349: 231–238

Weber GM, King VW, Clark RW, Hodson RG, Sullivan CV (2000) Morpho-physiological predictors of ovulatory success in captive striped bass (*Morone saxatilis*). Aquaculture 188(1–2):133–146

Weber GM, Lee CS (1985) Effects of 17α-methyltestosterone on spermatogenesis and spermiation in the grey mullet, *Mugil cephalus* L. J Fish Biol 26(1):77–84

Weber GM, Sullivan CV (2000) Effects of insulin-like growth factor-I on in vitro final oocyte maturation and ovarian steroidogenesis in striped bass, *Morone saxatilis*. Biol Reprod 63(4):1049–1057

Weber GM, Sullivan CV (2005) Insulin-like growth factor-I induces oocyte maturational competence but not meiotic resumption in white bass (*Morone chrysops*) follicles in vitro: evidence for rapid evolution of insulin-like growth factor action. Biol Reprod 72(5):1177–1186

Weil C, Crim LW (1983) Administration of LHRH analogues in various ways: effect on the advancement of spermiation in prespawning landlocked salmon, *Salmo salar*. Aquaculture 35(c):103–115

Withler FC, Morley RB (1968) Effects of chilled storage on viability of stored ova and sperm of sockeye and pink salmon. J Fish Res Board Can 25:2695–2699

Wohlfarth GW (1994) The unexploited potential of tilapia hybrids in aquaculture. Aquac Fish Manag 25(8): 781–788

Wong AC, Van Eenennaam AL (2008) Transgenic approaches for the reproductive containment of genetically engineered fish. Aquaculture 275(1–4):1–12

Wu C (1990) Retrospects and prospects of fish genetics and breeding research in China. Aquaculture 85(1–4):61–68

Yamamoto E (1999) Studies on sex-manipulation and production of cloned populations in hirame, *Paralichthys oliuaceus* (Temminck et Schlegel). Aquaculture 173(1–4):235–246

Yano A, Guyomard R, Nicol B, Jouanno E, Quillet E, Klopp C et al (2012) An immune-related gene evolved into the master sex-determining gene in rainbow trout, *Oncorhynchus mykiss*. Curr Biol 22:1423–1428

Yashouv A, Hefez A (1959) A key to tilapia species found in pond areas. Isr J Aquac Bamidgeh 11:36–42

Yaskowiak ES, Shears MA, Agarwal-Mawal A, Fletcher GL (2006) Characterization and multi-generational stability of the growth hormone transgene (EO-1α) responsible for enhanced growth rates in Atlantic Salmon. Transgenic Res 15(4):465–480

Yasui GS, Fujimoto T, Sakao S, Yamaha E, Arai K (2011) Production of loach (*Misgurnus anguillicaudatus*) germ-line chimera using transplantation of primordial germ cells isolated from cryopreserved blastomeres. J Anim Sci 89(8):2380–2388

Yoshizaki G, Okutsu T, Morita T, Terasawa M, Yazawa R, Takeuchi Y (2012) Biological characteristics of fish germ cells and their application to developmental biotechnology. Reprod Domest Anim 47: 187–192

Yoshizaki G, Tago Y, Takeuchi Y, Sawatari E, Kobayashi T, Takeuchi T (2005) Green fluorescent protein labeling of primordial germ cells using a nontransgenic method and its application for germ cell transplantation in salmonidae. Biol Reprod 73(1):88–93

Yu F, Xiao J, Liang X, Liu S, Zhou G, Luo K et al (2011) Rapid growth and sterility of growth hormone gene transgenic triploid carp. Chin Sci Bull 56(16): 1679–1684

Zajicek P, Goodwin AE, Weier T (2011) Triploid grass carp: triploid induction, sterility, reversion, and certification. N Am J Fish Manag 31:614–618

Zhang T, Lubzen E (2009) Cryopreservation of fish oocytes. In: Cabrita E, Robles V, Herráez P (eds) Methods in reproductive aquaculture: marine and freshwater species. CRC Press, Boca Raton, pp 251–264

Zhang X, Onozato H (2004) Hydrostatic pressure treatment during the first mitosis does not suppress the first cleavage but the second one. Aquaculture 240(1–4): 101–113

Zhu Z (1992) Generation of fast-growing transgenic fish: methods and mechanisms. In: Hew CL, Fletcher GL (eds) Transgenic fish. World Publishing, Singapore, pp 92–119

Zhu Z, Li G, He L, Chen S (1985) Novel gene transfer into the fertilized eggs of goldfish (*Carassius auratus* L. 1758). Z Angew Ichthyol 1:31–34

Zohar Y, Muñoz-Cueto JA, Elizur A, Kah O (2010) Neuroendocrinology of reproduction in teleost fish. Gen Comp Endocrinol 165(3):438–455

Zohar Y, Mylonas CC (2001) Endocrine manipulations of spawning in cultured fish: from hormones to genes. Aquaculture 197(1–4):99–136

Incorporation of Genetic Technologies Associated with Applied Reproductive Technologies to Enhance World Food Production

4

Robert A. Cushman, Tara G. McDaneld, Larry A. Kuehn, Warren M. Snelling, and Dan Nonneman

Abstract

Animal breeding and reproductive physiology have been closely related throughout the history of animal production science. Artificial insemination provides the best method of increasing the influence of sires with superior genetics to improve production traits. Multiple ovulation embryo transfer (MOET) provides some ability to increase the genetic influence of the maternal line as well. The addition of genetic technologies to this paradigm allows for improved methods of selecting sires and dams carrying the best genes for production and yield of edible products and resistance to diseases and parasites. However, decreasing the number of influential parents within a population also increases the risk of propagating a recessive gene that could negatively impact the species (Reprod Domest Anim 44:792–796, 2009; BMC Genomics 11:337, 2010). Furthermore, antagonistic genotypic relationships between production traits and fertility (Anim Prod Sci 49:399–412, 2009; Anim Genet 43:442–446, 2012) suggest that care must be taken to ensure that increasing the frequency of

Mention of trade names or commercial products in this article is solely for the purpose of providing specific information and does not imply recommendation or endorsement by the USDA. USDA is an equal opportunity provider and employer.

R.A. Cushman, Ph.D. (✉)
Reproduction Research Unit, U.S. Meat Animal Research Center, State Spur 18D, Clay Center, NE 68933-0166, USA
e-mail: bob.cushman@ars.usda.gov

T.G. McDaneld • L.A. Kuehn • W.M. Snelling
D. Nonneman
USDA-ARS, U.S. Meat Animal Research Center, Clay Center, NE, USA

G.C. Lamb and N. DiLorenzo (eds.), *Current and Future Reproductive Technologies and World Food Production*, Advances in Experimental Medicine and Biology 752, DOI 10.1007/978-1-4614-8887-3_4, genes with a positive influence on production does not negatively impact the fertility of the replacement females entering the herd.

Keywords
Genetic technology • Reproduction • Animal breeding • Genotypes • Multiple ovulation embryo transfer

Identification and Application of Genetic Parameters of Reproductive Traits

In contrast to the use of assisted reproductive technologies to rapidly propagate superior genetics for yields of edible food, the advancements in the use of genetic technologies to improve reproduction have been slow in domestic farm species, mostly due to the relatively low heritability of these traits (Cushman et al. 2008; Cammack et al. 2009). Among reproductive traits, those with the greatest heritability are associated with sexual maturity (Tables 4.1 and 4.2), probably because many of these traits depend on the animal attaining a certain age, body mass, or body composition (Gargantini et al. 2005; Johnston et al. 2009). The interaction between growth and reproductive traits has been demonstrated clearly in experiments with transgenic mice where intentional inactivation of specific genes that control growth and development results in decreased fecundity and decreased fertility (Elvin and Matzuk 1998; Matzuk and Lamb 2002). Conversely, selection for increased ovulation rate in cattle resulted in an increase in serum Insulin-Like Growth Factor-1 (IGF-1) concentrations (Echternkamp et al. 1990b), and, in sheep, was associated with polymorphisms in several genes in the transforming growth factor-beta family (Elvin et al. 2000; Galloway et al. 2000; Wilson et al. 2001; McNatty et al. 2005). Thus, the tight relationship between growth and development genes and reproductive success suggests that genetic technologies must be used with care to improve production efficiency without negatively impacting fertility.

Table 4.1 Heritability of reproductive traits in cattle

Trait	Heritability	References
Age at first ovulation	0.28	Mialon et al. (2001)
Age at first progesterone	0.38	Mialon et al. (2001)
Age at puberty	0.14 0.24	Snelling et al. (2012) Morris et al. (2000)
Reproductive tract score	0.30	Martin et al. (1992)
Yearling uterine horn diameter	0.20	Johnston et al. (2009)
Scrotal circumference	0.41 0.41	Morris et al. (2000) Lunstra et al. (1988)
Antral follicle count	0.44	Snelling et al. (2012)
Ovulation rate	0.07	Echternkamp et al. (1990a)
Follicle diameter	0.16	MacNeil et al. (2006)
Calving day	0.08	Minick Bormann and Wilson (2010)
Heifer pregnancy rate	0.11 0.21 0.28	Snelling et al. (2012) Doyle et al. (2000) Thallman et al. (1999)
Pregnancy rate	0.07	MacNeil et al. (2006)
Stayability	0.15	Doyle et al. (2000)

In a number of studies, age at puberty in beef heifers has been reported to have a moderate heritability (Martin et al. 1992; Morris et al. 2000; Mialon et al. 2001), and to be genetically correlated positively with postpartum interval to first ovulation (Mialon et al. 2000). In swine, age at first estrus is correlated with weaning to estrus interval and strength of estrus symptoms and is moderately heritable (Sterning et al. 1998; Knauer et al. 2010). This would suggest that e a group of genes influences the initiation of reproductive cycles both

Table 4.2 Heritability of reproductive traits in pigs

Trait	Heritability	References
Age at puberty	0.46 0.38	Kuehn et al. (2009) Rosendo et al. (2007b)
Age at first farrowing	0.22	Knauer et al. (2011)
Ovulation rate	0.28 0.34 0.45	Cassady et al. (2000) Rosendo et al. (2007a) Schneider et al. (2011)
Total number born	0.09	Schneider et al. (2012)
Number born alive	0.12	Schneider et al. (2011)
Number born dead	0.02	Schneider et al. (2011)
Litter birth weight	0.18	Schneider et al. (2011)
Piglet birth weight	0.44 0.36	Schneider et al. (2011) Rosendo et al. (2007b)
Stayability	0.14	Knauer et al. (2011)
Testes weight	0.35	Johnson et al. (1994)
Weaning to estrus interval	0.02	Schneider et al. (2011)

at puberty and during postpartum recovery. However, using age at puberty as a reproductive phenotype requires intensive observations and measurements, whether it is determined by detection of behavioral estrus, sequential blood sampling, or sequential ultrasonography. Furthermore, early age at puberty may increase the risk of a beef heifer being bred by a co-pastured bull before she is weaned, which could lead to trauma associated with a 12–14-month-old heifer trying to give birth. Therefore, caution must be used with the genomic applications for this trait in the beef industry. It most likely has more application for the dairy, swine, and sheep industries where management practices and physiology limit the chances of a very young female coming in contact with a fertile male.

In beef heifers, the use of sequential ultrasonography of the ovaries to detect the presence of a corpus luteum (CL) to determine age at puberty has been applied in research situations to identify chromosomal regions potentially affecting this trait (Johnston et al. 2009; Fortes et al. 2010). However, there is no reason to limit these exams to the identification of a CL. In heifers, the relative size of the reproductive tract (reproductive tract score) provides a good estimate of pubertal status; therefore, ultrasonography can also be used to determine the size of the reproductive tract (Martin et al. 1992). Holm et al. (2009) reported that yearling heifers with low reproductive tract scores (indicating small, immature tracts) had decreased pregnancy rates in their first breeding season. The conception rate in their second breeding season was also decreased, perhaps due to inherently poor development of the reproductive tract and not just inability to attain puberty. Thus, the application of ultrasonography to measure ovarian size and uterine horn diameter can provide additional phenotypes for genomic association studies to identify genetic markers of advanced reproductive tract development prior to the first breeding season. These could be the ideal markers for selecting replacement females at a very young age.

Antral follicle count, determined by ultrasonography, provides a good estimate of the number of microscopic follicles within the mammalian ovary, because the number of macroscopic follicles is correlated positively with the number of microscopic follicles (Cushman et al. 1999; Ireland et al. 2008). Antral follicle count is also predictive of response to exogenous follicle stimulating hormone (FSH) in MOET protocols in cattle (Cushman et al. 1999; Singh et al. 2004), and may be an indicator of fertility, although reports have been mixed (Starbuck-Clemmer et al. 2007; Cushman et al. 2009; Tessaro et al. 2011; Mossa et al. 2012). Furthermore, a positive correlation between antral follicle count and uterine horn diameter (Jimenez-Krassel et al. 2009) suggests a relationship to reproductive tract development. Finally, breed differences within cattle for antral follicle count (Alvarez et al. 2000) suggest a genetic component to this trait that could be exploited to identify genes associating with reproductive tract development and reproductive longevity.

Candidate genes associated with reproductive tract development in heifers might provide clues to genes that could be sequenced for functional polymorphisms in other domestic species such as pigs and sheep, where limitations of ultrasonography make it difficult to assess reproductive tract development in vivo. Functional polymorphisms within these candidate genes, identified from genomic and expressed cDNA sequence may open an avenue to select for reproductive tract development in species where direct evaluation of the tract is not feasible.

Selection Lines

Given that there are some reproductive traits with moderate heritability, a number of selection lines have been created by scientists to assess genetic parameters of reproductive traits in domestic farm animals. In 1980, a selection experiment was initiated at the U.S. Meat Animal Research Center for the purpose of increasing twinning rate in beef cows. This long-term study demonstrated that a lowly heritable reproductive trait (twinning) could be improved by extensive selection and that the resulting population could be used to identify quantitative trait loci that associated with reproductive traits. Serum IGF-1 concentrations were greater in the cows selected for an increased twinning rate than in contemporary herd mates (Echternkamp et al. 1990b, 2004), supporting a positive role for IGF-1 in follicular development and possibly fecundity.

In 1983, a herd of Droughtmaster cows at Australia's Commonwealth Scientific Industrial Research Organization Lansdown Research Station was used to establish two lines; the first was selected for high fertility and the second for low fertility. Estimated breeding values for all cows and bulls in the herd were computed from the results of 9,086 pregnancy records collected on 2,280 cows between the years 1964 and 1983. Three years of breeding within the high and low fertility line resulted in a 12 % greater pregnancy rate in the high fertility line compared to the low fertility line (Hetzel et al. 1989). Scrotal circumferences were greater at 9 and 18 months of age in bulls from the high fertility line than in bulls from the low fertility line (Mackinnon et al. 1990). While age at puberty was never measured, the authors suggested based on scrotal circumference differences in the bulls that the heifers in the high fertility line may have been reaching sexual maturity at an earlier age.

In 1984, a selection line in Angus cattle was established at New Zealand's Ruakura Agricultural Research Center to examine the effects of selection for a decreased or increased age at first observed behavioral estrus. After 5 years of selection, there was a 66-day difference in age at puberty between heifers selected for a decreased age at puberty or an increased age at puberty (Morris et al. 2000). In this study, heifer pregnancy rates were also improved by selecting for decreased age at puberty, demonstrating a connection between fertility and an early age at puberty in beef heifers and supporting the conclusions drawn from the Droughtmaster fertility selection line that age at puberty was a component of fertility in replacement beef heifers from an base population where age at puberty may have been limiting.

A selection line was initiated at the U.S. Ohio State University in 1989 to examine the influence of selecting for high or low serum IGF-1 concentrations on production traits in Angus cattle. Among females, there was a tendency for heifers in the high line to be 4 days younger at first calving; however, there was no change in age at puberty (Yilmaz et al. 2006). In contrast, Johnston et al. (2009) reported a moderate negative genetic correlation between serum IGF-1 concentrations and age at puberty (defined as detection of a CL by palpation) in Brahman heifers, suggesting that higher serum IGF-1 concentrations might be associated with an earlier onset of puberty. This may be related to follicle selection and ovulation, because immunizing beef heifers against growth hormone releasing factor increased the age at puberty (Schoppee et al. 1996), but did not change the number of antral follicles that were 6 mm or smaller (Cohick et al. 1996). Overall, the lack of a correlated response in age at puberty in the IGF-1 selection lines still suggests that other factors may be influencing the onset of reproductive cycles in replacement females. Thus, age at puberty may only be directly influencing fertility when it is limiting as in situations where many females from a population do not attain puberty until late in their first breeding season due to breed or nutritional status.

In the US dairy cattle population, intense selection for milk yield has clearly resulted in a decrease in fertility over the last several decades (Butler 1998). When serum concentrations of growth hormone, insulin, and IGF-1 were compared between modern Holsteins and a control unselected line from the Southern Experiment Station at the University of Minnesota, there was a decrease in insulin, increase in growth hormone, mixed responses in IGF-1 and decreases in luteal phase serum progesterone in Holsteins that had been selected for increased milk yield. The mixed response in IGF-1 reflected higher IGF-1 in the selected line in the first 2 weeks postpartum followed by lower IGF-1 levels during the breeding season, suggesting that the decrease in IGF-1 might negatively affect luteal function and fertility in Holstein cows (Bonczek et al. 1988; Lucy et al. 1998). This demonstrates the need for multi-trait selection and the potential of genetic markers of fertility for improving selection indices to avoid negative impacts of selection for yield traits. The dairy industry has already begun to estimate daughter pregnancy rates and sire conception rates as part of its selection criteria in order to reverse the decrease in fertility.

Although heritability for litter size is low, selection for ovulation rate, uterine capacity, or prenatal survival as component traits of litter size in pigs has improved litter size. A Large White and Landrace composite line selected for eight generations at the University of Nebraska to increase ovulation rate and embryonic survival resulted in an increase of four ova and nearly two piglets (Ruiz-Flores and Johnson 2001). Along with increased litter size, correlated responses included decreased age at puberty, increased number of pigs born alive and increased birth weights. Subsequent generations underwent a two-stage selection for increased ovulation rate followed by selection for increased litter size (Cassady et al. 2000). Plasma FSH was genetically correlated with ovulation rate and greater in gilts and boars of the selection line; therefore, selection for plasma FSH concentrations could be a noninvasive method to select for ovulation rate. Selection of Large White pigs for ovulation rate and prenatal survival for six generations at INRA resulted in an increase of 0.24 piglets/generation and 0.5 corpora lutea/generation in the ovulation rate line (Rosendo et al. 2007a). An unfavorable genetic correlation with age at puberty in both selection lines (Rosendo et al. 2007b). The genetic correlation between ovulation rate and age at puberty may depend upon whether the ovulation rate is measured at a constant age or at physiological maturity. In another study at the University of Nebraska a composite population was selected for high ovulation rate for nine generations followed by two generations of random selection and eight generations of increased litter size, decreased age at puberty or continued selection for ovulation rate (Lamberson et al. 1991). The response to selection for ovulation rate was about 3.7 ova and 1.4–1.8 piglets. The response to selection for reduced age at puberty was 15.7 days without a change in litter size. Selection by the swine industry for litter size and age at first farrowing, a reflection of age at puberty, has made significant advances in the number of pigs weaned and the number of parities a sow stays in the breeding herd (Saito et al. 2011; Foxcroft 2012).

Overall, selection lines in livestock provide proof of concept that while the heritability of most reproductive traits may be low, there are genes contributing to reproductive phenotypes and that use of genomic technologies could enhance reproductive management in domestic farm animals. Selection lines also contributed greatly to the ability of quantitative geneticists to determine the genetic parameters of reproductive and production traits and to identify antagonistic relationships between production traits and fertility. With the advent of genomic technologies, research scientists can begin to identify the genes influencing important reproductive traits and dissect the mechanisms by which they function using a combination of the information from transgenic mouse models and genomic research in domestic farm species.

Contributions from Transgenic Mouse Models

There have been studies with selection lines in mice as well. At one time, the decreased generation interval made this an attractive alternative to farm animals before transgenic technologies were developed. However, the advent of transgenic technologies increased the ability of investigators to understand how specific genes affected reproductive development and performance by constructing transgenic mice that over-expressed, under-expressed, or conditionally expressed specific genes of interest. While reproductive traits are lowly heritable, transgenic mouse models have identified a number of genes that influence fecundity and fertility in both females and males (Matzuk and Lamb 2002). These studies provide even greater support for the concept that there are genetic components to reproduction and that biotechnologies to harness and manipulate these genetic components will be possible for domestic farm animals in the future. Furthermore, researchers have used the results from these mouse models to identify candidate genes and investigate them more intensely in domestic farm animals.

Mice null for the anti-Müllerian hormone (AMH) gene had an increased rate of activation of primordial follicles that resulted in a depletion of the ovarian reserve at a younger age (Durlinger et al. 1999, 2002a, b). Follow up investigation in cattle suggested that AMH may control activation of primordial follicles in this species as well (Cushman et al. 2002; Gigli et al. 2005). However, a more recent study in sheep suggests that AMH does not control follicle activation in this species, because immunization of ewes against AMH did not halt follicular development until the early antral stage (Campbell et al. 2012). Interestingly, serum AMH concentration was identified as a potential biomarker of follicle number (Ireland et al. 2009) and ovulatory response to exogenous FSH in cattle (Rico et al. 2009). Polymorphisms in the AMH and AMH Type II receptor genes have been associated with premature ovarian failure (POF), polycystic ovarian syndrome, and age at menopause in women (Kevenaar et al. 2007, 2008); however, to date, no polymorphisms in these genes have been reported to associate with follicle number or reproductive longevity in domestic farm animals.

Growth and differentiation factor 9 (GDF9) and bone morphogenic protein 15 (BMP15) are two members of the transforming growth factor-beta family produced by the oocyte that were reported to influence early follicular development in mice. Dong et al. (1996) reported that transgenic mice that did not produce functional GDF9 had a lesion that halted follicular development at the primary stage, resulting in a focusing of research on this family of genes by reproductive biologists. In sheep, a mutation in the bone morphogenic protein receptor 1B (BMPr1B or ALK6) was demonstrated to be responsible for the increased ovulation rate in highly prolific Booroola Merino Sheep (Wilson et al. 2001). In total 5 mutations in BMP15 and 1 mutation in GDF9 have been associated with increased ovulation rate in prolific sheep (McNatty et al. 2005). This research, initiated in response to results from a transgenic mouse model, has provided a great amount of basic information on how these growth factors control early follicular development in ruminants.

Transgenic female mice that do not produce IGF-1 develop follicles to the early antral stage but fail to ovulate (Baker et al. 1996; Zhou et al. 1997; Kadakia et al. 2001). This is not alleviated by treatment with exogenous gonadotropins, suggesting that in early antral follicles, IGF-1 is necessary to stimulate production of gonadotropin receptors. Immunizing beef heifers against growth hormone releasing factor increased the age at puberty (Schoppee et al. 1996), but did not change the number of antral follicles that were 6 mm or smaller (Cohick et al. 1996). While indicus cattle generally have greater numbers of antral follicles detectable by ultrasound (Alvarez et al. 2000), a recent study reported no differences in the numbers of microscopic follicles in the ovaries of indicus cows compared to taurine cows (Silva-Santos et al. 2011). Taken together with the IGF-1 selection line results, these data suggest a role for IGF-1 in the secondary to antral follicle transition and ovulation.

Inactivation of the clock genes in murine models has been associated with different aspects of sub-fertility and disruption of the reproductive cycles. For example, in Period 1 (Per1) null mice there is not only an increase in the number of anestrous mice but also among those that are cycling there is an increase in the number of abnormally long estrous cycles (Pilorz and Steinlechner 2008). Clearly, further investigation of the clock genes in domestic farm animals is warranted as viable candidates influencing reproductive cycles. A role for clock genes in ovarian function in cattle has been reported. The amount of Circadian Locomotor Output Cycles Kaput (CLOCK) mRNA was greatest in small antral follicles and decreased as follicular development progressed (Shimizu et al. 2012). In contrast, Period 2 (Per2) expression was increased in the granulosa of preovulatory follicles. The treatment of bovine granulosa cells with CLOCK and Per2 siRNAs decreased granulosa cell proliferation and LH receptor expression, but only CLOCK siRNA decreased estradiol production and P450 aromatase expression (Shimizu et al. 2011). Period 1 was expressed in bovine oocytes and expression levels were not associated with differences in ovulation rate (Cushman et al. 2007a). Steroids have also been shown to increase the expression of Per1 in the rat uterus (He et al. 2007). Thus, these clock genes appear to be involved in follicular selection and proper uterine function, which could explain the alterations in reproductive cycles observed in transgenic mice that do not produce Per1.

This list of genes is not all-inclusive but demonstrates how some of the major candidate genes that are investigated for reproductive function in domestic farm species have been supported and even initiated by discoveries in transgenic mouse models. There are limitations to transgenic mouse models, especially when gene inactivating technologies are used. Before the development of conditional knockout models that allowed the inactivation of genes in specific tissues, dissecting whole animal effects from tissue specific effects was very difficult. However, even with conditional knockouts, a specific phenotype can still result from an indirect effect of alterations of whole animal physiology due to alterations of feedback loops that are not a direct effect at the level of the tissue of interest. Given the limitations of what can be concluded from transgenic mouse models, many of these genes have been demonstrated to be important regulators of reproductive function in domestic farm animals as well. Thus, the use of transgenic mouse models is an efficient application of research funding to begin to identify the genes controlling fertility.

Conversely, while these inactivating mutations are excellent models for beginning to understand how specific genes influence reproductive function, the frequency and relevance of naturally occurring mutations that result in loss of function remains questionable. A recent re-sequencing study suggested that human genomes contain about 100 genuine loss of function variants with approximately 20 genes completely inactivated (MacArthur et al. 2012). This result suggests a remarkable tolerance for gene inactivation, possibly due to redundancy of function within gene families, and implies that there may be very few major gene mutations in nature. Dominant variants causing inactivation of a gene affecting fertility cannot exist because the carriers could not reproduce. Recessive mutations that do not affect fertility of carriers, but render offspring that are homozygous for the recessive allele infertile may persist at low frequencies.

Human Genetic Disorders

Human disorders that result in altered reproductive development and fertility can also be used to identify genes that influence reproductive function in livestock. There are clearly familial patterns to some of these disorders demonstrating a genetic component. However, the fact that reproduction does occur to proliferate these disorders demonstrates that very few of them result in complete infertility. As stated previously, this is most likely due to the redundancy of many of the genes in these pathways, insuring that if one gene is not functional it does not result in a cessation of the propagation of the species.

There is a large network of genes involved in sexual differentiation and our knowledge of how these genes influence development of the female reproductive tract and subsequent fertility is limited. The Y-linked sex-determining region Y (SRY) gene is responsible for testis development and, in the absence of SRY, ovarian development progresses. Recently, Wingless-type MMTV integration site family, member 4 (WNT-4) has been implicated in female reproductive tract development and in mice lacking WNT-4 the ovarian reserve is greatly depleted (Vainio et al. 1999). Turner Syndrome (monosomy 45,X) is characterized by gonadal dysgenesis with primary amenorrhea. Individuals with 46,XY gonadal dysgenesis have Y chromosome but are phenotypically female. They have undeveloped streak gonads and are usually identified at puberty when secondary sexual characteristics do not develop. Use of the bovine high density chip in a DNA pooling study to identify genotypes associated with low fertility, defined as failing to produce a calf in the first two breeding seasons, identified a Y-chromosome signal in cows with low fertility (McDaneld et al. 2012). This could be due to unidentified freemartins in the DNA pools, but it is also possible that an XY gonadal dysgenesis could be causing some reproductive failure in heifers.

POF is defined as a depletion of follicles in the ovaries of women before the age of 40. Both genetic predisposition and autoimmune disorders have been demonstrated to be major factors causing POF (Cordts et al. 2011; Jasti et al. 2012), and a number of candidate genes have been sequenced in an attempt to explain the genetic causes of POF (Cordts et al. 2011). Polymorphisms in BMP15, FSHr, and the estrogen receptor were all identified in cases of POF. However, the relatively low frequencies of these polymorphisms provided inconsistent results across studies. Therefore, none of them have been pursued intensively as a genetic marker of POF. This line of research is a clear example of how the candidate gene approach can be as unsuccessful as or even more unsuccessful than genome-wide association studies.

Fragile X mental retardation 1 (FMR1) may be the most intensely studied genetic variant associated with POF. Fragile X syndrome is caused by a hyperexpansion and hypermethylation of a CGG repeat in the 5′ region of the FMR1 gene. Depending on the number of CGG repeats there are three alleles identified (Allen et al. 2004); the normal allele (less than 55 CGG repeats), the permutated allele (55–200 CGG repeats), and the full mutation (>200 CGG repeats); an increasing size of the CGG repeat was associated with an increased risk of POF (Bretherick et al. 2005). Further studies reported that the number of CGG repeats correlated positively with serum FSH concentrations and negatively with serum AMH concentrations in women presenting with a primary diagnosis of repeated pregnancy loss (Gleicher et al. 2009). However, in women undergoing in vitro fertilization, there was no difference between genotypes in serum AMH concentrations or the number of oocytes recovered (Gleicher et al. 2012). In a line of transgenic mice developed with an increased CGG repeat in the 5′ region of the FMR1 gene, the starting ovarian reserve was normal, however, these mice showed a greater rate of follicle depletion, suggesting that the lesion is more in maintaining the ovarian reserve than in establishing it.

Clearly, in combination with the transgenic mouse models, these human genetic disorders of sexual development and reproductive lifespan have provided a basis for understanding a number of genes influencing female reproductive tract development and fertility. Given that there are genes influencing these traits and that disruption of these genes can cause sub-fertility or infertility in women, it is feasible that genomic technologies can be applied to improve the efficiency of assisted reproductive technologies and fertility in domestic farm animals.

Advancements in Molecular Genetic Technologies for Agriculturally Important Species and Their Applications to Assisted Reproductive Technologies

Due to the obvious applications to biomedical sciences, funding for sequencing DNA was skewed toward the human and mouse genomes in initial applications of the technology. Therefore, much of the early work to identify critical regions of the genome in agriculturally important species was limited to association studies that used linkage analysis to identify broad regions of the genome that associated with production traits. Restriction fragment length polymorphisms (RFLP) and microsatellites were the first sequence variants used for such association analyses (Kappes 1999). However, the mouse and human genomes rapidly provided large volumes of sequence information that allowed researchers working with domestic farm species to develop primers and initiate efforts to sequence specific genes of interest in domestic food animals. Information from the transgenic mouse models helped to identify candidate genes and focus research toward understanding how those genes were involved in controlling reproduction in livestock.

As cost of sequencing has decreased, the quality of consensus sequences and the opportunities for individual animal sequencing have increased. While the cost of individual sequencing still greatly exceeds the value of market animals, it is becoming affordable to sequence influential animals to inform large numbers of descendents. Bioinformatic tools are necessary to distill the large volumes of data created by whole genome sequencing down to meaningful genotypes from sequence variants. However, there are a growing number of databases and software programs that facilitate summarizing these data, identifying and classifying novel sequence variants and developing biologically relevant networks from this information, especially when combined with low density genotyping to impute pedigrees and genotypes.

In production agriculture, use of genetic markers will continue to emphasize the identification of animals with superior production traits such as growth, carcass traits (White et al. 2007), milk production, disease resistance (Casas and Snowder 2008), and parasite resistance (Bishop 2012). However, they can also be used to manage or eliminate animals with undesirable traits or carriers of known defects such as osteopetrosis in Red Angus cattle (Meyers et al. 2010) or Factor XI in Holstein cattle (Ghanem and Nishibori 2009). Perhaps the most important concern to address will be insuring that selecting for genetic markers for production traits does not negatively impact reproductive performance, similar to the problem that selecting almost exclusively for milk production has caused for fertility in dairy cows (Butler 1998).

Genome-Wide Association Studies

In the absence of cost-effective, efficient technologies to develop transgenic models to understand gene function in domestic farm animals, investigators initiated studies to associate regions of the genome with specific traits to capture inherent variation for use in marker assisted selection schemes. In early studies, repetitive regions termed microsatellites were used for such analysis; however, researchers working with domestic species were quick to pick up on sequence information generated from mouse genetic models and the human genome project and develop cost-effective methods for generating sequence for specific genes of interest in cattle, pigs, and sheep. Continued improvement in sequencing technologies allowed the identification of a broad array of Single Nucleotide Polymorphisms (SNPs) and the development of SNP Chips for high-throughput genotyping in cattle and pigs (Wiedmann et al. 2008; Matukumalli et al. 2009; Ramos et al. 2009).

There are still limitations to the application of this technology for identifying genetic markers for fertility. Fortes et al. (2010) reported that the number of SNP with significant effects exceeded what was expected for age at puberty in Brahman cattle. The authors used pathway analysis to identify the key pathways associating with puberty. Snelling et al. (2012) used the data from a genome-wide association study in heifers from the Germplasm Evaluation population at the U.S. Meat Animal Research Center to calculate the genomic heritability for age at puberty and antral follicle count and calculate genetic correlations between these traits. Thus, while it may be difficult to identify genetic markers with a large influence on reproduction in domestic farm species, it may still be possible to develop customized low density SNP chips that will predict genetic merit of replacement females by accounting for the interactions of these different components of reproduction. This will allow the development of genetic selection indices to improve production traits while ensuring that fertility is not sacrificed.

The greatest limitation to whole genome association studies with these new genotyping technologies is the lack of large populations of animals with recorded reproductive phenotypes. As genotyping and sequencing costs decrease, the ability to impute genotypes across the entire world population of a species becomes more likely; however, the ability to collect intensive reproductive phenotypes such as age at puberty, postpartum interval, ovulatory follicle diameter, and antral follicle count is still extremely doubtful. It is simpler and more achievable to have a database that contains body weights at major life transition points and birth dates. The swine and poultry industries in many developed countries are vertically integrated, so the implementation of databases containing production data should be easier for these species. Comprehensive record keeping in the dairy industry is organized efficiently in many countries and information provided through sharing of data among countries provides large databases on several phenotypes that are associated with reproductive performance. While record keeping and databases in the beef industry have lagged behind, the breed associations are starting to compile this information as the value of genetic markers appears on the horizon.

The accuracy of a whole genome association is limited by the accuracy of the phenotype measured. Thus research populations like those at the U.S. Meat Animal Research Center where pedigreed animals can be intensively characterized for reproductive traits are a necessity. Genetic markers that are first discovered by broad phenotypes such as age at first calving or pregnancy rate in production populations can be validated in animals with a wide spectrum of accurately measured reproductive phenotypes. This aids in the fine-mapping of the functional polymorphisms that are influencing the specific reproductive traits, and there is no better genetic marker than a functional polymorphism.

Studies using pools of DNA from large numbers of animals that express the extremes of reproductive phenotypes such as age at first calving and pregnancy rate may be a viable method to decrease the cost of discovery (Huang et al. 2010; McDaneld et al. 2012; Snelling et al. 2012). Polymorphisms identified in these studies can then be validated in animals from research centers where both production and reproductive traits can be extensively characterized.

There are still issues as to who will curate the databases. It is most likely that each species will maintain its own database, but will the individual breed associations maintain individual databases? How will these data be handled across international borders? Furthermore, there is a need for quality assurance of the data that is entered into these databases or it becomes meaningless.

Marker Assisted Selection

Marker Assisted Selection is selecting animals within a population to increase the frequency of alleles that have been demonstrated to positively influence an economically relevant trait. No longer is the animal being selected on is phenotypic measurements, but instead based upon a sequence within its genome.

In production agriculture it is still likely that growth and production traits that have moderate heritability and carry a great deal of economic impact are going to be the traits where genomic tools are most easily marketed. Already a number of mutations in myostatin (GDF8), growth hormone, and the growth hormone receptor have been reported to influence production traits (Hammond et al. 1991; White et al. 2007; McCormack et al. 2009). While the true functionality of many of these polymorphisms remains questionable their association with increased carcass weights and growth traits suggest that packers will begin to pay for animals confirmed to carry these alleles and, thus, many producers will adopt these markers in the next decade.

The bigger concern for reproductive biologists is the potential negative impact that selection for these polymorphisms might have on reproductive performance. Johnston et al. (2009) reported that residual feed intake and meat color in Brahman steers had antagonistic genetic correlations with age at puberty in females. Antagonistic genetic correlations between production traits and reproductive traits have been reported (Collis et al. 2012). However, polymorphisms that had relatively large positive effects on production traits had relatively small negative effects on fertility. The difficulty is that without tangential selection for fertility, those small negative effects will be unchecked and gradually accumulate. A large number of transforming growth factor-beta family members are expressed in the male and female reproductive tract, and myostatin, for example, is produced by the granulosa cells of small antral follicles in cows (Skinner et al. 2008). While the polygenic effects observed in genome-wide association studies suggest very little evidence of a major gene effect on reproductive traits, this raises the question of what might happen when a number of these genes influencing growth traits become are under selection pressure through genomic selection? What happens when a loss of function variant of myostatin is combined with an allele in the growth hormone receptor that is also positive for growth? What if an IGF-1 polymorphism is added to that combination? At some point, these production trait polymorphisms could tip the scales against reproduction unless equal effort is focused on identifying positive alleles for reproductive function and balancing them in these populations. The easiest way to do this may be to incorporate sire lines and maternal lines into the beef and dairy industries as has been done for the swine and poultry industries.

Marker Assisted Management and Pharmacogenetics

Marker assisted management is an approach for managing individual animals based on their genetic information. Animals are grouped by specific genotypes and management decisions are optimized for those genotypes. For the reproductive biologist, pharmacogenetics is the applicable side of marker assisted management that has made headway in human medicine (Greb et al. 2005a). Pharmacogenetics explains variation in response to drug or hormone treatments based on an individual's genetic background. From the stand-point of reproductive management, the most obvious target genes for pharmacogenetic analysis are the receptors to the specific hormones that are being provided as part of a synchronization of estrus or a MOET protocol.

There is large variation among animals in the response to hormones for synchronization of estrus and MOET, and a great deal of research has focused on improving the timing and dosage of various hormones. However, very little research has focused on how genetic variation might influence an animal's individual response to hormonal stimulation. This has been most closely investigated in women, where a polymorphism in the FSH receptor is associated with decreased response to FSH (Greb et al. 2005a, b). Polymorphisms have been identified in the gonadotropin releasing hormone receptor (GnRHr) and the FSH receptor in cattle (Yang et al. 2010; Líron et al. 2011). However, intensive studies to investigate their influence on reproductive traits or response to exogenous hormones have not been forthcoming.

In cattle, the most intensively studied gene with a proposed influence on the response to exogenous hormones is the glutamate receptor ionophore A1 (GRIA1). Cows that were homozygous for this polymorphism had a decreased number of ovulations in response to exogenous FSH (Sugimoto et al. 2010). Interestingly, this polymorphism was also associated with a lower number of antral follicles detectable by ultrasound on day 9 and 10 of the estrous cycle. Immortalized hypothalamic cells transfected with the mutant form of the receptor released about 70 % less gonadotropin releasing hormone in response to glutamate treatment in vitro. There was a decreased LH surge after induced luteal regression in cows carrying the polymorphism and conception rate to artificial insemination was decreased as well. This suggests that beyond being a useful pharmacogenetic marker for response to FSH, this polymorphism might also associate with fertility. Further dissection of the molecular pathways in cows carrying this polymorphism could provide information on the genes and pathways influencing fertility.

A polymorphism in the growth hormone gene in miniature Brahman cattle has been associated with a decreased number of small (<5 mm) antral follicles detectable by ultrasonography and decreased serum concentrations of IGF-1 (Hammond et al. 1991; Chase et al. 1998). The polymorphism was reported to produce a form of growth hormone that had approximately 60 % of normal activity when binding to the growth hormone receptor. This is further support for IGF-1 stimulating secondary follicles to grow to antral follicles as has been reported for the IGF-1 deficient mouse (Baker et al. 1996). However, there is no decrease in fertility in the miniature Brahman, and based on the phenotype of a decrease in stature, this polymorphism does not appear to occur outside the Brahman breed.

Similar functional polymorphisms in the receptors for GnRH, LH, or FSH could influence the variation observed in responses to exogenous hormones during synchronization of estrus or MOET protocols. If the polymorphism is in the hormone, as in the case of the GH polymorphism in the miniature Brahman, it can be assumed that the animal carrying the polymorphism would have a comparable response to the exogenous hormone to wild-type animals. However, a polymorphism in LH that affects its binding to the LHr could influence the response to GnRH in an ovulation synchronization protocol. If the function of the receptor itself is affected as in the case of the GRIA1 polymorphism, than a larger dose of exogenous hormone might compensate for the decreased effect. However, this situation could require a pharmaceutical intervention that targets the second messenger system.

These functional polymorphisms and others that could be discovered provide a great deal of opportunity for understanding the biological pathways that control these traits. In the dissection of the second messenger system responses to these less effective hormones and receptors is the potential to discover new sites for pharmaceutical interventions that may improve reproductive management. Thus, the impact of understanding these polymorphisms is not just in tailoring hormone doses to the individual genotype but also in better understanding the biological pathways. These functional polymorphisms could be part of larger physiological pathways where a number of other functional polymorphisms could be identified, such as in the network of growth factors and developmental genes that control the development and function of the reproductive tract. This could aid in producing panels of markers to maintain fertility while selecting to improve production traits.

Paternity Panels

It may seem at this point that using genetic markers to determine paternity is so simple that it should not even be considered. However, there are still cases of record-keeping errors in the dairy industry where this technology can provide correct sire identification. It also has great applications in situations where herd sires are still being used, such as major sections of the beef industry and with sheep, goats, and buffalo. Paternity panels do not require a great deal of genetic information for those species that are still lagging behind in genome sequence as well. In beef cattle in multi-sire pasture situations, even among a group of bulls that have passed a breeding soundness exam, paternity testing has demonstrated that one or two bulls sire the majority of calves and this dominance is moderately repeatable (Drake et al. 2011). While it is interesting to ponder whether this is due to fertility, libido, or physical dominance, it does not really matter. Herd sires that are not producing offspring are an economic drain on the operation and the use of paternity testing to gain this information, and cull these sires is worth the cost of the technology.

Functional Genomics

Functional genomics is the large scale analysis of mRNAs and proteins to understand how the change in the expression of certain genes is involved in key physiological processes. Understanding the function of genes and the pathways that they are involved in is as important to improving assisted reproductive technologies as understanding the sequence variation that exists among individuals. Clearly, the relative level of methylation of a gene can cause very different levels of transcription of that gene in two individuals with identical sequence resulting in vastly different phenotypes. Polymorphisms that affect the location or sequence of microRNA binding sites can also have dramatic effects on the level of translation of a transcript such that two animals with the exact same coding sequence could still produce vastly different amount of the protein even though analysis of mRNA expression does not identify differences in the transcript levels.

As sequencing technologies have become less costly and allowed less expensive annotation of the genome of domestic farm species, microarray chips for evaluating large scale gene expression have become more available. However, at the same time, this decrease in sequencing cost has increased the opportunity for RNA-sequencing (RNA-seq) experiments. RNA-sequencing involves extracting the RNA from a tissue sample and generating a cDNA library from that pool of RNA. Instead of hybridizing the cDNA to a microarray, the cDNA is submitted for sequencing to identify the transcripts that are present and their relative levels. The advantages of RNA-seq is that it is not limited to the specific probes that are on the chip unlike a microarray, and that it allows for the identification of novel SNP in the coding regions of genes (Wilhelm and Landry 2009; Nagalakshmi et al. 2010). While this does not necessarily improve the chances of identifying a functional mutation, it adds to the number of SNP within the coding region that can be investigated for functional properties, because a truly functional mutation is the best genetic marker.

Annotation of protein databases are lagging behind gene databases for domestic species because they are dependent on the quality of the consensus sequence (Kayser et al. 2004; Soares et al. 2012). This continues to make global proteomics studies difficult. However, there is no reason to think that as DNA sequencing technologies continue to become less expensive and the annotations of the genomes of the domestic farm species become better that proteomic databases will not also improve. Thus, the identification of protein-based biomarkers of physiological status will aid in the reproductive management of domestic farm species, if they can be implemented at a cost that makes them feasible to production agriculture. The advent of multiplex arrays for measuring multiple proteins in a single biological sample at very low concentrations may aid in reducing the cost (Leng et al. 2008).

However, recent studies have begun to demonstrate the potential use of mRNA levels in blood cells or other tissues as biomarkers of physiological status (Han et al. 2006; Cushman et al. 2007b; Arigami et al. 2012). Real-time RT-PCR has facilitated the processing of large numbers of mRNA samples in a relatively short time and with a comparable cost to immunoassays (estimated current costs in the United States of $5.00 a sample to extract RNA and $2.00 to analyze a sample for a total of about $9.00 if run in duplicate). Thus, there are reasons for using mRNA as a biomarker instead of protein.

There are many factors to consider when choosing between protein and mRNA as a potential biomarker. The handling of samples requires less care when processing them for proteomic analysis than for transcriptomic analysis. However, due to their applicability to human health issues the costs of immunoassays are rising to a point where they are not feasible for those working with domestic farm species. The combination of the increase in sequence data for farm species and the decrease in the cost to synthesize oligonucleotide primers is making it easier and cheaper to obtain primers than antibodies. Thus, as long as one can show that gene expression in a specific tissue differs consistently with physiological state, then mRNA can be used as a feasible biomarker.

The combination of increased sequence and the understanding of specific genes and how they function derived from genome-wide association studies, global transcriptomic and proteomic studies provide further information for developing biomarkers of fertility, pregnancy status (Han et al. 2006), and pubertal status in domestic farm species. Because of the large environmental component to many of these traits, biomarkers measured close to the time that animals go to breeding that act as an indicator of fertility can be more useful than genetic markers that predict overall merit.

Epigenetics

Epigenetics is defined as changes in gene expression caused by mechanisms other than changes in DNA sequence. This may be due to maternal diet or uterine crowding during pregnancy or could be due to the culture media that cells are in during in vitro fertilization or somatic cell nuclear transfer.

A few examples of potential epigenetic effects on reproduction have begun to surface in domestic farm species. Martin et al. (2007) reported that the daughters of cows that were fed a protein supplement during the last trimester of pregnancy conceived earlier in their first breeding season and had overall greater pregnancy rates, suggesting that changes in maternal nutrient status during late pregnancy were influencing the reproductive performance of the daughters. Similar effects have been seen with maternal nutrition and the ovarian reserve of daughters. Sullivan et al. (2009) reported that heifers that were born to dams that were fed a diet high in protein during the second trimester had a decreased number of antral follicles at 2 years of age. Furthermore, decreased energy to pregnant heifers during the first two trimesters of pregnancy resulted in a decreased number of antral follicles detectable by ultrasonography in daughters during the first year of life (Ireland et al. 2011).

Taken together these data indicate that modifying the way that pregnant females are fed may be a method to alter gene function and improve production traits of their offspring. The ability to custom feed animals or their dams and influence genes that alter reproductive function could have awesome implications for production agriculture. However, caution must be maintained until the proper studies are performed to understand the mechanisms by which these trans-generational influences on reproductive performance may be occurring (Funston et al. 2012).

Genetically Modified Organisms

The concept of genetically modified organisms is a source of concern for many consumers. Is the day coming when we will have genetically modified farm animals? We already have genetically modified salmon, but is there a need for genetically modified cattle, pigs or sheep even with the predicted increase in world population and predicted food shortage? These are not easy questions to answer, but if there is a need, the most likely candidates for genes to genetically modify are those associated with production traits and disease resistance. The concern will be that modifying these genes could negatively impact reproductive performance.

A recent survey reported that consumers had a greater awareness of cloning as it pertains to animals than other assisted reproductive technologies such as artificial insemination (Brooks and Lusk 2011). Less than half (43 %) of those surveyed were unwilling to consume meat from a cloned animal. The combination of cloning with genetic technologies can create founder animals to propagate the genetic modifications in the population at greater efficiency and without introducing cloned animals into the food chain. Consumer acceptance might depend upon the types of genetic modifications that were made. If the food was made safer by genetic modification or healthier (Zhang et al. 2010; Hu et al. 2011), this might help increase consumer acceptance while altering the function of specific receptors to increase growth or ovulation rate might be less acceptable.

It is clear that to construct these genetically modified food animals, reproductive physiologists and technologists will be involved. The need for individuals with the skills to collect embryos, perform in vitro fertilization, perform somatic cell nuclear transfer, maintain and transfect cell lines, and transfer embryos will only increase in this era. There will be the issue of what the manipulations in vitro of cell lines or embryos could do to the epigenome. This could have negative or positive effects. As our understanding of the epigenome grows there may be targeted modifications that can be performed in vitro to enhance performance of the offspring rather than just the negative impacts of culture systems on performance of the offspring. Perhaps a better understanding of how culture media causes large calf syndrome could actually be harnessed to improve birth weights without negatively impacting the health of the offspring or increasing the risk of dystocia.

Thoughts for Feeding the World

A world food shortage has been predicted by the year 2050. Food animals use surplus grains, forages, and by-products to produce protein in the form of milk and protein as well as hides and fibers. Manure will become an increasingly important source of nutrients for crops. For much of the world, the affordability of these products will remain a major issue. During the last century, advancements in assisted reproductive technologies such as synchronization of estrus and MOET have contributed to the ability to propagate superior genetics. In the last 30 years, research to identify the genes involved in attainment of puberty, development of the reproductive tract, and follicular development has increased our knowledge of the biological pathways and aided in developing pharmaceutical interventions to better control reproductive function in domestic farm species. In the coming years, the identification of functional polymorphisms will provide even better estimates of the biological functions of these genes. However, proving that a polymorphism is truly functional is still the most difficult aspect of this research, and where more of the research dollars will need to be focused in the future.

Both reproductive technologies and genomic technologies have lagged behind for species like the goat and the buffalo that are of major importance to agriculture outside the United States (Womack 2005; Singh et al. 2009). These species will become more important in coming years and will require greater research focus. This may create the best opportunities for collaborations across borders as researchers in countries where these species originate draw from the experience of researchers who have worked to establish the DNA libraries, database, and bioinformatic systems using the more common domestic farm species, thereby extending these technologies from more developed countries to less developed countries. In return, novel forms of genes identified in this these exotic breeds may aid in improving the efficiency of production in developed countries.

As sequencing technologies continue to decrease in cost, the only limitations to their use to improve reproductive technologies are the creativity and technical training of the reproductive biologists. Thus, they will need to be more broadly trained across disciplines to assist in research projects that will become more all-encompassing to address production efficiency at the whole animal level and improve sustainability (Britt et al. 2008). This may be the only way to continue to improve production efficiency to meet the ever growing world demand and make food affordable for as many people as possible.

References

Allen EG, He W, Yadav-Shah M, Sherman SL (2004) A study of the distributional characteristics of FMR1 transcript levels in 238 individuals. Hum Genet 114:439–447

Alvarez P, Spicer LJ, Chase CC Jr, Payton ME, Hamilton TD, Stewart RE, Hammond AC, Olson TA, Wettemann RP (2000) Ovarian and endocrine characteristics during an estrous cycle in Angus, Brahman, and Senepol cows in a subtropical environment. J Anim Sci 78:1291–1302

Arigami T, Uenosono Y, Ishigami S, Hagihara T, Haraguchi N, Matsushita D, Yanagita S, Nakajo A, Okumura H, Hokita S, Natsugoe S (2012) Expression of stanniocalcin 1 as a potential biomarker of gastric cancer. Oncology 83:158–164

Baker J, Hardy MP, Zhou J, Bondy C, Lupu F, Bellve AR, Efstratiadis A (1996) Effects of an Igf1 gene null mutation on mouse reproduction. Mol Endocrinol 10:903–918

Bishop SC (2012) Possibilities to breed for resistance to nematode parasite infections in small ruminants in tropical production systems. Animal 6:741–747

Bonczek RR, Young CW, Wheaton JE, Miller KP (1988) Responses of somatotropin, insulin, prolactin, and thyroxine to selection for milk yield in Holsteins1,2. J Dairy Sci 71:2470–2479

Bretherick KL, Fluker MR, Robinson WP (2005) FMR1 repeat sizes in the gray zone and high end of the normal range are associated with premature ovarian failure. Hum Genet 117:376–382

Britt JH, Aberle ED, Esbenshade KL, Males JR (2008) Animal science departments of the future. J Anim Sci 86:3235–3244

Brooks KR, Lusk JL (2011) U.S. consumers attitudes toward farm animal cloning. Appetite 57:483–492

Butler WR (1998) Review: effect of protein nutrition on ovarian and uterine physiology in dairy cattle. J Dairy Sci 81:2533–2539

Cammack KM, Thomas MG, Enns RM (2009) Review: reproductive traits and their heritabilities in beef cattle. Prof Anim Sci 25:517–528

Campbell BK, Clinton M, Webb R (2012) The role of anti-Müllerian hormone (AMH) during follicle development in a monovulatory species (sheep). Endocrinology 153:4533–4543

Casas E, Snowder GD (2008) A putative quantitative trait locus on chromosome 20 associated with bovine pathogenic disease incidence. J Anim Sci 86:2455–2460

Cassady JP, Johnson RK, Ford JJ (2000) Comparison of plasma FSH concentration in boars and gilts from lines selected for ovulation rate and embryonal survival, and litter size and estimation of (co)variance components for FSH and ovulation rate. J Anim Sci 78:1430–1435

Chase CC Jr, Kirby CJ, Hammond AC, Olson TA, Lucy MC (1998) Patterns of ovarian growth and development in cattle with a growth hormone receptor deficiency. J Anim Sci 76:212–219

Cohick WS, Armstrong JD, Whitacre MD, Lucy MC, Harvey RW, Campbell RM (1996) Ovarian expression of insulin-like growth factor-I (IGF-I), IGF binding proteins, and growth hormone (GH) receptor in heifers actively immunized against GH-releasing factor. Endocrinology 137:1670–1677

Collis E, Fortes MR, Zhang Y, Tier B, Schutt K, Barendse W, Hawken R (2012) Genetic variants affecting meat and milk production traits appear to have effects on reproduction traits in cattle. Anim Genet 43:442–446

Cordts EB, Christofolini DM, Dos Santos AA, Bianco B, Barbosa CP (2011) Genetic aspects of premature ovarian failure: a literature review. Arch Gynecol Obstet 283:635–643

Cushman RA, DeSouza JC, Hedgpeth VS, Britt JH (1999) Superovulatory response of one ovary is related to the micro- and macroscopic population of follicles in the contralateral ovary of the cow. Biol Reprod 60:349–354

Cushman RA, Wahl CM, Fortune JE (2002) Bovine ovarian cortical pieces grafted to chick embryonic membranes: a model for studies on the activation of primordial follicles. Hum Reprod 17:48–54

Cushman RA, Allan MF, Jones SA, Rupp GP, Echternkamp SE (2007a) Localization of Period 1 mRNA in the ruminant oocyte and investigations of its role in ovarian function. Anim Reprod Sci 99:93–105

Cushman RA, Allan MF, Kuehn LA (2007b) Achievements of research in reproduction sciences. In: Rosati A, Tewolde A, Mosconi C (eds) Animal production and animal science worldwide. WAAP book of the year 2007. Wageningen Academic, Wageningen, pp 59–66

Cushman RA, Allan MF, Kuehn LA (2008) Characterization of biological types of cattle: indicator traits of fertility in beef cows. Rev Bras Zootec 37:116–121

Cushman RA, Allan MF, Kuehn LA, Snelling WM, Cupp AS, Freetly HC (2009) Evaluation of antral follicle count and ovarian morphology in crossbred beef cows: Investigation of influence of stage of the estrous cycle, age, and birth weight. J Anim Sci 87:1971–1980

Dong J, Albertini DF, Nishimori K, Kumar TR, Lu N, Matzuk MM (1996) Growth differentiation factor-9 is required during early ovarian folliculogenesis. Nature 383:531–535

Doyle SP, Golden BL, Green RD, Brinks JS (2000) Additive genetic parameter estimates for heifer pregnancy and subsequent reproduction in Angus females. J Anim Sci 78:2091–2098

Drake DJ, Weber KL, Van Eenennam AL (2011) What are herd bulls accomplishing in multiple sire breeding pastures? In: Proceedings, applied reproductive strategies in beef cattle, p 305–319, Joplin, MO, 31 Aug–1 Sept

Durlinger ALL, Kramer P, Karels B, de Jong FH, Uilenbroek JTJ, Grootegoed JA, Themmen APN (1999) Control of primordial follicle recruitment by Anti-Müllerian hormone in the mouse ovary. Endocrinology 140:5789–5796

Durlinger ALL, Gruijters MJG, Kramer P, Karels B, Ingraham HA, Nachtigal MW, Uilenbroek JTJ, Grootegoed JA, Themmen APN (2002a) Anti-Müllerian hormone inhibits initiation of primordial follicle growth in the mouse ovary. Endocrinology 143:1076–1084

Durlinger ALL, Visser JA, Themmen APN (2002b) Regulation of ovarian function: the role of anti-Mullerian hormone. Reproduction 124:601–609

Echternkamp SE, Gregory KE, Dickerson GE, Cundiff LV, Koch RM, Van Vleck LD (1990a) Twinning in cattle: II. Genetic and environmental effects on ovulation rate in puberal heifers and postpartum cows and the effects of ovulation rate on embryonic survival. J Anim Sci 68:1877–1888

Echternkamp SE, Spicer LJ, Gregory KE, Canning SF, Hammond JM (1990b) Concentrations of insulin-like growth factor-I in blood and ovarian follicular fluid of cattle selected for twins. Biol Reprod 43:8–14

Echternkamp SE, Roberts AJ, Lunstra DD, Wise T, Spicer LJ (2004) Ovarian follicular development in cattle selected for twin ovulations and births. J Anim Sci 82:459–471

Elvin JA, Matzuk MM (1998) Mouse models of ovarian failure. Rev Reprod 3:183–195

Elvin JA, Yan C, Matzuk MM (2000) Oocyte-expressed TGF-b superfamily members in female fertility. Mol Cell Endocrinol 159:1–5

Fortes MR, Reverter A, Zhang Y, Collis E, Nagaraj SH, Jonsson NN, Prayaga KC, Barris W, Hawken RJ (2010) Association weight matrix for the genetic dissection of puberty in beef cattle. Proc Natl Acad Sci U S A 107:13642–13647

Foxcroft G (2012) Reproduction in farm animals in an era of rapid genetic change: will genetic change outpace our knowledge of physiology? Reprod Domest Anim 47(Suppl 4):313–319

Funston RN, Summers AF, Roberts AJ (2012) Alpharma Beef Cattle Nutrition Symposium: implications of

nutritional management for beef cow-calf systems. J Anim Sci 90:2301–2307
Galloway SM, McNatty KP, Cambridge LM, Laitinen MPE, Juengel JL, Jokiranta TS, McLaren RJ, Luiro K, Dodds KG, Montgomery GW, Beattie AE, Davis GH, Ritvos O (2000) Mutations in an oocyte-derived growth factor gene (*BMP15*) cause increased ovulation rate and infertility in a dosage-sensitive manner. Nat Genet 25:279–283
Gargantini G, Cundiff LV, Lunstra DD, van Vleck LD (2005) Genetic relationships between male and female reproductive traits in beef cattle. Prof Anim Sci 21:195–199
Ghanem ME, Nishibori M (2009) Genetic description of factor XI deficiency in holstein semen in Western Japan. Reprod Domest Anim 44:792–796
Gigli I, Cushman RA, Wahl CM, Fortune JE (2005) Evidence for a role for anti-Mullerian hormone in the suppression of follicle activation in mouse ovaries and bovine ovarian cortex grafted beneath the chick chorioallantoic membrane. Mol Reprod Dev 71:480–488
Gleicher N, Weghofer A, Barad DH (2009) A pilot study of premature ovarian senescence: I. Correlation of triple CGG repeats on the FMR1 gene to ovarian reserve parameters FSH and anti-Mullerian hormone. Fertil Steril 91:1700–1706
Gleicher N, Weghofer A, Kim A, Barad DH (2012) The impact in older women of ovarian FMR1 genotypes and sub-genotypes on ovarian reserve. PLoS One 7:e33638
Greb RR, Behre HM, Simoni M (2005a) Pharmacogenetics in ovarian stimulation—current concepts and future options. Reprod Biomed Online 11:589–600
Greb RR, Grieshaber K, Gromoll J, Sonntag B, Nieschlag E, Kiesel L, Simoni M (2005b) A common single nucleotide polymorphism in exon 10 of the human follicle stimulating hormone receptor is a major determinant of length and hormonal dynamics of the menstrual cycle. J Clin Endocrinol Metab 90:4866–4872
Hammond AC, Elsasser TH, Olson TA (1991) Endocrine characteristics of a miniature condition in Brahman cattle: circulating concentrations of some growth-related hormones. Proc Soc Exp Biol Med 197:450–457
Han H, Austin KJ, Rempel LA, Hansen TR (2006) Low blood ISG15 mRNA and progesterone levels are predictive of non-pregnant dairy cows. J Endocrinol 191:505–512
He PJ, Hirata M, Yamauchi N, Hattori MA (2007) Up-regulation of Per1 expression by estradiol and progesterone in the rat uterus. J Endocrinol 194:511–519
Hetzel DJ, Mackinnon MJ, Dixon R, Entwistle KW (1989) Fertility in a tropical beef herd divergently selected for pregnancy rate. Anim Prod 48:73–81
Holm DE, Thompson PN, Irons PC (2009) The value of reproductive tract scoring as a predictor of fertility and production outcomes in beef heifers. J Anim Sci 87:1934–1940
Hu ZL, Ramos AM, Humphray SJ, Rogers J, Reecy JM, Rothschild MF (2011) Use of genome sequence information for meat quality trait QTL mining for causal genes and mutations on pig chromosome 17. Front Genet 2:43
Huang W, Kirkpatrick BW, Rosa GJ, Khatib H (2010) A genome-wide association study using selective DNA pooling identifies candidate markers for fertility in Holstein cattle. Anim Genet 41:570–578
Ireland JL, Scheetz D, Jimenez-Krassel F, Themmen AP, Ward F, Lonergan P, Smith GW, Perez GI, Evans AC, Ireland JJ (2008) Antral follicle count reliably predicts number of morphologically healthy oocytes and follicles in ovaries of young adult cattle. Biol Reprod 79:1219–1225
Ireland JJ, Zielak-Steciwko AE, Jimenez-Krassel F, Folger J, Bettegowda A, Scheetz D, Walsh S, Mossa F, Knight PG, Smith GW, Lonergan P, Evans AC (2009) Variation in the ovarian reserve is linked to alterations in intrafollicular estradiol production and ovarian biomarkers of follicular differentiation and oocyte quality in cattle. Biol Reprod 80:954–964
Ireland JJ, Smith GW, Scheetz D, Jimenez-Krassel F, Folger JK, Ireland JL, Mossa F, Lonergan P, Evans AC (2011) Does size matter in females? An overview of the impact of the high variation in the ovarian reserve on ovarian function and fertility, utility of anti-Mullerian hormone as a diagnostic marker for fertility and causes of variation in the ovarian reserve in cattle. Reprod Fertil Dev 23:1–14
Jasti M, Warren BD, McGinnis LK, Kinsey WH, Petroff BK, Petroff MG (2012) The autoimmune regulator (Aire) prevents premature reproductive senescence in female mice. Biol Reprod 86:1–9
Jimenez-Krassel F, Folger JK, Ireland JL, Smith GW, Hou X, Davis JS, Lonergan P, Evans AC, Ireland JJ (2009) Evidence that high variation in ovarian reserves of healthy young adults has a negative impact on the corpus luteum and endometrium during estrous cycles in cattle. Biol Reprod 80:1272–1281
Johnson RK, Eckardt GR, Rathje TA, Drudik DK (1994) Ten generations of selection for predicted weight of testes in swine: direct response and correlated response in body weight, backfat, age at puberty, and ovulation rate. J Anim Sci 72:1978–1988
Johnston DJ, Barwick SA, Corbet NJ, Fordyce G, Holroyd RG, Williams PJ, Burrow HM (2009) Genetics of heifer puberty in two tropical beef genotypes in northern Australia and associations with heifer- and steer-production traits. Anim Prod Sci 49:399–412
Kadakia R, Arraztoa JA, Bondy C, Zhou J (2001) Granulosa cell proliferation is impaired in the Igf1 null ovary. Growth Horm IGF Res 11:220–224
Kappes SM (1999) Utilization of gene mapping information in livestock animals. Theriogenology 51:135–147
Kayser JP, Vallet JL, Cerny RL (2004) Defining parameters for homology-tolerant database searching. J Biomol Tech 15:285–295
Kevenaar ME, Themmen APN, Rivadeneira F, Uitterlinden AG, Laven JSE, van Schoor NM, Lips P, Pols HAP, Visser JA (2007) A polymorphism in the AMH type II receptor gene is associated with age at

menopause in interaction with parity. Hum Reprod 22:2382–2388

Kevenaar ME, Laven JS, Fong SL, Uitterlinden AG, de Jong FH, Themmen AP, Visser JA (2008) A functional anti-Mullerian hormone gene polymorphism is associated with follicle number and androgen levels in polycystic ovary syndrome patients. J Clin Endocrinol Metab 93:1310–1316

Knauer MT, Cassady JP, Newcom DW, See MT (2010) Estimates of variance components for genetic correlations among swine estrus traits. J Anim Sci 88:2913–2919

Knauer MT, Cassady JP, Newcom DW, See MT (2011) Phenotypic and genetic correlations between gilt estrus, puberty, growth, composition, and structural conformation traits with first-litter reproductive measures. J Anim Sci 89:935–942

Kuehn LA, Nonneman DJ, Klindt JM, Wise TH (2009) Genetic relationships of body composition, serum leptin, and age at puberty in gilts. J Anim Sci 87:477–483

Lamberson WR, Johnson RK, Zimmerman DR, Long TE (1991) Direct responses to selection for increased litter size, decreased age at puberty, or random selection following selection for ovulation rate in swine. J Anim Sci 69:3129–3143

Leng SX, McElhaney JE, Walston JD, Xie D, Fedarko NS, Kuchel GA (2008) ELISA and multiplex technologies for cytokine measurement in inflammation and aging research. J Gerontol A Biol Sci Med Sci 63:879–884

Líron JP, Prando A, Ripoli MV, Rogberg-Munoz A, Posik DM, Baldo A, Peral-Garcia P, Giovambattista G (2011) Characterization and validation of bovine gonadotropin releasing hormone receptor (GNRHR) polymorphisms. Res Vet Sci 91:391–396

Lucy MC, Weber WJ, Baumgard LH, Seguin BS, Koenigsfeldl AT, Hansen LB, Chester-Jones H, Crooker BA (1998) Reproductive endocrinology of lactating dairy cows selected for increased milk production. J Dairy Sci 81(Suppl 1):246

Lunstra DD, Gregory KE, Cundiff LV (1988) Heritability estimates and adjustment factors for the effects of bull age and age of dam on yearling testicular size in breeds of bulls. Theriogenology 30:127–136

MacArthur DG, Balasubramanian S, Frankish A, Huang N, Morris J, Walter K, Jostins L, Habegger L, Pickrell JK, Montgomery SB, Albers CA, Zhang ZD, Conrad DF, Lunter G, Zheng H, Ayub Q, DePristo MA, Banks E, Hu M, Handsaker RE, Rosenfeld JA, Fromer M, Jin M, Mu XJ, Khurana E, Ye K, Kay M, Saunders GI, Suner MM, Hunt T, Barnes IH, Amid C, Carvalho-Silva DR, Bignell AH, Snow C, Yngvadottir B, Bumpstead S, Cooper DN, Xue Y, Romero IG, Wang J, Li Y, Gibbs RA, McCarroll SA, Dermitzakis ET, Pritchard JK, Barrett JC, Harrow J, Hurles ME, Gerstein MB, Tyler-Smith C (2012) A systematic survey of loss-of-function variants in human protein-coding genes. Science 335:823–828

Mackinnon MJ, Hetzel DJ, Corbet NJ, Bryan RP, Dixon R (1990) Correlated responses to selection for cow fertility in a tropical beef breed. Anim Prod 50:417–424

MacNeil MD, Geary TW, Perry GA, Roberts AJ, Alexander LJ (2006) Genetic partitioning of variation in ovulatory follicle size and probability of pregnancy in beef cattle. J Anim Sci 84:1646–1650

Martin LC, Brinks JS, Bourdon RM, Cundiff LV (1992) Genetic effects on beef heifer puberty and subsequent reproduction. J Anim Sci 70:4006–4017

Martin JL, Vonnahme KA, Adams DC, Lardy GP, Funston RN (2007) Effects of dam nutrition on growth and reproductive performance of heifer calves. J Anim Sci 85:841–847

Matukumalli LK, Lawley CT, Schnabel RD, Taylor JF, Allan MF, Heaton MP, O'Connell J, Moore SS, Smith TP, Sonstegard TS, Van Tassell CP (2009) Development and characterization of a high density SNP genotyping assay for cattle. PLoS One 4:e5350

Matzuk MM, Lamb DJ (2002) Genetic dissection of mammalian fertility pathways. Nat Cell Biol 4(Suppl):s41–s49

McCormack BL, Chase CC Jr, Olson TA, Elsasser TH, Hammond AC, Welsh TH Jr, Jiang H, Randel RD, Okamura CA, Lucy MC (2009) A miniature condition in Brahman cattle is associated with a single nucleotide mutation within the growth hormone gene. Domest Anim Endocrinol 37:104–111

McDaneld TG, Kuehn LA, Thomas MG, Snelling WM, Sonstegard TS, Matukumalli LK, Smith TP, Pollak EJ, Keele JW (2012) Y are you not pregnant: identification of Y chromosome segments in female cattle with decreased reproductive efficiency. J Anim Sci 90:2142–2151

McNatty KP, Galloway SM, Wilson T, Smith P, Hudson NL, O'Connell A, Bibby AH, Heath DA, Davis GH, Hanrahan JP, Juengel JL (2005) Physiological effects of major genes affecting ovulation rate in sheep. Genet Sel Evol 37(Suppl 1):S25–S38

Meyers SN, McDaneld TG, Swist SL, Marron BM, Steffen DJ, O'Toole D, O'Connell JR, Beever JE, Sonstegard TS, Smith TP (2010) A deletion mutation in bovine SLC4A2 is associated with osteopetrosis in Red Angus cattle. BMC Genomics 11:337

Mialon MM, Renand G, Krauss D, Menissier F (2000) Genetic variability of the length of postpartum anoestrus in Charolais cows and its relationship with age at puberty. Genet Sel Evol 32:403–414

Mialon MM, Renand G, Krauss D, Menissier F (2001) Genetic relationship between cyclic ovarian activity in heifers and cows and beef traits in males. Genet Sel Evol 33:273–287

Minick Bormann J, Wilson DE (2010) Calving day and age at first calving in Angus heifers. J Anim Sci 88:1947–1956

Morris CA, Wilson JA, Bennett GL, Cullen NG, Hickey SM, Hunter JC (2000) Genetic parameters for growth, puberty, and beef cow reproductive traits in a puberty selection line. New Zeal J Agr Res 43:83–91

Mossa F, Walsh SW, Butler ST, Berry DP, Carter F, Lonergan P, Smith GW, Ireland JJ, Evans AC (2012) Low numbers of ovarian follicles ≥3 mm in diameter are associated with low fertility in dairy cows. J Dairy Sci 95:2355–2361

Nagalakshmi U, Waern K, Snyder M (2010) RNA-Seq: a method for comprehensive transcriptome analysis. Curr Protoc Mol Biol Chapter 4:Unit 4.11.1–13

Pilorz V, Steinlechner S (2008) Low reproductive success in Per1 and Per2 mutant mouse females due to accelerated ageing. Reproduction 135:559–568

Ramos AM, Crooijmans RP, Affara NA, Amaral AJ, Archibald AL, Beever JE, Bendixen C, Churcher C, Clark R, Dehais P, Hansen MS, Hedegaard J, Hu ZL, Kerstens HH, Law AS, Megens HJ, Milan D, Nonneman DJ, Rohrer GA, Rothschild MF, Smith TP, Schnabel RD, Van Tassell CP, Taylor JF, Wiedmann RT, Schook LB, Groenen MA (2009) Design of a high density SNP genotyping assay in the pig using SNPs identified and characterized by next generation sequencing technology. PLoS One 4:e6524

Rico C, Fabre S, Medigue C, di Clemente N, Clement F, Bontoux M, Touze JL, Dupont M, Briant E, Remy B, Beckers JF, Monniaux D (2009) Anti-Mullerian hormone is an endocrine marker of ovarian gonadotropin-responsive follicles and can help to predict superovulatory responses in the cow. Biol Reprod 80:50–59

Rosendo A, Druet T, Gogue J, Bidanel JP (2007a) Direct responses to six generations of selection for ovulation rate or prenatal survival in Large White pigs. J Anim Sci 85:356–364

Rosendo A, Druet T, Gogue J, Canario L, Bidanel JP (2007b) Correlated responses for litter traits to six generations of selection for ovulation rate or prenatal survival in French Large White pigs. J Anim Sci 85:1615–1624

Ruiz-Flores A, Johnson RK (2001) Direct and correlated responses to two-stage selection for ovulation rate and number of fully formed pigs at birth in swine. J Anim Sci 79:2286–2297

Saito H, Sasaki Y, Koketsu Y (2011) Associations between age of gilts at first mating and lifetime performance or culling risk in commercial herds. J Vet Med Sci 73:555–559

Schneider JF, Rempel LA, Rohrer GA, Brown-Brandl TM (2011) Genetic parameter estimates among scale activity score and farrowing disposition with reproductive traits in swine. J Anim Sci 89:3514–3521

Schneider JF, Rempel LA, Rohrer GA (2012) GWAS of swine farrowing traits Part I: genetic and genomic parameter estimates. J Anim Sci 90(10):3353–3359

Schoppee PD, Armstrong JD, Harvey RW, Whitacre MD, Felix A, Campbell RM (1996) Immunization against growth hormone releasing factor or chronic feed restriction initiated at 3.5 months of age reduces ovarian response to pulsatile administration of gonadotropin-releasing hormone at 6 months of age and delays onset of puberty in heifers. Biol Reprod 55:87–98

Shimizu T, Hirai Y, Murayama C, Miyamoto A, Miyazaki H, Miyazaki K (2011) Circadian Clock genes Per2 and clock regulate steroid production, cell proliferation, and luteinizing hormone receptor transcription in ovarian granulosa cells. Biochem Biophys Res Commun 412:132–135

Shimizu T, Hirai Y, Murayama C, Miyamoto A, Miyazaki H, Miyazaki K (2012) Expressions of the circadian genes Per2, Bmal1, Clock and Cry1 during the different stages of follicular development and their regulation by FSH in bovine granulosa cells from small follicles. Livest Sci 145:292–297

Silva-Santos KC, Santos GM, Siloto LS, Hertel MF, Andrade ER, Rubin MI, Sturion L, Melo-Sterza FA, Seneda MM (2011) Estimate of the population of preantral follicles in the ovaries of Bos taurus indicus and Bos taurus taurus cattle. Theriogenology 76:1051–1057

Singh J, Dominguez M, Jaiswal R, Adams GP (2004) A simple ultrasound test to predict the superstimulatory response in cattle. Theriogenology 62:227–243

Singh B, Chauhan MS, Singla SK, Gautam SK, Verma V, Manik RS, Singh AK, Sodhi M, Mukesh M (2009) Reproductive biotechniques in buffaloes (Bubalus bubalis): status, prospects and challenges. Reprod Fertil Dev 21:499–510

Skinner MK, Schmidt M, Savenkova MI, Sadler-Riggleman I, Nilsson EE (2008) Regulation of granulosa and theca cell transcriptomes during ovarian antral follicle development. Mol Reprod Dev 75:1457–1472

Snelling WM, Cushman RA, Fortes MR, Reverter A, Bennett GL, Keele JW, Kuehn LA, McDaneld TG, Thallman RM, Thomas MG (2012) Physiology and Endocrinology Symposium: how single nucleotide polymorphism chips will advance our knowledge of factors controlling puberty and aid in selecting replacement beef females. J Anim Sci 90:1152–1165

Soares R, Franco C, Pires E, Ventosa M, Palhinhas R, Koci K, Martinho de Almeida A, Varela Coelho A (2012) Mass spectrometry and animal science: protein identification strategies and particularities of farm animal species. J Proteomics 75:4190–4206

Starbuck-Clemmer MJ, Hernandez-Fonseca H, Ahmad N, Seidel G, Inskeep EK (2007) Association of fertility with numbers of antral follicles within a follicular wave during the oestrous cycle in beef cattle. Reprod Domest Anim 42:337–342

Sterning M, Rydhmer L, Eliasson-Selling L (1998) Relationships between age at puberty and interval from weaning to estrus and between estrus signs at puberty and after the first weaning in pigs. J Anim Sci 76:353–359

Sugimoto M, Sasaki S, Watanabe T, Nishimura S, Ideta A, Yamazaki M, Matsuda K, Yuzaki M, Sakimura K, Aoyagi Y, Sugimoto Y (2010) Ionotropic glutamate receptor AMPA 1 is associated with ovulation rate. PLoS One 5:e13817

Sullivan TM, Micke GC, Greer RM, Irving-Rodgers HF, Rodgers RJ, Perry VE (2009) Dietary manipulation of Bos indicus x heifers during gestation affects the reproductive development of their heifer calves. Reprod Fertil Dev 21:773–784

Tessaro I, Luciano AM, Franciosi F, Lodde V, Corbani D, Modina SC (2011) The endothelial nitric oxide synthase/nitric oxide system is involved in the defective

quality of bovine oocytes from low mid-antral follicle count ovaries. J Anim Sci 89:2389–2396
Thallman RM, Cundiff LV, Gregory KE, Koch RM (1999) Germplasm evaluation in beef cattle—Cycle IV: post-weaning growth and puberty of heifers. J Anim Sci 77:2651–2659
Vainio S, Heikkila M, Kispert A, Chin N, McMahon AP (1999) Female development in mammals is regulated by Wnt-4 signalling. Nature 397:405–409
White SN, Casas E, Allan MF, Keele JW, Snelling WM, Wheeler TL, Shackelford SD, Koohmaraie M, Smith TP (2007) Evaluation in beef cattle of six deoxyribonucleic acid markers developed for dairy traits reveals an osteopontin polymorphism associated with post-weaning growth. J Anim Sci 85:1–10
Wiedmann RT, Smith TP, Nonneman DJ (2008) SNP discovery in swine by reduced representation and high throughput pyrosequencing. BMC Genet 9:81
Wilhelm BT, Landry JR (2009) RNA-Seq-quantitative measurement of expression through massively parallel RNA-sequencing. Methods 48:249–257
Wilson T, Wu X-Y, Juengel JL, Ross IK, Lumsden JM, Lord EA, Dodds KG, Walling GA, McEwan JC, O'Connell AR, McNatty KP, Montgomery GW (2001) Highly prolific Booroola sheep have a mutation in the intracellular kinase domain of bone morphogenic protein 1B receptor (ALK-6) that is expressed in both oocytes and granulosa cells. Biol Reprod 64:1225–1235
Womack JE (2005) Advances in livestock genomics: opening the barn door. Genome Res 15:1699–1705
Yang WC, Li SJ, Tang KQ, Hua GH, Zhang CY, Yu JN, Han L, Yang LG (2010) Polymorphisms in the 5′ upstream region of the FSH receptor gene, and their association with superovulation traits in Chinese Holstein cows. Anim Reprod Sci 119:172–177
Yilmaz A, Davis ME, Simmen RCM (2006) Analysis of female reproductive traits in Angus beef cattle divergently selected for blood serum insulin-like growth factor I concentration. Theriogenology 65:1180–1190
Zhang S, Knight TJ, Reecy JM, Wheeler TL, Shackelford SD, Cundiff LV, Beitz DC (2010) Associations of polymorphisms in the promoter I of bovine acetyl-CoA carboxylase-alpha gene with beef fatty acid composition. Anim Genet 41:417–420
Zhou J, Kumar TR, Matzuk MM, Bondy C (1997) Insulin-like growth factor I regulates gonadotropin responsiveness in the murine ovary. Mol Endocrinol 11:1924–1933

Impacts of Reproductive Technologies on Beef Production in the United States

5

Carl Dahlen, Jamie Larson, and G. Cliff Lamb

Abstract

Estimations of world population growth indicate that by the year 2050 we will reach nine billion habitants on earth. These estimates impose a tremendous challenge in the current agricultural systems as food supply will need to increase by 100 % in the next 40 years (Food and Agriculture Organization of the United Nations 2009). Beef will be a primary protein source that will assist in meeting the requirements for a portion of the protein in diets of this expanding global populace. Beef is a high-quality protein that contains all essential amino acids for the human body and also contains additional essential nutrients such as iron, zinc, B vitamins, riboflavin, selenium, choline, and conjugated linoleic acid (CLA). Adopting reproductive technologies at greater rates than currently used is a viable method to dramatically enhance production efficiency of beef cattle enterprises.

Artificial insemination (AI), estrous synchronization and fixed-time AI (TAI), semen and embryo cryopreservation, multiple ovulation and embryo transfer (MOET), in vitro fertilization, sex determination of sperm or embryos, and nuclear transfer are technologies that are used to enhance the production efficiency of beef operations. In many cases, the development of these technologies is responsible for significant changes to traditional livestock production practices. However, adoption of these

C. Dahlen
Department of Animal Sciences, North Dakota State University, Fargo, ND 58108-6050, USA
e-mail: Carl.Dahlen@ndsu.edu

J. Larson
Department of Animal and Dairy Sciences, Mississippi State University, Mississippi State, MS 39762, USA
e-mail: JLarson@ads.msstate.edu

G.C. Lamb (✉)
North Florida Research and Education Center, University of Florida, Marianna, FL 32446, USA
e-mail: gclamb@ufl.edu

G.C. Lamb and N. DiLorenzo (eds.), *Current and Future Reproductive Technologies and World Food Production*, Advances in Experimental Medicine and Biology 752,
DOI 10.1007/978-1-4614-8887-3_5,

technologies appears to has not grown at the same rate in the United States as other formidable beef producing nations. For example, sales of beef semen for AI increased from 3.3 to 11.9 million units between 1993 and 2011 in Brazil, whereas that in the United States has increased from 2.9 to 3.8 million units during the same period. The significant increases in adoption of reproductive technologies in developing countries is likely as a result of the development of practical estrous synchronization and TAI systems that have allowed beef producers the opportunity to eliminate detection of estrus in their AI programs with a high degree of success. In the United States, slow adoption rates of these technologies may result in a future loss of international market share of beef products as other nations take advantage not only of the additional kilogram of beef that can be produced but also the improved quality of beef that can be realized through incorporation of reproductive technologies and resultant genetic improvement. However, current difficulties the US producers have with the incorporation of applied reproductive technologies, such as TAI, MOET, and sex semen, must not be the reason to overlook and incorporate more traditional reproductive technologies such as castration, breeding season management, or weaning. In many cases, beef producers in the United States fail to incorporate these more traditional technologies, which results in a reduction in production efficiency of the US beef industry. This chapter will focus on both traditional and more developed reproductive technologies that will play a role in enhancing future production efficiencies of the US beef cattle production system.

Keywords
Artificial insemination • Beef cattle • Fertilization • Cryopreservation • Reproductive technology

Castration of Male Cattle

Removal of testis from bull calves is likely the most frequently used reproductive technique in US beef cattle production systems. A 2007 survey indicated that 59.2 % of cattle operations castrated at least a portion of their bull calves and this represented 77.1 % of the total calves owned by survey respondents (NAHMS 2008). In addition, large regional variations exist with 43.9 % of producers in the South Central US castrating bulls compared with 92.3 % of operation in the Western Region. Sales price of steer calves are greater than those of bull calves and 91.6 % of feedlot operators reported that castration and dehorning at least 4 weeks prior to arrival at the feedlot was an extremely effective or very effective for reducing sickness and death loss (APHIS 2012). The main reason for this thought is that when calves are not castrated prior to feedlot entry, they will be castrated shortly thereafter. Calves recovering from castration have lower intake and gains compared with calves already healed.

From a standpoint of weight gain, feed efficiency, and final feedlot weight, intact males actually perform better than castrated males (Worrell et al. 1987). In an era where maximizing efficiencies is such a high priority this would be a logical place to change our production systems. However, castration was implemented on a large scale for good reason and is not likely to change in the immediate future.

At its very basic level, castration is a method to foster-selective breeding. Castration offers producers the opportunity to decide which animals may best be suited to sire future calf crops and those which are better suited to the commodity beef market. Reasons for some cattle being suited for mating within a herd and others not being suitable for such tasks are vast. For one operation, it may be a desire to have a certain phenotype/genotype combination for their bull customers, and for others it may be as simple as not wanting to keep bull calves from within a herd to avoid issues with inbreeding. Another major concern with maintaining bulls rather than steers is the aggressive tendencies of bulls. This aggressive behavior is in response to sexual stimuli as well as a desire to maintain social dominance. The sexual stimuli would be relevant if pens of bulls were in the same vicinity of a feedlot as pens of heifers not receiving an estrus-suppressant (such as melengestrol acetate, MGA). Behavioral characteristics of bulls cause concerns over safety of humans and of other cattle, as well as maintenance of fences and other equipment.

Perhaps the overriding issue with why castration is prevalent in the US beef production systems is consumer preference. Tenderness and consistency of products are very important attributes for consumers choosing to purchase beef. Compared with that from bulls, meat from steers contains more fat resulting in greater quality grades (Calkins et al. 1986), as well as increased tenderness, juiciness, and flavor ratings of the longissimus muscle along with a brighter color (Carroll et al. 1975). This preference for meat from steer carcasses has been highlighted by prices received when selling finished cattle. A large disparity in prices exists among steers and intact bulls sold on a value-based marketing system. Therefore, from a management standpoint, bulls should be castrated early in life to not only reduce pain and stress association with castration but also to avoid development of secondary sex characteristics which would decrease value of finished steers, and to take advantage of the improved marbling and tenderness compared with castrating later in life (Worrell et al. 1987).

The US swine industry and the beef industries in several other countries have approved the use of immunological castration techniques, whereby male livestock are vaccinated against their own gonadotropin-releasing hormone. The technique includes an initial and booster dose of the vaccine and the end results are temporary reductions in testicular size and concentrations of testosterone for vaccinated bulls compared with unvaccinated bulls (Janett et al. 2012). In addition, bulls vaccinated against GnRH had similar performance and improved meat quality (Amatayakul-Chantler et al. 2012). Reductions in physical activity (perhaps related to sexual aggression and hierarchy maintenance) in vaccinated bulls may explain the similar performance measures in spite of the reduced concentrations of endogenous testosterone compared with unvaccinated bulls (Janett et al. 2012). As producers look for alternative castration techniques to optimize animal welfare and sensitivity of the public increases toward management practices such as castration, immunocastration may become a management practice utilized to a greater extent in the US beef production system.

Controlling Estrus in Nonbreeding Females

Reports indicate that in many cases preweaned heifer calves on a pasture may be pregnant at the end of a grazing season. The frequency of pregnant heifers entering feedlots has ranged from 4 to 17 % of all heifers, depending on the management at the heifer source of origin. In addition, some pens of heifers have been harvested where the pregnancy rate approaches 20 % (Laudert 1988). While pregnant heifers gain similarly compared with open heifers, a portion of this weight is being partitioned toward the developing fetus and the overall feed efficiency is reduced in those pregnant heifers (Jim et al. 1991). Producers have the option of continuing to feed heifers that are pregnant or to administer an abortifactant to heifers that are found to be pregnant. Retaining pregnant heifers and adding excess body condition may result in major problems associated with calf survival and heifer health when they calve in feedlot pens.

Removal of ovaries (spaying) in heifers is a practice that is conducted on only a small proportion of females but, when successful, eliminates the opportunity for pregnancy in females. Spaying heifers offers feedlot operators the ability to maintain mixed-sex pens or to house pens of spayed heifers in the vicinity of a pen of cull bulls. In addition, spaying reduces the incidence of estrus activity. Estrus activity of intact heifers causes a temporary reduction in feed intake and associated performance and estrus activity near the time of harvest may cause an increased incidence of dark cutting beef in intact heifers compared with spayed heifers and steers (Scanga et al. 1998). The process of spaying removes the heifer's natural source of estrogen. Therefore, implanting spayed heifers with a steroid implant is an important management strategy to increase feed intake and feed efficiency (Garber et al. 1990).

In lieu of spaying, feeding MGA is an approved method of controlling estrus in feedlot heifers. The label for MGA claims a suppression of estrus, an increase in gain, and an improvement in efficiency for feedlot heifers. According to a 1999 survey of feedlot operators, 79 % of feedlot heifers in the United States received MGA when in feedlots (NAHMS 2000). A portion of the remaining heifers were likely prohibited from receiving MGA, because they were destined to specific markets that mandated they do not receive MGA. Although a small increase in daily gain was observed when a low dose of MGA was fed to steers (Moseley et al. 2003), it is still not approved for use when feeding MGA to steers. Therefore, feedlot operators are prohibited from feeding MGA to mixed-sex pens of cattle.

Implementation of a Breeding Season

In the United States, 55 % of operations surveyed had no set breeding season, whereas 34 % of operations has a single defined breeding season and 12 % of operations had two defined breeding seasons (NAHMS 2009a). Implementation of defined breeding seasons significantly impacts profitability of beef operations by matching cattle to available resources, refining nutrient delivery to groups of cattle, concentrating labor resources, and by increasing the sales price of calves.

Many traditional commercial breeding seasons are designed to place a young, growing calf on pasture forages that are at their peak of quality and availability. Grass breaks dormancy earlier in southern US compared with locations further north and thus breeding seasons are typically earlier in the calendar year the further south an operation is located. Matching growing calves to high quality growing forages allows for maximal gains through forage intake and through milk produced via forage intake of the dam during lactation.

A defined breeding season also allows for delivery of proper nutrients to cows during key times of gestation. At the time of lowest nutrient requirements (during the second trimester of pregnancy and non-lactating), a 590 kg beef cow requires 4.9 kg of total digestible nutrients (TDN) and 0.7 kg of crude protein (CP) on a daily basis (NRC 1996). At the time of greatest nutrient requirements (cow nursing a calf, producing 9.1 kg milk per day), the same 590 kg cow would require 7.2 kg TDN and 1.3 kg of CP; an increase of 47 % for TDN and 87 % for CP. Having groups of cattle at similar stages of gestation allows producers to set precise targets for nutrient delivery and to manage feeding cattle to their requirements rather than to their appetite. Without a specific breeding season, producers would have difficulty implementing a precision-based nutrition system without continually overfeeding a portion of their herd and underfeeding another portion. The concept also pertains to allocation of pasture resources to optimize forage utilization.

Another impact of a defined breeding season is that labor resources are concentrated. When no breeding season is established, a producer has the opportunity to monitor cows in a herd for calving every day of the year. By implementing a defined breeding season, calving activity is concentrated into a period of time slightly longer than the breeding season (given the natural variation in length of gestation). This phenomenon of more calves being born over a shorter period of time also concentrates the need for labor to monitor and assist cows during calving. The concentrated labor may also increase attentiveness of producers as they monitor cows regularly during calving and ultimately lead to greater calf survival (and coincident profit potential) by offering prompt assistance to heifers or cows experiencing dystocia.

As the number of calves in an auction lot increases, the price received subsequently increases; this is true in both sale barn (Leupp et al. 2008) and in video auction lots (Seeger et al. 2011). Upon arrival at a sale barn, groups of calves are sorted into relatively similar cohort groups and presented in the sale accordingly (unless producers live in a region where calves are sold individually through the ring). A defined breeding season that concentrates calving increases the uniformity of calves, resulting in fewer groups of cohorts being sold at a single time. Fewer cohort lots result in a greater number of calves in each lot and a greater sales price.

Actual dates of breeding season vary by region, and tradition (cited by 43 % of survey respondents) plays a major role in deciding when a breeding season should be held (NAHMS 2009a). Additional factors of feed cost, cow performance, calf performance, and environmental conditions need to be considered when deciding which breeding season is appropriate for a particular herd (Grings et al. 2005).

Breeding Soundness Examinations

Bulls

Breeding soundness examinations (BSE) are targeted at identifying bulls that will likely not sire calves if mated to fertile cows. A BSE includes a physical evaluation, measurement of scrotal circumference, and an evaluation of semen motility and morphology (Society of Theriogenology 1993). Currently, 71 % of operations have a BSE conducted on bulls that are newly purchased or leased and the proportion increases with increasing herd size (Fig. 5.1; NAHMS 2009a). Recommendations are to perform a BSE on every bull annually prior to the initiation of the breeding season. Specific timing of the examination is vague, but it should be conducted far enough in advance of bull turnout to access additional bulls if any bulls are classified as "unsatisfactory," or to provide sufficient time for a second examination if bulls are classified as "deferred." Timing of the BSE is important. Many yearling bulls are sold in winter (January and February) after passing a BSE. However, in northern climates bulls may experience testicles that become frozen after being turned into a pen or pasture at their new home may not pass another BSE administered at a later stage, closer to the breeding season. Other injuries or illnesses that occur throughout the

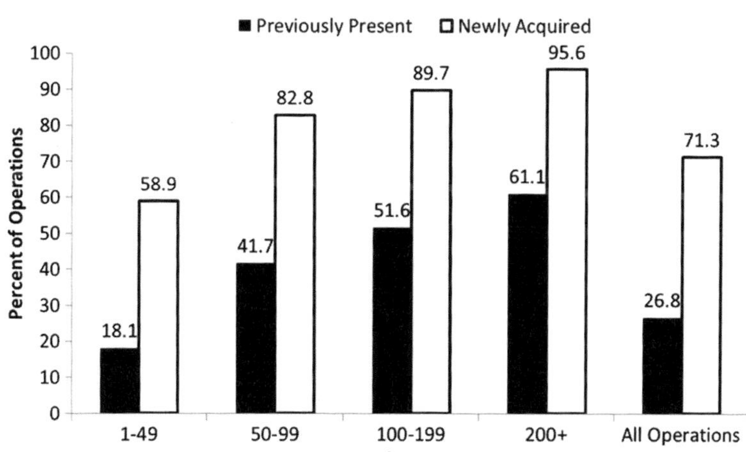

Fig. 5.1 Proportion of operations that performed semen tests on newly acquired bulls and on bulls previously present on their operations for at least 2 years (NAHMS 2009a)

Table 5.1 Description of reproductive tract scores (RTS) in beef heifers

RTS	Uterine horn diameter (mm)	Ovarian structures
1	<10.0	No palpable follicles
2	10.0–14.9	8 mm follicles
3	15.0–19.9	8–10 mm follicles
4	20.0–24.9	>10 mm follicles, CL possible
5	≥20.0	>10 mm follicles, CL present

Adapted from Dahlen et al. (2003)

year may be detected during a BSE of mature bulls, yet only 26.8 % of operations evaluate the semen of their mature herd bulls on a yearly basis (Fig. 5.1).

When cows were bred to bulls in either the satisfactory or deferred categories (termed "questionable" prior to 1993), a greater proportion of females bred to bulls classified as satisfactory breeders (46.6 %) were pregnant at the end of the breeding season compared with females bred to bulls classification as deferred (36.5 %; Farin et al. 1989). No differences, however, were observed among BSE classification in number of times bulls mounted females, number of services, or percentage of females serviced. Thus, a breeding soundness exam is an indication of potential fertility and not an indication of libido. Having a high libido in herd sires is essential to ensure high fertility in the female herd. In a multi-sire breeding pasture setting, 7.3 % of all bulls failed to sire a single calf (Drake et al. 2011). Whether this issue is one of libido, social hierarchy, or other factors not detectable by a BSE is unknown; bulls failing a BSE sire very few calves (Magee 2005) and thus should be removed from the herd.

Females

The purpose of breeding soundness exams in heifers is to identify the proportion of females that are likely cyclic prior to the breeding season and to identify the individual animals that will likely not become pregnant and remove them from the herd. Components of a breeding soundness exam for heifers include the palpation of the uterus and ovaries to determine the size of the uterine horns and the structures present in the ovaries (Table 5.1). A pelvic measurement may also be determined to provide an additional selection tool to identify heifers that have a small pelvis relative to measure of heifer weight and frame size and, thus, have a greater risk of experiencing dystocia.

As reproductive tract scores increase from 1 to 5, body weight, pelvic area, and the proportion of heifers observed in estrus increased (Patterson et al. 2000), and the proportion of heifers becoming pregnant after breeding also increased (Anderson et al. 1991). Assigning reproductive tract scores to heifers allows producers to gauge potential estrus response to estrous synchronization protocols essentially by estimating the proportion of females in a herd that are cyclic. Heifers that have multiple estrous cycles prior to first breeding are more likely to become pregnant compared with those exposed to mating on their first cycle (Byerley et al. 1987). In addition to fertility during the immediate breeding season, reproductive tract scores were found to be positively associated with calf weaning weights and fertility the following breeding season and, thus, a potential indicator of lifetime cow productivity (Holm et al. 2009).

A method of determining breeding soundness in cows has not been refined to the extent previously reviewed for heifers. As reproductive development progresses in heifers, the uterus and ovaries increase in size. The presence of a corpus luteum (CL) in an ovary is an indication that a female is cyclic, but is usually only palpable from day 5 to 18 (67 % of the time). Since mature cows have already undergone growth and development of their uterus, and the major component of postpartum recovery is uterine involution, the palpation of a cow reproductive tract for size would likely not be significantly informative except in cases of uterine infection. However in reproductive management experiments, a method frequently used by scientists is to collect two blood samples 10 days apart. If concentrations of progesterone are high (>1 ng/mL) at either of the two samples a female is defined as being cyclic, since concentrations of progesterone are below 1 ng/mL in females that do not have a CL (Perry et al. 1991). Similarly, the same concept could be applied to palpation of a CL (i.e., palpation of the ovary twice, 10 days apart to determine the presence or absence of a CL). This could be further refined by conducting the examinations with ultrasonography (Ribadu et al. 1994). The utility of a single point in time breeding soundness exam for mature cows is limited, however, and the requirement to handle cows twice to palpate for the presence of a CL further limits the use of breeding soundness exams in postpartum beef cows.

Determination of Pregnancy

Determination of pregnancy status is a way to identify nonpregnant cows and subsequently decide how to best manage nonpregnant females in a beef cattle operation. Pregnancy diagnosis may be the single most effective reproductive management tool available to producers to enhance production efficiency of their operations, especially as input costs, such as feed, fuel, and fertilizer, continue to rise. If a nonpregnant cow is maintained over an extended period of time, a significant feed cost is incurred with no calf to market to offset the expense of feeding her for 1 extra year. Yet, surprisingly, fewer than 20 % of beef producers in the United States perform a pregnancy diagnosis on their cow herd annually (NAHMS 2009a).

The rationale behind individual producers' decisions of when and how to utilize pregnancy determination in their operations varies. If heifers are maintained for spring breeding, producers may diagnose pregnancy status as soon as possible after breeding in order to market open heifers in a favorable market for yearling cattle. Similarly, mature cows may be diagnosed for pregnancy status to ensure that nonpregnant cows are sold during favorable market conditions for cull cows, or producers may wish to identify nonpregnant cows and place them on high grain diets prior to marketing. This practice can add value to cull cows by targeting a "white fat" cow market (Schnell et al. 1997). In either case, producers may use knowledge gained during pregnancy examinations to optimize the use of existing resources and to improve profitability of their operations.

Table 5.2 Comparison of different methods of pregnancy detection in beef cattle

	Method of pregnancy detection		
Item	Palpation	Ultrasound	Blood tests
Minimum fetal age detected	35–45[a]	25–30[a]	28–32
Accurate fetal aging	Yes	Yes	No
Identification of twins	No	Yes	No
Evaluate fetal viability	No[b]	Yes	No
Determine sex of fetus	No	Yes	No
Veterinarian required[c]	Yes	Yes	No
Immediate answer	Yes	Yes	No
Does experience impact accuracy?	Yes	Yes	No
Price	Medium	Higher	Low

[a]Each veterinarian has a comfort level regarding the gestational age they are comfortable detecting
[b]If the fetus is old enough, some movement may be felt using palpation per rectum
[c]Regulations requiring a veterinarian or allowing layperson technicians vary by state

Three major methods of pregnancy detection are currently available and suitable for beef cattle producers: palpation per rectum, ultrasound, and pregnancy-associated glycoproteins (PAGs). Given the variety of production systems and operation goals that modern beef producers have, each of the method of pregnancy determination has a place in the industry. For a comparison of attributes among methods of pregnancy diagnosis described, see Table 5.2.

Palpation per rectum is the most common and inexpensive method of pregnancy diagnosis. Early pregnancy diagnosis may be safely made by sensing layers of the fused chorion/allantois and the amnion slipping between your fingers when the uterus is gently pinched through the rectal wall (Romano et al. 2007). As pregnancy progresses the developing fetus may be palpated with the fetus progressively increasing in size, and beginning around 90 days after mating the placentomes can be palpated.

As veterinarians become more adept at transrectal ultrasonography and the equipment becomes less expensive, ultrasonography is becoming a more popular method of pregnancy diagnosis and has several benefits. By measuring the fetal size (crown-rump length or biparietal distance), accurate fetal age can be determined in early pregnancy. In addition, because the operator has the opportunity to visualize the fetal characteristics, fetal viability, fetal sex, and number of fetuses may all be determined (Lamb et al. 2003).

In recent years, the development of pregnancy testing of blood samples to identify PAGs has generated less expensive opportunities for beef producers to diagnose pregnancy in their operations. From day 28 after cattle are mated blood samples may be collected by producers and mailed to laboratories that are contracted with one of the three primary companies for analysis of PAGs. Barring mislabeled blood tubes, missing ID tags, and multiple animals with the same ID number, test results will be returned to the producer to describe the pregnancy status of each female. Based on data reported by the commercial companies providing these tests, they report that the tests generally yield results that are greater than 99 % accurate when a cow is diagnosed as not pregnant (false-negative), whereas the false-pregnant (false-positive) rate for these tests is approximately 5 %. A primary concern among producers using PAG analyses is that an immediate diagnosis is not available, whereas with the use of rectal palpation and transrectal ultrasonography a diagnosis is immediately available while the female is in the working facility. Once PAG results are obtained the cattle producer must gather the cattle to sort off the nonpregnant females. However, if large animal veterinarians become less accessible, less willing to provide pregnancy diagnosis services, or not sufficiently skilled to provide the services, the use of PAGs is a viable alternative to palpation per rectum and transrectal ultrasonography.

Artificial Insemination

Artificial insemination (AI) is not a new technology. The developmental research that preceded our modern techniques dates back to Russia in the late 1800s and early 1900s (Foote 2002). The accidental discovery that glycerol has properties to protect and maintain semen viability though freezing set the stage for the development of the AI industry as we know it today.

For beef producers a major opportunity exists to increase the genetic potential of their herd through the use of AI. With AI, the most genetically superior sires are available to a large number of producers rather than being confined to the cows that are on a single pasture. In addition, the accuracy of expected progeny differences (EPDs) of young sires with no progeny (typical of most natural service sires) is less than that of sires with a large number of offspring (typical of "proven" AI sires; Harris and Newman 1994). One of the primary advantages of using AI is that semen from sires with EPDs and accuracies far superior to most natural service sires is available. High accuracy of EPDs in proven AI sires allow producers more confidence that the advertised performance and phenotypic characteristics of offspring will be realized compared with offspring from low accuracy natural service sires. The risk of unexpected performance is simply much greater when using low accuracy natural service sires (Pruzzo et al. 2003). Coincidently, expected returns should be adjusted for the risk of using a natural service sire that is "overrated" or that simply does not live up to its genetic potential (Pruzzo et al. 2003). In addition, improving the accuracy of sire breeding value predictions may increase the overall rate of genetic change on beef operations (Betz 2007) and improved rate of genetic change can lead to subsequent improvements in overall profitability (Harris and Newman 1994).

Estrus or Ovulation Synchronization

Synchronizing the estrous cycle with the use of exogenous hormones (estrus synchronization and ovulation synchronization) has been developed and incorporated into beef production systems primarily to facilitate the use of AI for over 40 years. A primary factor that limits the use of AI is the labor required not only to perform AI but to detect estrus in females and ensure they are inseminated at the appropriate time. It is now possible to expect to achieve pregnancy from AI in more than 50 % of the herd during the first week of the breeding season (Lamb et al. 2006; Larson et al. 2006). The success of estrus synchronization in increasing the proportion of pregnancies derived from AI will increase the rate of genetic improvement through mating with genetically superior AI sires. However, other benefits have become evident including the potential to alter the calving season and increase uniformity of calves (Dziuk and Bellows 1983; Rodgers et al. 2012). Estrus synchronization protocols, particularly those which include a progestin, may induce cyclicity in noncyclic females (Thompson et al. 1999; Lamb et al. 2001; Stevenson et al. 2003). These mentioned advantages to utilize estrous synchronization have enhanced its use in beef operations and is usually used in conjunction with AI.

Several limitations of the initial estrous synchronization protocols contributed to slow incorporation into beef operations. Some of the first protocols devised did not improve fertility in pre- or peri-pubertal heifers or cows with postpartum anestrus. These protocols also failed to control follicular waves leading to asynchronous follicular maturation. This lack of a tight synchrony resulted in females exhibiting estrus over a period of days rather than a period of hours (Lamb et al. 2010). Estrus detection was still necessary, albeit to a lesser extent than without the ES protocol, resulting pregnancy rates were still less than ideal. As research led to more effective protocols, some of the limitations have been reduced. More recent goals have been to develop protocols that focus on control of follicular growth and development of the corpus luteum (CL) as well as to maintain protocols as being convenient and cost-effective. A primary advancement in efforts to reduce labor requirements is to move to a single fixed-time AI (TAI) which eliminates the need for estrus detection and to limit the number of times each female must be handled to 3.

Today producers can select from many estrous synchronization protocols based on their needs and goals. Protocols which require detection of estrus are available and they tend to be some of the least expensive options. Protocols also exist which incorporate some estrus detection along with a clean-up TAI. These protocols have a reduced cost because a female detected in estrus receives one less injection while all females in the herd are still given the opportunity to become pregnant via the clean-up TAI. The third category of protocols are the TAI protocols and these, although they tend to be the most expensive, result in the greatest number of pregnancies generated on the first day of the breeding season with no estrus detection required.

Currently only 7.6 % of beef operations in the United Sates utilize AI as a reproductive management tool (NAHMS 2009a), whereas pregnancies from AI are present in 72.5 % of dairy females (NAHMS 2009b). When queried as to their reluctance to utilize AI, over 53 % of operations cited labor concerns or complicated estrous synchronization protocols as primary reasons for not implementing this reproductive technology (NAHMS 2009a). Research projects addressing these key areas of producer concern have been developed and improvements in the actual protocols and their subsequent ability to effectively synchronize estrous and ovulation have been made (Lauderdale 2009).

While many options exist for synchronization of estrus and ovulation, a short list of protocols was developed based on available research data and field use by the Beef Cattle Reproduction Leadership Team (http://beefrepro.unl.edu/). This group is composed of representatives from the AI and pharmaceutical industries, veterinarians, and reproductive physiologists from the Beef Reproduction Task Force with active research programs in this area. The primary objectives of the Task Force are to: (1) improve the understanding of the physiological processes of the estrous cycle, the procedures available to synchronize estrus and ovulation and the proper application of these systems and (2) improve the understanding of methods to assess male fertility and how it affects the success of AI programs. Annually the Task Force reviews and updates estrous synchronization protocols that are recommended for use in heifers (Fig. 5.2) and cows (Fig. 5.3). In addition, these protocols are distributed to members of the AI and pharmaceutical industries to be published in their publications.

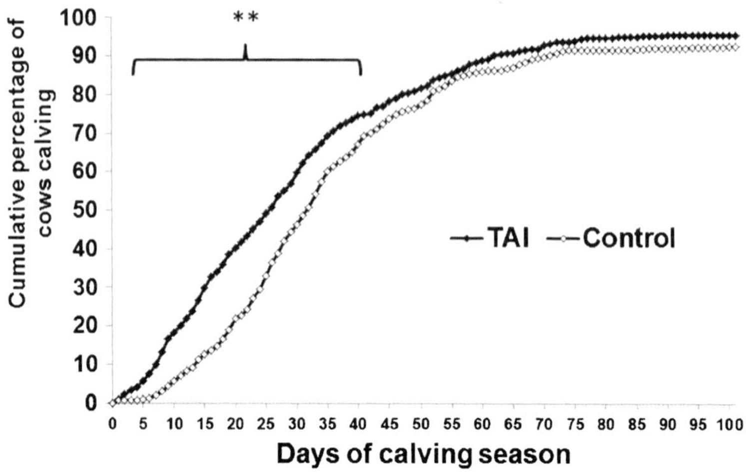

Fig. 5.4 Survival analysis of the percentage of cows calving by day during the calving season (Rodgers et al. 2012). **Cumulative calving percentage differs ($P<0.05$) between TAI (cows were exposed to the CO-synch + CIDR estrous synchronization protocol at the initiation of the breeding season) and control (cows were only exposed to natural service during the breeding season) treatments

Economic Advantages of Artificial Insemination and Estrus Synchronization

Incorporation of estrus synchronization and AI has potential to influence economic efficiency of cow/calf enterprises (Sprott 1999). Modeling exercises demonstrated a potential increased return of $25–$40 per calf born from AI breeding for producers who decide to dedicate the time and effort required to successfully implement an AI protocol (Johnson 2005). In addition, 72 % of respondents to a survey administered at the *Applied Reproductive Strategies in Beef Cattle* workshops estimated the additional value of calves from AI breeding compared with natural service breeding to be over $20, whereas 48 % of respondents estimated the additional value at over $50 (Johnson et al. 2011). Hedonic modeling of data generated from sales of the Show-Me Replacement Heifer, Inc. revealed a premium of $18.69/hd for heifers pregnant with a calf from AI breeding, and a $24.30/hd discount for heifers that were due to calve outside a 30 days window (Parcell et al. 2006). The $18.69 economic advantage for AI pregnancies may need to be adjusted upward, as using estrous synchronization and AI may result in a greater proportion of cows calving within the first 30 days of a breeding season (Larson et al. 2006).

Early models assessed economic benefits derived from incorporating estrus synchronization and TAI into an operation with an outcome of heavier weaning weights in calves (Johnson and Jones 2005). In this analysis, the use of estrous synchronization produced calves approximately 10 days older than calves born from natural mating. Based on an assumed daily growth rate of 0.91 kg/day, it was determined that calves born from ES gained 9.1 more kilograms. Profitability depends on several factors including cost of estrous synchronization and AI and selling price per kilogram. In an analysis conducted which investigates the incorporation of ES and AI compared to natural mating in a cow/calf production setting, 84 % of cows exposed to ES/TAI weaned a calf compared to 78 % of cows in the natural mating group (Rodgers et al. 2012). Calving distribution also differed (Fig. 5.4), resulting in the mean calving day from initiation of the calving season to be 26.8 days for cows exposed to ES/TAI and 31.3 days for cows exposed to natural mating (Rodgers et al. 2012). According to these data, not only are more calves weaned per cow exposed to estrus synchronization and TAI, on average, but calves may be older at weaning and have the opportunity to gain more weight.

This increase in weaning weight may have the greatest potential to offset the cost of estrus synchronization and TAI systems. Although the improvement in genetics is a significant and long-term improvement, many producers have a desire for an immediate recovery of costs. The increase in total pounds produced because more cows exposed to ES/TAI produced a weaned calf is significant. When Johnson and Jones (2005) determined a 9.1 kg increase in weaning weight, the increase was attributed to calves being born earlier in the calving season and improved genetic growth potential due to the selection of sires. Using the estimated 9.1 kg increase in weaning weight (Johnson and Jones 2005) in conjunction with a 5-year mean selling price as of 2011, a net return of $16.23 would be realized for the calves derived from an AI system primarily through an increase in weaning weight of exposed females (Rodgers et al. 2012). Weaning weight of cows exposed is a major driver of profit but it also influences the total kilograms of beef available for consumers and the efficiency with which that weight was produced. It is clear that the benefits of estrus synchronization in combination with AI will continue to be realized and incorporated into beef production systems, with a subsequent improvement in efficiency of beef cattle operations.

Sex-Sorted Semen

The technology that has been developed to sort spermatozoa by the presence of either a Y or X chromosome has the potential to alter the efficiency of beef production. Depending on the production goals of an operation, the availability of either more bull or heifer calves creates the opportunity for more profitability. Males are preferred over females when feeding animals for the production of beef. Steers are more efficient at converting feed to muscle which equates to more efficient production of beef. Many producers focus on the generation of replacement females, and in these operations a benefit may be realized for more heifer calves.

Processes to generate sex-sorted spermatozoa are fairly inefficient and costly, which has limited its use. Spermatozoa are passed through a flow cytometer single file, and with upward of seven billion sperm in an ejaculate, it may take three or four times longer to process an ejaculate for sex-sorted semen compared to conventionally processed semen. Damage incurred during the sorted process and/or fewer spermatozoa per dose result in decreased fertility with sex-sorted spermatozoa. A producer can expect pregnancy rates of approximately 70 % of that normally achieved with the use of conventional semen (DeJarnette et al. 2009). An economic evaluation is needed to determine whether an increase in the preferred sex of offspring is outweighed by the increased cost of the semen and the likely decreased fertility.

In commercial beef cattle operations, sexed semen provides the opportunity to use a small number of elite cows to generate replacements while mating the remainder of the cows to terminal sires. However, the most common use of sexed semen in the beef industry is to increase the number of desired sex animals in purebred operations. Generating more bull calves from a superior herd sire to produce bulls for the commercial sector is an important consideration. Similarly, deriving more daughters from a purebred maternal line would also be advantageous to certain purebred breeders. Therefore, although sexed semen may not be utilized extensively throughout the beef industry, sexed semen will continue to provide beef producers an opportunity to alter management practices that will enhance beef production efficiency.

Embryo Transfer

Incorporating embryo transfer (ET) into beef production systems is the fastest way to change the genetic base of a herd using existing females. Females of poor or even average genetic potential have the opportunity via ET to serve as a surrogate to carry a calf of exceptional genetic merit.

Table 5.3 Bovine in vivo-derived embryo activity in 2011

Continents	Flushes	Transferrable embryos	Number of transferred embryos			
			Fresh	Frozen	Total	Percentage[a]
Africa	1,438	9,401	4,056	2,469	6,525	1.14
Asia	15,444	124,362	24,026	51,697	75,723	13.23
Europe	23,480	108,712	41,040	69,381	110,421	19.29
North America	54,837	362,781	109,197	139,418	248,615	43.44
South America	12,174	68,187	36,953	26,054	63,007	11.01
Oceania	10,755	59,419	32,921	35,130	68,051	11.52
2010 total	104,651	732,000	243,885	291,279	590,561	100.00
2011 total	118,128	732,862	248,193	324,149	572,342	

Adapted from IETS (2012)
[a]Percentage of total embryos collected in 2010

In vivo embryo production through superovulation of donor females and in vitro production following ovum pickup (OPU) both allow a single female to generate a substantially greater number of offspring than she would be capable of producing in conventional systems. When ET technology is coupled with the use of spermatozoa from genetically superior sires and possibly the use of sex-sorted semen, genetic improvement can increase exponentially within a herd. The use of estrous synchronization protocols to achieve donor–recipient synchrony decreased the number of available recipients necessary. The advancements of cryopreservation also decreased the number of recipients necessary as well as relaxed the timing requirements of embryo transfer making it more feasible and efficient for many producers and has increased the use of the technology (Hasler 2003). In spite of these advantages, many procedures used in embryo transfer are expensive and inefficient which limits the practical application for beef producers. Perhaps the most promising aspect of these technologies as they relate to food security is the ability to transport embryos, rather than live animals, to areas where improved genetics would rapidly increase production of beef.

Much of the data regarding the use of embryo transfer is collected and reported by members of the International Embryo Transfer Society (IETS). The data retrieval committee is composed of members from most of the countries where embryo transfer technologies are practiced. This allows for the discussion of the technology and its role in food security on a global scale.

According to IETS (2011, 2012), there were 732,862 in vivo-derived embryos collected across the world in 2011 which was an increase from the 702,000 embryos collected in 2009 and the 732,000 collected in 2010. After increasing from 534,000 in 2009 to 591,000 in 2010, the number of transferred in vivo-derived embryos decreased to 572,342 in 2011. Over the last few years, the general trend is increasing for the number of in vivo-derived embryos being transferred in all continents except Africa. Although the potential to increase efficient beef production as a result of embryo transfer exists, efforts must be made to overcome current obstacles with the technology and to facilitate increased adoption. Part of that goal may be achieved with the use of frozen–thawed embryos. More frozen in vivo-derived embryos were transferred in 2011 (309,806 frozen and 238,194 fresh) compared to fresh and this trend has been consistent since the 1990s when freezing technologies have advanced. In most continents, except South America, more frozen embryos are transferred than fresh embryos (Table 5.3).

Even though the use of in vivo-derived embryos continues to increase, a major limitation has been the lack of successful superovulation in donor females. Although research continues in the development of superovulation protocols as well as techniques to predict which donor females may respond well to superovulation (Hasler 2003; Betteridge 2006), this remains an inhibitor of using in vivo-derived embryos. As number of embryos per flush increases the overall cost per embryo produced will likely decrease. Since cost

Table 5.4 Bovine in vitro-produced embryos in 2011

		Number of transferred embryos			
Continents	Transferrable embryos	Fresh	Frozen	Total	Percentage[a]
Africa	0	0	0	0	0.00
Asia	62,418	4,086	6,699	10,785	2.88
Europe	8,034	8,034	3,419	11,453	3.06
North America	48,474	17,850	2,930	20,780	5.56
South America	325,349	307,278	15,879	323,157	86.44
Oceania	9,196	6,679	1,015	7,694	2.06
2010 total	450,549	315,715	23,970	339,685	100.00
2011 total	453,471	343,927	29,942	373,869	

Adopted from IETS (2012)
[a]Percentage of total embryos collected in 2010

of the technology is one of the reasons that producers have been hesitant to incorporate ET, finding methods to improve superovulatory response and coincident number of transferrable embryos per flush would likely increase use of the technology.

To avoid this potential disadvantage of poor response to superovulation, the use of in vitro-produced embryos is increasing. Proper facilities and expertise are required, but when females can be subjected to frequent sessions of transvaginal ultrasonically guided OPU, oocytes can then be subjected to in vitro fertilization and culture resulting in more transferrable embryos. These embryos are more likely to be transferred fresh because their viability decreases with cryopreservation to a greater extent than in vivo-derived embryos (Palasz and Mapletoft 1996). Until this hurdle can be mediated, the ability to transport and store these embryos will be limited and will thus limit its use on a global scale to improve overall efficiency of production. It is possible to culture in vitro-produced embryos in the oviduct of sheep, and these embryos survive cryopreservation as well as their in vivo counterparts (Galli et al. 2003), giving promise to increased potential in this area.

The use of in vitro-produced embryos continues to increase as well. The number of transferable embryos produced worldwide was 453,471 in 2011 compared to 377,000 in 2009 (IETS 2011, 2012). This increase was led by South America, primarily Brazil. While the United States produced 40,602 embryos, in vitro, Brazil produced 318,116 in 2011 (IETS 2012). Furthermore, while the United States transferred 30,427 in vitro-produced embryos, Brazil transferred 318,119 in 2011 (IETS 2012). There were 373,836 in vitro-produced embryos transferred worldwide in 2011, which was an increase of 10 % from 2010 (IETS 2012). In contrast to in vivo-derived embryos, most in vitro-produced embryos are transferred fresh. Only 7 % (23,970 embryos) of the in vitro-produced embryos transferred in 2010 were frozen (IETS 2011). In 2010, all in vitro-derived embryos that were produced in the United States were generated via OPU (IETS 2011). The 6,876 reported OPU sessions in 2010 (up from 4,885 in 2009) resulted in 109,615 oocytes and 34,969 transferrable embryos (IETS 2011). Brazil has an extensive system for generating in vivo-produced embryos via OPU but Argentina is increasing the pace it generates embryos as well. Eighty-five percent of transfers of in vivo-derived embryos occurred in Brazil in 2011 (IETS 2012). The number of in vitro-produced embryos that are frozen continues to increase as cryopreservation techniques advance (Table 5.4).

When assessing all embryos (in vivo and in vitro, fresh and frozen), 841,540 were transferred in 2009, 990,993 in 2010, and 921,836 in 2011 (IETS 2011, 2012). With the rapid pace of improved genetics garnered with the use of embryo transfer, efficiency of beef production in herds incorporating the technology will rapidly improve as well.

When considering food security and the benefits of technologies such as embryo transfer, it is import to assess the transport of genetics in the form of embryos. While countries such as the United States, Brazil, Argentina, France, Japan, and Australia lead the countries that flush donor females, produce embryos via in vitro fertilization, and transfer embryos, it is the export of embryos from these countries to others that are in desperate need of improved genetics that might stand to have the greatest influence on world food security. However, it is believed that data regarding imports are grossly underreported. So while we know more about the embryos being exported, we know less about where they are going and if they are eventually transferred into recipient females. In 2011, 18,189 beef embryos were exported and the United States exported 5,076 of them (IETS 2012). However, without data on the importation of embryos, it is nearly impossible to determine the genetic progress being made in countries without established infrastructures and production capabilities.

A discussion of how embryo transfer technologies may impact food security of the world is not complete without mentioning the role these technologies play in research. Significant gains in understanding the development of the early embryo, embryo mortality and viability, and the interaction of the embryo and the endometrium of the uterus have all been possible as a result of studies conducted using in vivo and in vitro methods. One could quickly realize how significant embryo survival is in the efficiency of livestock production. The result of improved efficiency of production is more efficient production of food.

Conclusion

The term "reproductive technology" encompasses a wide variety of interventions that beef producers can implement on their operations. On one end of the spectrum some reproductive techniques discussed grew out of a simple desire to control animal behavior or minimize calving losses. On the other end of the spectrum some of the techniques discussed would not be possible without the years of technical research it has taken to give us a better understanding of biology and subsequent ability to manipulate natural processes. However, all of the reproductive technologies discussed can play a significant role in improving the efficiency, production capability, and sustainability of the US beef industry.

In most cases the land available for grazing is being utilized and portions of existing land are actually being transitioned from grazing land to crop land or even being overtaken by expanding urban development. The net result is either an unchanging or decrease in land available to support grazing beef herds. Therefore, the long-term sustainability of the US beef production system likely depends on the intensity of management and production level achieved with existing resources (Galyean et al. 2011). Reproductive technologies allow us to maximize the potential of existing resources which will be imperative if our industry is to rise to the challenge of providing an affordable nutritious protein source to an expanding global population for generations to come.

References

Amatayakul-Chantler S, Jackson JA, Stenger J, King V, Rubio LMS, Howard R, Lopez E, Walker J (2012) Immunocastration of *Bos indicus* × Brown Swiss bulls in feedlot with gonadotropin-releasing hormone vaccine Bopriva provides improved performance and meat quality. J Anim Sci 90:3718–3728

Anderson KJ, Lefever DG, Brinks JS, Odde KG (1991) The use of reproductive tract scoring in beef heifers. Agri-Practice 12:19–26

APHIS, Veterinary Service: Centers for Epidemiology and Animal Health (2012) Importance of pre-arrival management practices to operators of U.S. feedlots. Info Sheet: Safeguarding American Agriculture

Betteridge KJ (2006) Farm animal embryo technologies: achievements and perspectives. Theriogenology 65: 905–913

Betz GCM (2007) Using the rate of genetic change and the population structure of cattle to better target genetic progress. In: Proceedings of 39th beef improvement federation symposium, Fort Collins, CO, pp 103–109

Byerley DJ, Staigmiller RB, Berardinelli JG, Short RE (1987) Pregnancy rates of beef heifers bred either on puberal or third estrus. J Anim Sci 65:645–650
Calkins CR, Clanton DC, Berg TJ, Kinder JE (1986) Growth, carcass and palatability traits of intact males and steers implanted with zeranol or estradiol early and throughout life. J Anim Sci 62:625–631
Carroll FD, Rollins WC, Wagnon KA, Loy RG (1975) Comparison of beef from bulls and des implanted steers. J Anim Sci 41:1008–1013
Dahlen CR, Lamb GC, Zehnder CM, Miller LR, DiCostanzo A (2003) Fixed-time insemination in peri-puberal, light-weight replacement beef heifers synchronized with PGF2alpha and GnRH. Theriogenology 59(8):1827–1837
DeJarnette JM, Nebel RL, Marshall CE (2009) Evaluating the success of sex-sorted semen in US dairy herds from on farm records. Theriogenology 71:49–58
Drake DJ, Weber KL, Van Eenennaam AL (2011) What are herd bulls accomplishing in multiple sire breeding pastures? In: Proceedings of applied reproductive strategies in beef cattle. Joplin: MO
Dziuk PJ, Bellows RA (1983) Management of reproduction in beef cattle, sheep and pigs. J Anim Sci 57(suppl 2):355–379
Farin PW, Chenoweth PJ, Tomky DF, Ball L, Pexton JE (1989) Breeding soundness, libido and performance of beef bulls mated to estrus synchronized females. Theriogenology 32(5):717–725
Food and Agriculture Organization of the United Nations (2009) The State of Food and Agriculture. Rome Italy. Accessed online: http://www.fao.org/docrep/012/i0680e/i0680e.pdf
Foote RH (2002) The history of artificial insemination: selected notes and notables. J Anim Sci 80:1–10
Galli C, Duchi R, Crotti G, Turini P, Ponderato N, Colleoni S, Laqutina I, Lazzari G (2003) Bovine embryo technologies. Theriogenology 59:599–616
Galyean ML, Ponce C, Schutz J (2011) The future of beef production in North America. Anim Front 1(2):29–36
Garber MJ, Roeder RA, Combs JJ, Eldridge L, Miller JC, Hinman DD, Ney JJ (1990) Efficiency of vaginal spaying and anabolic implants on growth and carcass characteristics in beef heifers. J Anim Sci 68:1469–1475
Grings EE, Short RE, Klement KD, Geary TW, MacNeil MD, Haferkamp MR, Heitschmidt RK (2005) Calving system and weaning age effects on cow and preweaning calf performance in the Northern Great Plains. J Anim Sci 83:2671–2683
Harris DL, Newman S (1994) Breeding for profit: synergism between genetic improvement and livestock production (a review). J Anim Sci 72:2178–2200
Hasler JF (2003) The current status and future of commercial embryo transfer in cattle. Anim Reprod Sci 79:245–264
Holm DE, Thompson PN, Irons PC (2009) The value of reporductive tract scoring as a predictor of fertility and production outcomes in beef heifers. J Anim Sci 87: 1934–1940
IETS (2011) IETS 2011 statistics and data retrieval committee report. Pages 14-23 in Embryo transfer newsletter. International Embryo Transfer Society, Champaign, IL
IETS (2012) IETS 2012 statistics and data retrieval committee report. Pages 16-26 in Embryo transfer newsletter. International Embryo Transfer Society, Champaign, IL
Janett F, Gerig T, Tschuor AC, Amatayakul-Chantler S, Walker J, Howard R, Bollwein H, Thun R (2012) Vaccination against gonadotropin-releasing factor (GnRF) with Bopriva significantly decreases testicular development, serum testosterone levels, and physical activity in pubertal bulls. Theriogenology 78:182–188
Jim GK, Ribble CS, Guichon PT, Thorlakson BE (1991) The relative economics of feeding open, aborted, pregnant feedlot heifers. Can Vet J 10:613–617
Johnson SK (2005) Possibilities with today's reproductive technologies. Theriogenology 64:639–656
Johnson SK, Jones R (2005) Costs and comparisons of estrus synchronization systems. In: Proceedings of applied reproductive strategies in beef cattle workshop, North Platte, NE, pp 103–115
Johnson SK, Funston RN, Hall JB, Kesler DJ, Lamb GC, Lauderdale JW, Patterson DJ, Perry GA, Strohbehn DR (2011) Multi-state Beef Reproduction Task Force provides science-based recommendations for the application of reproductive technologies. J Anim Sci 89:2950–2954
Lamb GC, Stevenson JS, Kesler DJ, Garverick HA, Brown DR, Salfen BE (2001) Inclusion of an intravaginal progesterone insert plus GnRH and prostaglandin F2α for ovulation control in postpartum suckled beef cows. J Anim Sci 79:2253–2259
Lamb GC, Dahlen CR, Brown DR (2003) Symposium paper: reproductive ultrasonography for monitoring ovarian structure development, fetal development, embryo survival, and twins in beef cows. Prof Anim Sci 19:135–143
Lamb GC, Larson JE, Geary TW, Stevenson JS, Johnson SK, Day ML, Ansotegui RP, Kesler DL, DeJarnette JM, Landblom DG (2006) Synchronization of estrus and artificial insemination of replacement beef heifers using gonadotropin-releasing hormone, prostaglandin $F_2\alpha$, and progesterone. J Anim Sci 84:3000–3009
Lamb GC, Dahlen CR, Larson JE, Marquezini G, Stevenson JS (2010) Control of the estrous cycle to improve fertility for fixed-time artificial insemination in beef cattle: a review. J Anim Sci 88(13 suppl):E181–E192
Larson JE, Lamb GC, Stevenson JS, Johnson SK, Day ML, Geary TW, Kesler DJ, DeJarnette JM, Schrick FN, DiCostanzo A, Arseneau JD (2006) Synchronization of estrus in suckled beef cows before detected estrus and artificial insemination and timed artificial insemination using gonadotropin-releasing hormone, prostaglandin $F_2\alpha$, and progesterone. J Anim Sci 84:332–342
Lauderdale JW (2009) ASAS centennial paper: contributions in the Journal of Animal Science to the development

of protocols for breeding management of cattle through synchronization of estrus and ovulation. J Anim Sci 87:801–812
Laudert SB (1988) Incidence of pregnancy in feedlot heifers at slaughter. Kansas Agric Exp Stn Rep Prog 539:112
Leupp JL, Lardy GP, Daly R, Wright CL, Paterson JA (2008) Factors influencing price of North Dakota, South Dakota and Montana feeder calves. NDSU beef cattle and range research report, pp 46–49
Magee D (2005) Breeding soundness evaluation of bulls. In: Proceedings of applied reproductive strategies in beef cattle, College Station, TX
Moseley WM, Meeuwse DM, Boucher JF, Dame KJ, Lauderdale JW (2003) A dose-response study of melengestrol acetate on feedlot performance and carcass characteristics of beef steers. J Anim Sci 81(11): 2699–2703
National Animal Health Management Service (NAHMS) (2000) Part I: baseline reference of feedlot management practices, 1999. National Animal Health Management Service, Fort Collins, CO, pp 46–47
National Animal Health Management Service (NAHMS) (2009a) Part II. Reference of beef cow-calf management practices in the United States, 2007–08. National Animal Health Management Service, Fort Collins, CO, pp 5–23
National Animal Health Management Service (NAHMS) (2009b) Part IV. Reference of dairy cattle health and management practices in the United States, 2007–08. National Animal Health Management Service, Fort Collins, CO, p 27
National Animal Health Monitoring Service (NAHMS) (2008) Part I. Reference of beef cow-calf management practices in the United States, 2007–08. National Animal Health Management Service, Fort Collins, CO, pp 37–40
National Research Council (NRC) (1996) Nutrient requirements of beef cattle, 7th edn. National Academy Press, Washington, DC
Palasz AT, Mapletoft RJ (1996) Cryopreservation of mammalian embryos and oocytes: recent advances. Biotechnol Adv 14:127–149
Parcell JL, Dhuyvetter KC, Patterson DJ, Randle R (2006) The value of heifer and calf characteristics in bred heifer price. Prof Anim Sci 22:217–224
Patterson DJ, Wood SL, Randle Randle RF (2000) Heifer programs that add value to the beef industry: procedures that support reproductive management of replacement beef heifers. J Anim Sci 77:1–15
Perry RC, Corah LR, Cochran RC, Beal WE, Stevenson JS, Minton JE, Simms DD, Brethour JR (1991) Influence of dietary energy on follicular development, serum gonadotropins, and first postpartum ovulation in suckled beef cows. J Anim Sci 69:3762–3773
Pruzzo L, Cantet RJC, Fioretti CC (2003) Risk-adjusted expected returns for selection decisions. J Anim Sci 81:2984–2988
Ribadu AY, Ward WR, Dobson H (1994) Comparative evaluation of ovarian structures in cattle by palpaltion per rectum, ultrasonography and plasma progesterone concentration. Vet Rec 135:452–457
Rodgers JC, Bird SL, Larson JE, DiLorenzo N, Dahlen CR, DiCostanzo A, Lamb GC (2012) An economic evaluation of estrous synchronization and timed artificial insemination in suckled beef cows. J Anim Sci 90:4055–4062
Romano JE, Thompson JA, Kraemer DC, Westhusin ME, Forrest DW, Tomaszweski MA (2007) Early pregnancy diagnosis by palpation per rectum: influence on embryo/fetal viability in dairy cattle. Theriogenology 67:486–493
Scanga JA, Belk KE, Tatum JD, Grandin T, Smith GC (1998) Factors contributing to the incidence of dark cutting beef. J Anim Sci 76:2040–2047
Schnell TD, Belk KE, Tatum JD, Miller RK, Smith GC (1997) Performance, carcass, and palatability traits for cull cows fed high-energy concentrate diets for 0, 14, 28, 42, or 56 days. J Anim Sci 75:1195–1202
Seeger JT, King ME, Grotelueschen DS, Rogers GM, Stokka GS (2011) Effect of management, marketing, and certified health programs on the sale price of beef calves sold through livestock video auction service from 1995 through 2009. J Am Vet Med Assoc 239:451–466
Society of Theriogenology (1993) Guidelines for the bull breeding soundness evaluation. In: Guidelines for uniform beef improvement programs, 9th edition, 2010. Beef Improvement Federation, North Carolina State University, Raleigh, pp 24–27
Sprott LR (1999) Management and financial considerations affected the decision to synchronize estrus in beef females. J Anim Sci 77:1–10
Stevenson JS, Lamb GC, Johnson SK, Medina-Britos MA, Grieger DM, Harmoney KR, Cartmill JA, El-Zarkouny SZ, Dahlen CR, Marple TJ (2003) Supplemental norgestomet, progesterone, or melengestrol acetate increases pregnancy rates in suckled beef cows after timed inseminations. J Anim Sci 81:571–586
Thompson KE, Stevenson JS, Lamb GC, Grieger DM, Löest CA (1999) Follicular, hormonal, and pregnancy responses of early postpartum suckled beef cows to GnRH, norgestomet, and PGF2α. J Anim Sci 77: 1823–1832
Worrell MA, Clanton DC, Calkins CR (1987) Effect of weight at castration on steer performance on the feedlot. J Anim Sci 64:343–347

Impact of Reproductive Technologies on Dairy Food Production in the Dairy Industry

6

Jeffrey S. Stevenson

Abstract

Reproductive technologies drive the efficiency of managing dairy cows because the lactation cycle of the dairy cow depends on regular calving to renew lactation yields. Achieving timely pregnancies to allow calving every 12–14 months, therefore, is critical in modern dairy production. To meet the demands to produce sufficient milk for fluid and dairy products, various technologies are applied to enhance efficiencies on the dairy farm. Artificial insemination (AI), embryo transfer, ultrasonographic and chemical detection of pregnancy, various monitors that detect or predict estrus, and handheld communication and testing devices allow managers to retrieve information to make cow-side decisions about health and reproductive status. Genomic testing of young potential sires or young heifers is now possible and can provide information about their genetic merit years before any progeny tests can be completed. In many countries, the challenge faced by dairy producers is their ability to afford these technologies in the face of rising feed and labor costs and volatile milk prices received at the farm gate. Government policies often place obstacles, trade barriers, and unfunded mandates that preclude operations from making a modest profit. Unlike nearly all other manufacturing industries, agriculture producers have little control over the price received for their products. Therefore, dairy production is vulnerable to many uncontrolled factors including climate, government policy, economic conditions, and skilled labor shortages. It is clear that the impact of emerging and current reproductive technologies is critical to the management of dairy cattle to produce sufficient milk to meet consumer demands for quality fluid and dairy products.

J.S. Stevenson, Ph.D. (✉)
Department of Animal Sciences and Industry, Kansas State University, Manhattan, KS 66506-0201, USA
e-mail: jss@k-state.edu

G.C. Lamb and N. DiLorenzo (eds.), *Current and Future Reproductive Technologies and World Food Production*, Advances in Experimental Medicine and Biology 752,
DOI 10.1007/978-1-4614-8887-3_6, © Springer Science+Business Media New York 2014

Keywords

Artificial insemination • Assisted reproductive technologies • Estrus-detection aids • Genomics • Pregnancy diagnosis • Recombinant bovine somatotropin • Timed artificial insemination programs

Introduction

Demand for protein in the diet increases in emerging economies of nations where poverty has prevailed. Demand for protein in diets can be met by the quality of milk proteins. Milk and dairy products supply abundant amounts of protein, calcium, phosphorus, magnesium, riboflavin, and vitamin B_{12}. These are some of the more expensive nutrients that are required in the human diet. Dairy products supply about 21 % of the protein and 76 % of the calcium in a typical US diet.

One pint of milk supplies about 30 % of the daily adult protein requirement and milk protein quality ranks third, exceeded only by human milk and egg protein, in fulfilling amino acid requirements of the human. One pint of milk provides all of the minimum daily requirements of essential amino acids needed by a woman and more than a man's requirements, except for methionine. Milk protein consists of 78 % casein, 18 % noncasein proteins, and 3 % nonprotein nitrogen compounds. Lactose, naturally found in milk, stimulates the absorption of calcium and promotes the growth of desirable types of bacteria in the intestines. Milk fat is a very complex material, very similar to human milk fat, containing more than 140 individual fatty acids. The five most important fatty acids, by weight, are oleic, palmitic, stearic, myristic, and butyric. The fatty acids are 59 % (by weight) saturated and 41 % unsaturated.

Per capita consumption of fluid milk and dairy products in the United States increased from 240 to 267 kg per person from 1970 to 2010 (NMPF 2012). Increasing proportions of manufactured products are produced while fluid milk consumption generally has declined since 1970. Although approximately 22 % of milk production is consumed as fluid products (whole, reduced fat, skim, flavored milks, and all cream products), 78 % is manufactured into various dairy products. The major US manufactured dairy product is American cheese.

During the past several decades in the United States as well as in other countries, the historical trends of increasing milk yield and declining fertility of dairy cows have resulted in dramatic changes in herd management (Bello et al. 2012). One obvious change is that of fewer commercial dairy operations, each of which concentrates a greater number of cows. For example, the numbers of US dairy farms have decreased from 329,680 in 1980 to 131,509 in 1992 to 53,127 in 2010, whereas average herd size increased from 19 cows in 1970 to more than 172 cows by 2010 (NMPF 2012). In 1980, only 4 % of the US dairy farms were milking more than 100 cows, whereas by 2010, 25.2 % of dairy farms were milking 100 or more cows. In 1980, 33 % of the milk was produced on dairy farms of more than 100 cows, whereas in 2010, 85 % of milk was produced on farms with 100 or more cows. Furthermore, no dairy herds were milking more than 500 cows in 1980, but by 2010, 6.2 % of herds milked 500 or more cows accounting for 61 % of the total milk output in the United States (NMPF 2012).

It is unclear what adaptations in management and management effects have changed relative to the increase in size of dairy operations. Although no data are readily available, it seems likely that the average number of cows managed per employee has increased. Moreover, the labor force has changed culturally from predominantly immigrant Americanized Europeans to predominantly Hispanics, with the need for bilingual managers with a working knowledge of Spanish. In 2010, 1 in every 9 Americans were Hispanic, accounting for the largest minority in the United States. Currently, Hispanic employees (of which approximately 63 % are Mexican) represent approximately 75 % of the hired labor force on dairy farms in California (Bello et al. 2012). Further management changes in recent decades include greater usage of confinement housing whereby cows are housed almost exclusively on concrete after first calving. Diet composition has also changed from a mainly roughage diet to one that contains more concentrates than forage. Recently, in some areas of the United States, more dairies are converting to predominantly pasture-fed operations in an effort to reduce feed costs while also realizing less milk production per cow.

It is clear that reproductive biotechnologies have become important drivers in management of day-to-day dairy operations, particularly in larger herds. Breeding strategies have moved from predominantly natural service to an overwhelming prevalence of artificial insemination (AI) and ovulation synchronization programs (Caraviello et al. 2006). For example, in applying AI changes in semen packaging, dose, and using inseminates of sex-sorted semen has facilitated genetic progress by making more progeny-tested sires available for use in the dairy industry (Foote 1996). Detection of estrus by means of measuring correlated traits such as increased activity or body temperature are available in addition to once-used heat mount detectors and permanent rump-mounted, pressure-sensitive electronic detectors that record mounts received by the cow in estrus (Roelofs et al. 2010). Pregnancy diagnosis occurs by means of transrectal palpation, transrectal ultrasonography, or by chemical tests as early as 28 days post-insemination (NAHMS 2007). Other assisted reproductive techniques include embryo transfer, in vitro maturation of oocytes collected from ovaries of valuable cows at slaughter, in vitro fertilization (IVF) of those oocytes followed by in vitro culture, and transfer to recipient surrogate cows for gestation (Rodriguez-Martinez 2012). Applying recombinant DNA-derived bovine somatotropin (bST) enhances milk yields and has been demonstrated to improve pregnancy rates of dairy cows in some studies. This chapter will review the direct impact of these technologies on reproductive management of dairy cattle and subsequent milk production to provide adequate supplies of dairy fluid milk and dairy foods in the diet of consumers.

Artificial Insemination

Preservation of semen for AI is well established and provides semen of good quality for commercial applications. The number of doses of bovine semen produced worldwide during the last 50 years is greater than 250 million (Rodriguez-Martinez 2012). Genetic progress in cattle can be increased up to 50 % by application of AI to replace natural mating. The increase in milk yield per cow is a result of genetic selection for yield of milk and milk components. Increase in milk production of dairy cattle is a remarkable success story that is the result of transferring superior genes of progeny-tested sires through AI (Fig. 6.1). The first national sire evaluations were calculated in 1936. The Animal Improvement Programs Laboratory (AIPL) of the U.S. Department of Agriculture conducts research to discover, test, and implement improved genetic evaluation techniques for economically important traits of dairy cattle. Data are collected and research is directed at genetic improvement of yield traits (e.g., milk, fat, and protein) and nonyield traits that affect health and profitability (longevity, udder conformation, fertility, calving, and disease resistance) of dairy cows. Three times annually the AIPL produces sire summaries of multiple traits obtained from herd records of dairy producers participating in the Dairy Herd Information Association (DHIA). Six dairy record processing centers collate data and provide information on progeny daughters of bulls from which sire summaries are calculated and provided by the AIPL.

Despite the remarkable progress made in all other technologies, AI remains the most important single reproductive technology adopted by dairy producers. Historically, fresh semen was used by breeders from nondescript bulls of unknown genetic merit as early as the 1930s. The art and science of producing frozen semen was perfected in the 1950s with addition of glycerol

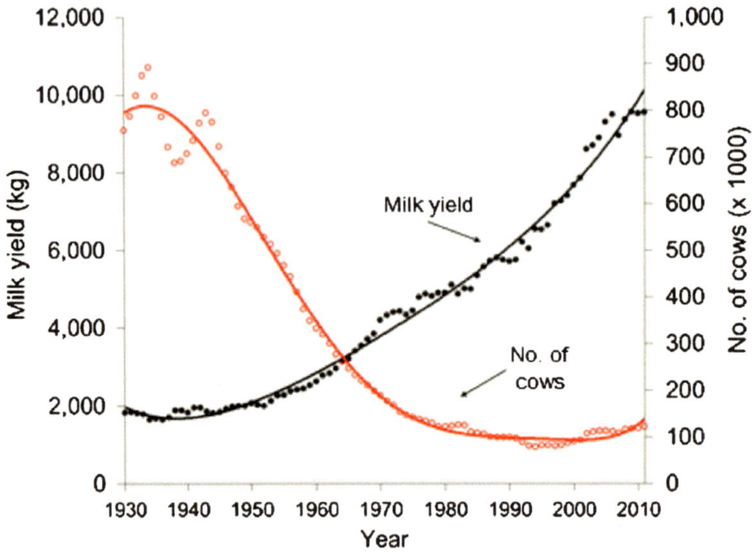

Fig. 6.1 A historical summary of milk yield per cow and numbers of dairy cows in Kansas from 1930 to 2011, which mirrors similar trends for all 50 US states. Artificial insemination became vogue in the 1950s with the advent of frozen semen and its adoption by dairy producers. Genetic progress in milk yield per cow is evident by the increasing yields after the 1960s when daughter-dam comparisons were replaced by the more sophisticated animal model that included genetic contributions of all female relatives of dairy sires in addition to their progeny

and other diluents for long-term storage of sperm packaged in plastic straws at liquid nitrogen temperatures (−196 °C). This progress led to the rapid growth of the AI industry in dairy cattle populations. By 1970, more than seven million cows were inseminated artificially.

In spite of the advantages of using AI, not all cows and heifers receive AI. A survey was conducted in 103 Alta Genetics progeny-test herds (averaging 613 cows), located in Wisconsin (26), California (12), New York (11), Minnesota (10), Michigan (7), Washington (6), Pennsylvania (6), Iowa (5), Idaho (5), Texas (4), Ohio (4), and other states (7; Caraviello et al. 2006). The results showed that 56 % of the herds used AI for all breeding services. Moreover, a national survey sponsored by the National Association of Animal Breeders (NAAB) revealed that, depending on herd size, 55–63 % of dairy heifers were serviced using AI (Fricke 2004). A more recent NAAB survey showed that 62–68 % of dairy heifers receive at least one AI service (Fricke 2004).

As recent as 2007, more than 72 % of all dairy operations were using AI compared with 29 % of natural mating (Fig. 6.2). Today, AI accounts for 85 % of all Holstein births according to the Holstein Association USA. Of 9.2 million dairy cows in the United States, approximately 90 % are Holstein with six other dairy breeds (Jersey, Brown Swiss, Red and White, Guernsey, Ayrshire, and Milking Shorthorn) making up the remainder. Surveys of dairy producers report between 55 and 96 % used natural service bulls in part for impregnating their milk cows (Stevenson 2009). Of herds processed at the Dairy Records Management System in Raleigh, N.C. during 1999–2002, only 1 % of herds used natural service bulls exclusively, whereas 26 % used only AI. This indicates that approximately 73 % of herds use a combination of both AI breeding and natural service bulls. Not only are pregnancy outcomes better for cows inseminated artificially than for cows exposed to natural service bulls, but return on investment is greater for AI (Stevenson 2009).

Semen historically was packaged in 1-mL glass ampules but today is packaged in either 0.25- or 0.5-mL plastic straws that can be more easily stored in liquid nitrogen storage containers specifically manufactured to store frozen semen on the farm. The straw technology reduces

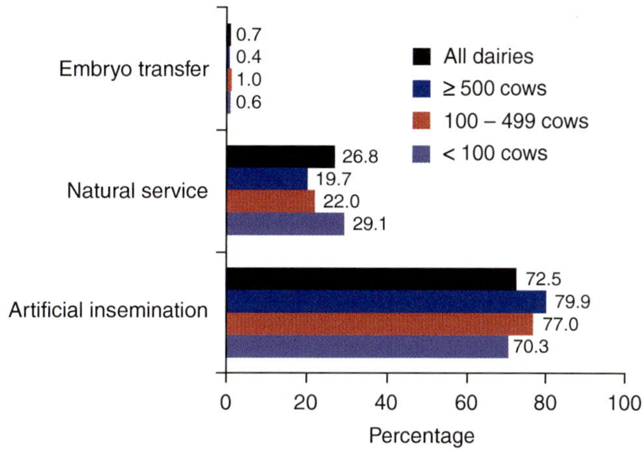

Fig. 6.2 Percentage of pregnancies achieved by embryo transfer, natural service (bull bred), and artificial insemination in the US dairy operations of various herd size during 2006–2007 (*source*: NAHMS 2007)

shipping weights and storage space in addition to be more easily handled by AI technicians in the field. Different suppliers of bovine semen for AI use both sizes of straws for packaging their frozen semen. Meta-analyses applied to fixed- or random-effect models of 15 different studies indicated that the average odds of having a greater pregnancy outcome with the 0.25-mL straw were 3–4 % greater than for the 0.5-mL straw (Stevenson et al. 2009). Based on these odds ratios, the expected proportion of difference in pregnancy outcome translated into a difference of only 0.74 %.

Sex-biased or gender-sorted semen became available in the US market place since 2003. Altering the sex ratio in favor of heifer calves has led to a greater abundance of replacement heifers in dairy operations. Sorting sperm based on their DNA content was successfully done in 1987 by flow cytometry allowing relatively pure populations of X- and Y-chromosome bearing sperm to be collected (Moore and Thatcher 2006) with typical biased gender ratios of 95 %. Disadvantages of applying sex-sorted semen are its cost and reduced fertility. Because the technology of sorting sperm is a slow process (150–200 straws of sexed per machine per day), commercial doses are packaged at a lesser concentration (e.g., two million sperm per straw) compared with conventional semen packaged in doses of ten million or more sperm per straw. Pregnancy outcomes are generally 70–90 % of what can be achieved with conventional semen (DeJarnette et al. 2009). The impact of this technology is yet to be revealed. Because of cost of producing sex-sorted semen and poorer fertility associated with its application, most of it is used in heifers because they are the most fertile females on a dairy operation. Sex-sorted semen was used for inseminations in 11.4 % of heifers and 3.5 % of cows.

Assisted Reproductive Technologies

During the past 30 years, other artificial reproductive technologies in addition to AI have been developed including superovulation and embryo transfer, in vitro production of embryos, cloning, and transgenesis (Moore and Thatcher 2006). Inducing multiple ovulations by applying reproductive hormones during the estrous cycle, coupled with AI, embryo collection, and embryo transfer, allows multiple offspring to be produced from genetically superior females to be transferred as embryos into recipient surrogate cows. Using this technology allows one to exploit the genetic merit of superior cows in the same way AI extends the use of superior genes of the bull. By comparison, however, using AI is much more efficient because a single ejaculate of a bull may supply 200–300 conventional doses of semen, whereas a single embryo collection of a cow may produce only 2–4 offspring. Semen from bulls can be collected several times per week, whereas the embryos from each cow can be collected but once every other week. Impact of producing pregnancies by embryo transfer in the US dairy industry is very small (Fig. 6.2) compared with AI. Part of this limited use is associated with costs of approximately $100 per embryo and transfer costs of $25–$50 per embryo, which is further complicated by the needed expertise to perform the procedures. Comparable costs of $15 per dose of conventional semen and less time and skill associated with the AI procedure makes AI much more common place in producing pregnancies in dairy operations.

In vitro production of embryos led to the first live calf produced by IVF in 1981 (Moore and Thatcher 2006). The first live calf resulting from an oocyte matured in vitro (IVM) and then fertilized in vitro was reported in 1986. Three years later, the first live calf was produced by in vitro (IVM-IVF). This technology allows for production of viable embryos from oocytes collected from slaughter house ovaries of culled cows and the generation of large numbers of embryos. Ovum pick-up (OPU) was first reported in 1988, in which oocytes were aspirated from ovaries by transvaginal ultrasonography. This technique allows collection of oocytes for later IVM-IVF procedures from genetic superior females and at any age or reproductive status. These procedures allow for production of embryos and have the potential to substantially increase the reproductive lifetime of superior females.

Not all of quirks of these in vitro techniques, however, have been resolved. For example, only 20–50 % of IVF oocytes develop into viable embryos. Compared with AI, IVF results in reduced pregnancy outcomes, increased embryonic deaths, especially during the first 30 days of pregnancy, and prolonged gestation and increased birth weights (8–50 % larger). Larger birth weights are attributed to the so-called large offspring syndrome requiring cesarean section (Moore and Thatcher 2006). Application of this technology to produce pregnancies on dairy operations is very small and likely included in the embryo transfer category (Fig. 6.2).

Producing multiple copies of an individual, either naturally (identical twins) or artificially (cloning) is reported for dairy cattle (Moore and Thatcher 2006). Clones can be produced by nuclear transfer of a nucleus from one cell into the cytoplasm of a second enucleated cell or into an enucleated oocyte. This technology became commercially available for the dairy industry, but was limited to a number of cells within an embryo. The main application of cloning to the dairy industry is for expanding the use of genetically superior cows or bulls. This application would not be limited by age, injury, or health of the donor. Cloning technology is quite inefficient and very costly at present. Only 10 % of cloned embryos transferred to surrogates are carried to full term (less than 1 % of cloned embryos originally constructed) because of excessive abortions during gestation. Gestation is often prolonged and cloned calves also suffer from the large offspring syndrome. If once perfected, cloning will be of value to the dairy industry only if cloned animals are genetically superior.

Production of transgenic animals was first introduced in 1980. Transgenesis technologies allow improvements not currently possible by traditional breeding schemes. Beneficial traits can be added from other species. Such benefits include transferring selected genes from one species to another, such as was done with transfer of a bacterial gene to cattle that when expressed in the mammary gland produced resistance to certain strains of mastitis (Moore and Thatcher 2006). Pharmaceutical companies have used the dairy cow as a biopharmaceutical production unit: a transgenic cow is produced that contains a gene for a particular drug or protein that is expressed and later harvested in the milk of the transgenic cow. For example, protein C, a blood-clotting protein has been synthesized by the mammary glands of transgenic pigs and secreted into their milk for use by hemophiliac patients. Efficiencies of producing transgenic animals (<1 %) hinders the propagation of many cows, because the likelihood of expression of a particular gene in a particular organ (e.g., mammary gland) is very small. Production of transgenic cows are expensive, nonetheless, transgenesis has great potential because the cow serves as a very efficient factory when producing large quantities of milk from which the transgenic product can be harvested and subsequently purified.

Estrus-Detection Systems

Artificial insemination is the most viable option to produce pregnancies in cows; however, implementation of a successful AI program on dairy farms requires accurate and efficient detection of estrus. A number of significant physiological changes have been reported to occur during the periestrual period (Lewis and Newman 1984; Roelofs et al. 2010) enabling detection of estrual behavior and other correlated traits in cows. Some of these changes include physical activity, vaginal cytology and pH, electrical resistance of vaginal mucus and genital tissues, body temperatures, pulse and heart rates, blood flow, pheromones or odors, blood metabolites, and hormones, stage of the estrus cycle, milk yield, and feed intake.

Cows are traditionally monitored for visual signs of estrus such as "standing to be mounted" by a herd mate. In order to be effective in identifying when to inseminate, one must visually observe cows for estrual behavior at least twice daily. To reduce labor and improve efficiencies associated with detection of estrus other methods have been developed. Automation of estrus-detection efforts in dairy cows is well documented (Firk et al. 2002). Technologies developed to take advantage of other physiological correlates of estrus include pedometry or activity monitors, pressure-sensitive, rump-mounted radiotelemetric devices, temperature sensors, and milking inline chemical sensors. Although technologically simple, such products as tail paint or tail chalk applied to the tail head is the most common form of estrus-detection aid in US dairy operations. When marked cows are mounted from the rear, some of the chalk or paint is rubbed off indicating that the painted cow possibly stood in estrus while mounted by a herd mate. Validation of accurate rubs by concentrations of progesterone in milk or subsequent pregnancy after insemination showed that accuracy of chalk or paint as estrus-detection devices varied widely from 33 to 90 % (Stevenson 1999). It seems from many studies that 5–30 % of cows or heifers inseminated are not really in estrus (Roelofs et al. 2010).

Various heat mount single-use devices that are affixed to the tail head of cattle have been used as detector aids (Stevenson and Phatak 1999). These devices or patches are made to change in color once pressure is applied by the weight of the mounting animal on the potentially estrual female that stands in estrus to accept the bull or mounting herd mate. One of the newest patches is made like a Lottery or scratch ticket that reveals a color once rubbed by the mounting female.

Increased walking activity that is associated with estrus (Roelofs et al. 2010) led to the development of pedometry as a means of detecting estrus as early as the 1970s. Pedometers affixed to the leg quantified cow movement or counted the number of steps taken by the cow. Increase in physical activity of the cow provided 70–80 % accuracy of heat detection. Cows housed in free stalls were approximately 2.75 times more active during estrus than when not in estrus. Further, relatively little within-cow variation in activity occurred from day to day when cows were not in estrus. Therefore, activity monitors could be excellent predictors of sexual behaviors associated with estrus (Roelofs et al. 2010).

The latest version of electronic estrus-detection aids that appeared in the early part of this decade was the neck-mounted activity tags containing a microprocessor and a three dimensional accelerometer. The accelerometer allows accurate measurement of cow movement. The activity tag monitors specific estrus-related movement and its intensity resulting in heat detection accuracies up to 90 % (Roelofs et al. 2010). By 2010, the best-selling system in the world with approximately one million estrus-detection tags sold, demonstrated that dairy farmers were willing to invest in technologies that provide a real solution to detection of estrus. These systems are effective management tools in the AI program because their use will increase estrus-detection rates. Increasing estrus-detection rates result in increased AI submission rates and more potential pregnancies. At least four

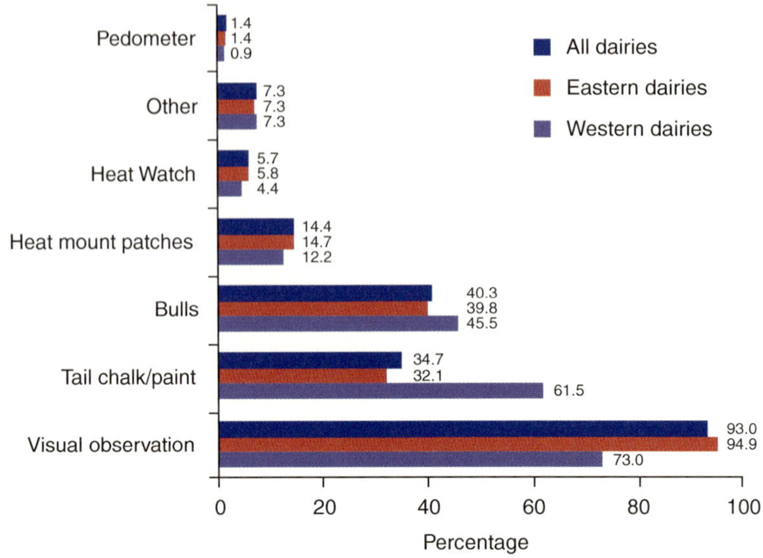

Fig. 6.3 Percentage of the US dairy operations by region using various estrus-detection methods during 2006–2007 (*source*: NAHMS 2007)

patents have been issued describing some type of transponder system that is capable of detecting movement or motion that includes the ability to be interrogated in the parlor or send signals via radiotelemetry.

Electronic mounting sensors are pressure-sensitive devices that are applied to the rump of the cow and are activated by a mounting herd mate. Accuracy and efficiency of these systems are quite high because they are associated with specific sexual behavior and are functional 24 h per day (Roelofs et al. 2010). The downside to these systems is the labor associated with applying the sensor patches and the maintenance to keep sensor patches affixed to cows until pregnancy occurs. Video systems also have been investigated to capture activity in free stall barns (Bruyère et al. 2012), but cost and practicality of such systems require further study.

Devices have been patented that claim to detect ovulation by changes in body temperature via an implanted temperature sensor or an implanted vaginal telemetry system that apparently detects tissue impedance, temperature, and activity of the animal. Radio-telemetered monitoring of vaginal or ear skin temperature was accomplished with more than 80 % accuracy of detection and only few false positives (Roelofs et al. 2010).

Impact and use of the most common estrus-detection aids by region in the United States are illustrated (Fig. 6.3). Since the NAHMS survey was made in 2007, at least a dozen companies are offering activity monitor technology for use in dairy operations and will likely have a larger impact in dairy operations seeking for low-cost methods of identifying cows in estrus. Regardless of methods applied, visual detection of estrus has been the gold standard management practice upon which a successful AI program is built.

AI-Breeding Management Programs

Artificial insemination after detected estrus not induced by hormones for first-service and repeat breeding services for replacement heifers and lactating cows exceeds 50 % of all dairy US dairy operations (Fig. 6.4). Natural service (use of bulls) was the second most common practice used at first services for the majority of heifers and cows (33.2 and 21.7 % of operations, respectively). For repeat services, use of AI ranged from 39.6 to 46.5 %, respectively. Bulls were used for the second or greater service for heifers on 35.1 % of operations and for cows on 22.2 % of operations. At least

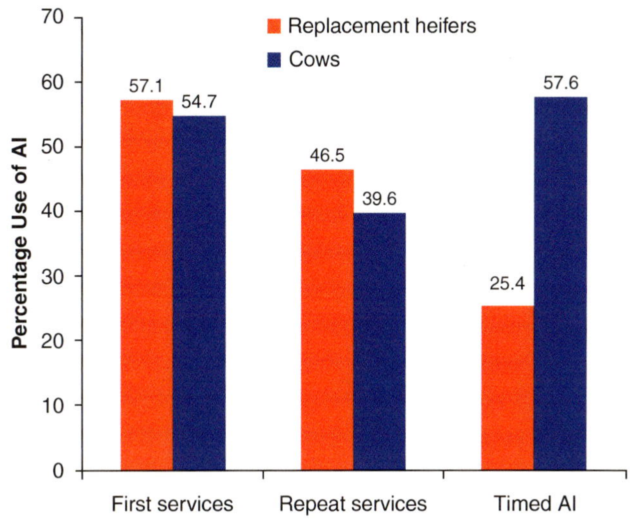

Fig. 6.4 Percentage of the US dairy operations that apply artificial insemination (AI) in replacement heifers and lactating dairy cows and incorporate timed AI programs during 2006–2007 (*source*: NAHMS 2007)

50 % of all US dairy operations apply some method of fixed time insemination without visual detection of estrus to manage reproduction in cows, but more operations used timed AI in cows than in heifers (Fig. 6.4).

For operations with pregnancies conceived by AI, the majority of AI services were performed by the owner-operators on 51 % of operations and by a professional AI technician on 40.7 % of operations (NAHMS 2007). In 95.9 % of operations, the person responsible for the majority of the AI services also had received some kind of formal AI training.

For operations with pregnancies conceived by AI, and for cows in which AI was unsuccessful, on average, cows received AI three to six times on 70.9 % of all dairy operations before cows were designated for a different status (e.g., moved to a bull pen, sold, etc.; NAHMS 2007). More than 72 % of pregnancies were conceived by AI—either after detected estrus or timed AI (Fig. 6.2).

Hormone-based AI breeding programs have become common place in the dairy industry since the late 1990s. Although discovery in the early 1970s that $PGF_{2\alpha}$ is the natural uterine luteolysin in cattle, its use on dairy operations was originally limited to inducing estrus for first AI services or to induce estrus in cows diagnosed not pregnant but having a palpable corpus luteum. Injections of $PGF_{2\alpha}$ successfully regress the corpus luteum, with the majority of estrus activity occurring between 2 and 5 days after treatment with $PGF_{2\alpha}$. Early in its application, two injections of $PGF_{2\alpha}$ were given 12–14 days apart to allow cows to be inseminated after the first injection and follow through with the second injection for all noninseminated cows. Waiting 12–14 days before reinjecting noninseminated cows allowed for some of the remaining cows to have a $PGF_{2\alpha}$-sensitive corpus luteum (not responsive to the first injection) or for the remaining cows to develop a new $PGF_{2\alpha}$-sensitive corpus luteum. The latter group of cows was most likely those that responded (corpus luteum regression) to the first $PGF_{2\alpha}$ injection but were not detected in estrus.

The large variation in the interval to estrus after $PGF_{2\alpha}$ was not understood until the application of transrectal ultrasonography to monitor growth patterns of individual follicles during the mid-1980s. It was discovered that follicles emerge together every 8–10 days (follicle wave) with cows have two or three follicular waves per estrous cycle. Each wave comprises successive phases referred to as recruitment, selection, deviation, dominance, and atresia. Within a wave of FSH-recruited antral follicles, one follicle is selected and undergoes deviation at approximately 8.5 mm in size and continues to grow while subordinate follicles cease to grow and become atretic. As a result, injection of $PGF_{2\alpha}$ in the presence of a dominant, estrogen-active follicle, for example, on day 7 of the estrous cycle, results in an earlier occurrence of estrus than an injection given at day 11, when a dominant follicle is absent. At day 11, a second wave of follicles begins to emerge during the estrous cycle, a process that requires several days before a new dominant follicle is selected. As a result, several days will pass after the injection of $PGF_{2\alpha}$ before the new dominant follicle matures to produce sufficient estrogen to induce estrus once the corpus luteum has regressed.

Based on the understanding of normal follicular wave dynamics of the estrous cycle, all brought about by the technology of transrectal ultrasonography, it becames clear that synchronizing follicle growth must be coupled with induced regression of the corpus luteum to better synchronize the occurrence of estrus and subsequent ovulation.

In the late 1960s, the discovery of hypothalamic substances that control the secretory function of the anterior pituitary led to the synthesis of gonadotropin-releasing hormone (GnRH) and methods of radioimmunoassay to detect biological substances in blood and tissue. These discoveries led to a shared Nobel Prize in Physiology or Medicine for three different scientists in 1977 (Yalow 1978). In the early 1970s, studies demonstrated that injection of GnRH-induced release of LH and FSH (Britt et al. 1981) and later was effective to induce ovulation of bovine ovarian follicles 10 mm or greater in diameter during approximately 64 % of the estrous cycle (Moore and Thatcher 2006).

It was not until the mid to late 1990s that combinations of GnRH and $PGF_{2\alpha}$ were applied to dairy heifers and cows to develop various successful timed AI programs used on dairy operations today. Various refinements of the programs are ongoing, but the basic timed AI program known as Ovsynch (injection of GnRH 7 days before and 48–56 h after an injection of $PGF_{2\alpha}$, with timed AI occurring 12–16 h after the second GnRH injection) is the basis for the majority of the timed AI programs, particularly in lactating cows (Moore and Thatcher 2006). A recent variant of the Ovsynch was to reduce the interval between the first GnRH injection and $PGF_{2\alpha}$ from 7 to 5 days to facilitate better managed follicular maturation for improved fertility. This move necessitated the inclusion of a second injection of $PGF_{2\alpha}$ up to 24 h after a first injection of $PGF_{2\alpha}$ to ensure luteolysis occurred in cows that ovulated after the first GnRH and formed a new corpus luteum. Otherwise, a large proportion of the new younger induced corpus luteum does not regress in response to one injection of $PGF_{2\alpha}$ administered 5 days after the first GnRH injection.

Variations of the Ovsynch program also are applied to cows diagnosed not pregnant in dairy operations (Resynch-Ovsynch; Fricke 2002). Dairy producers may choose to initiate the Ovsynch program (administer the first injection of GnRH) 5 or 7 days before the not-pregnant diagnosis is conducted in cows to allow resumption of the program in all nonpregnant cows with an injection of $PGF_{2\alpha}$ at the time of nonpregnant diagnosis and complete the program with timed AI 3 days later. Injection of GnRH 5 or 7 days before the pregnancy status is known has no negative effects in the pregnant cows. Alternatively, the 5- or 7-day program may be initiated on the day of the not-pregnant diagnosis.

Controlled internal drug release (CIDR) inserts containing progesterone are used in some of these timed AI programs since its market availability in 2002. The CIDR is either applied intravaginally for 7 days with an injection of $PGF_{2\alpha}$ upon insert removal or in combination with Ovsynch (inserted when the first GnRH is administered and removed 7 days later at the time of $PGF_{2\alpha}$ injection). About one-third of dairy operations (32.4 %) used CIDR inserts during the 2006–2007 period. Their greatest use was to treat anestrous females (65.7 %), cows with ovarian cysts (43.5 %), and to synchronize estrus (34.3 %) in females (NAHMS 2007).

The majority of large US dairy operations (>500 cows) in one 103-herd survey (Caraviello et al. 2006) used timed AI programs almost exclusively to inseminate their cows. More than 85 % of those large herds were using timed AI at first services after calving and 77 % for resynchronizing repeat services. Further, another study reported the mean percentage of 231,288 cows inseminated after a timed AI program was 43.4 % and differed among four regions of the United States.

Pregnancy Diagnosis

Pregnancy exams are important diagnostic tools used on dairy operations to identify pregnancy status of inseminated or mated females. Because the majority of US dairies operate a continuous breeding season and cows are calving year-round, it is critical to determine pregnancy status and identify nonpregnant cows so that can be managed and reinseminated in a timely manner to reduce the rebreeding and calving intervals. Additional benefits of pregnancy exams include detection of uterine or ovarian abnormalities, diagnosis of twins, and estimation of conception and calving dates for females with unobserved natural service by bulls.

In the 1800s, palpation per rectum was first identified as a method of pregnancy diagnosis. Traditional diagnosis of pregnancy involves transrectal palpation of the conceptus (detection of fluid-membrane slip, or amniotic vesicle, or both) between 35 and 40 days since last insemination or mating. About two-thirds of all dairy operations (67 %) performed pregnancy exams monthly or more frequently during 2006–2007 (NAHMS 2007). The majority of large (>500 cows) operations (75 %) performed pregnancy exams weekly or every 2 weeks, whereas 50.2 % of smaller operations (<100 cows) performed exams on a monthly basis. Palpation of the uterus and its contents per rectum is the method used most routinely to determine pregnancy status (Fig. 6.5).

Transrectal ultrasonography is becoming more accepted because portable, battery-powered units are available on the market. About one-fourth of all dairy operations surveyed (Fig. 6.5) use this technology for pregnancy diagnosis. Evidence of a viable embryo can be detected as early as 20 days in heifers, but accuracy and efficiency are best in routine herd exams for cows that are at least 28–30 days since last insemination. Use of ultrasonography requires skill and training. Common errors include a false positive diagnosis of uterine intraluminal fluid that is associated with estrus and not with the presence of a conceptus. Ultrasound on-farm use is also diagnostic in assessing follicular structures (differentiating between follicular and luteal cysts), functional corpora lutea, twins, embryonic loss, and fetal sexing (Fricke 2002). A major advantage of applying ultrasound is earlier and accurate detection of pregnancy, which has great impact on reproductive management of dairy herd.

Pregnancy-specific proteins are produced by the conceptus during early pregnancy. One class of these proteins is called pregnancy-associated glycoproteins (PAGs). Although application of blood tests to identify PAGs is done on a minority

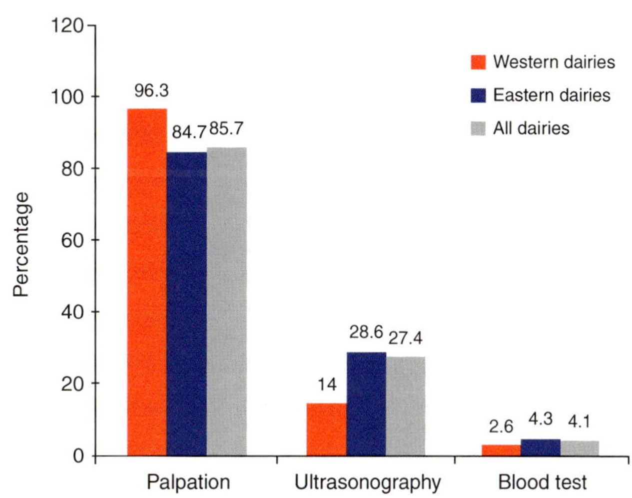

Fig. 6.5 Percentage of the US dairy operations applying various methods to diagnose pregnancy during 2006–2007 (*source*: NAHMS 2007)

of dairy operations (Fig. 6.5) in the 2007 survey, its use has increased greatly since then. At least three companies in North America are marketing the blood tests routinely performed in veterinary clinics with a 24–36-h turnaround. Incorporating these blood tests in reproductive programs usually involve collecting a blood sample at 28–32 days since last insemination. The blood sample can be collected when a resynchronization program is initiated (first GnRH injection of the Ovsynch protocol is administered) so the protocol can be continued 7 days later when the not-pregnant status of the female is confirmed. Alternately, the Resynch-Ovsynch is simply initiated after the blood test results become available. These blood tests are inexpensive, equally accurate as transrectal ultrasonography diagnoses made by skilled technicians and do not require expensive equipment or skilled training.

Because pregnancy is diagnosed earlier by ultrasonography or by blood tests before some natural spontaneous embryo loss occurs, later verification of the viable pregnancy is essential. Losses of embryos identified as viable on day 28 can be up to 10–15 % during the next 4–6 weeks. If pregnancy was detected per rectum at 35–45 days after last AI, then these losses would never have been detected; the cow would have been diagnosed not pregnant without any knowledge that she was pregnant for some time before the diagnosis. For these reasons, it is recommended that when early pregnancy diagnoses are made that they be verified later, perhaps by 80–100 days since AI, and again before terminating the lactation to begin the dry period.

Recombinant Bovine Somatotropin

One of the first potential biotechnology products for animal production was bST. Discovered more than 70 years ago, bST was reported to increase milk yield in lactating dairy cows regardless of stage of lactation (Bauman 1978). It was later synthesized and produced for commercial use through recombinant DNA technology similar to procedures used to produce human insulin. By 1978, research in the technology of bST involved hundreds of scientists publishing more than 1,000 studies involving more than 20,000 dairy cows. Treating cows with long-acting recombinant bST (once every 14 days) not only increases milk production on commercial dairy farms today, but it was observed in a few studies to have positive effects on pregnancy outcomes of well-managed cows. First treatment with bST at first services after calving following a timed AI program (Ovsynch) or upon detected estrus resulted in increased pregnancy rates and reduced early embryonic death between days 31 and 45 after AI (Moore and Thatcher 2006). The mechanism of bST action includes improved fertilization rate, accelerated embryo development, improved embryo quality, and increased conceptus growth at the time pregnancy is recognized approximately 15–16 days after AI. Use of bST has waxed and waned since it first became publicly available in the United States in 1994 (Caraviello et al. 2006) in part because of misunderstandings of consumers about the science and safety of such products and reluctance of dairy processors to market milk products coming from dairy operations using bST.

Management Software Applications

Dairy record program software has been available for many years through various dairy record processing centers associated with the DHIA. The first DHI organizations were established in 1905 to provide a standardized method of collecting data from individual cows. Approximately 39 % of the 53,127 US dairy operations participate in the DHI testing program representing 47 % of the 9.2 million dairy cows (NMPF 2012). Originally, computer printouts were mailed to dairy producers within a few days of the monthly herd test summarizing milk yield, and various milk components (e.g., fat, protein, somatic cells, lactose, milk urea nitrogen) that were analyzed at DHI-affiliated laboratories. Daily yield of milk and milk components and other information such as calving and breeding dates, calf identification, pregnancy diagnoses, feed components and costs, and culling information are collated and entered into the computer software to produce the monthly printed reports. Today, nearly 100 % of the US dairies use on-farm computers that allow the same information to be collected, entered, and used in daily management routines in addition to measured milk and milk component data that occurs monthly. Lists of cows due to dry off, to calve, to receive various health tests or immunizations, pregnancy exams, hoof trimming, bST injections, etc., can be printed for organizing daily chores. These options also include printing out schedules for hormone injections to manage custom timed AI programs. Another advantage of computerized records is to share that information with herd consultants and veterinarians who assist in closely monitoring client herd reproductive, production, and health issues. Consultants can easily obtain this information by attaining a backup of each dairy's computer records, or with permission of the dairy producer, obtain a download from one of the dairy record processing centers.

Not only are computers common place but many dairy managers use hand-held devices (including smart phones) that communicate with the farm computer to reveal needed information about cows as they work in cow pens. With the advent of electronic identification chips that are attached to ear tags, cows can be identified by blue-tooth wands to retrieve various information about individual cow's daily treatment. Milking equipment companies are now investing in in-line milk monitoring equipment that can measure progesterone concentrations in the milk of cows during the milking process and produce elegant profiles of milk progesterone to assist managers in determining reproductive status (cycling or pregnancy status) of individual cows. Use of computers, computer-generated reports, and access to electronic data at the finger tips of dairy managers has impacted not only production but increased efficiency of labor and management to provide better animal care.

Technologies of Emerging Application

During the last 25 years, much has been learned about the molecular makeup of living organisms. The term genomics refers to the study of nucleic acids (e.g., DNA and RNA) within cells. In 1977, nucleic acids were first sequenced and now the entire bovine genome has been sequenced to reveal approximately three billion nucleotides with approximately 1 % coding for functional genes, of which 40 % or more have unknown functions. This technology is now being applied in addition to traditional progeny testing to reveal potential young sires that express superior genes coding for important economic traits for breeding and future selection purposes.

Genetic merit of a bull or cow was determined traditionally by pedigree, performance, and progeny tests. The three traditional sources of information all originate as performance data. A lactation record is "performance" in the genetic evaluation of the cow that made the record, progeny data for her sire and dam, and pedigree data for her son or daughter. Today, a genetic map of what traits they inherited from each of their parents is available and summarized into genomic evaluations. The genomic evaluation provides more information so the results are more reliable. Having a more accurate estimate of genetic merit increases the potential for more rapid genetic progress. For sires that have progeny with type and production data, their genomic estimates are a combination of their genotypic data, parent average, and progeny performance data.

With the release of genomic evaluations in 2009, information about a cow or bull's genetic potential that would have previously taken years to obtain by progeny testing can be revealed at a young age. This means young sires have genetic information equivalent to proven sires and can be implemented and marketed accordingly. Today, 3K- and 6K-chip tests are available and economically feasible for commercial dairy producers, allowing for greater genetic improvement on the female side of the pedigree. Compared with the trend line for all active AI sires, the genetic merit of genomic-proof young sires is increasing a much faster rate and the difference between the top 25 % of all active sires and that of the genomic-proof sires is ever widening (Fig. 6.6). It is recommended, therefore, that limiting use of frozen semen from only progeny-tested sires will result in lagging behind the leading edge of genetic improvement on average.

Future prospects for improving reproductive performance of dairy cows through genetic selection based on early genomic tests are real. Although milk yield is often implicated as the

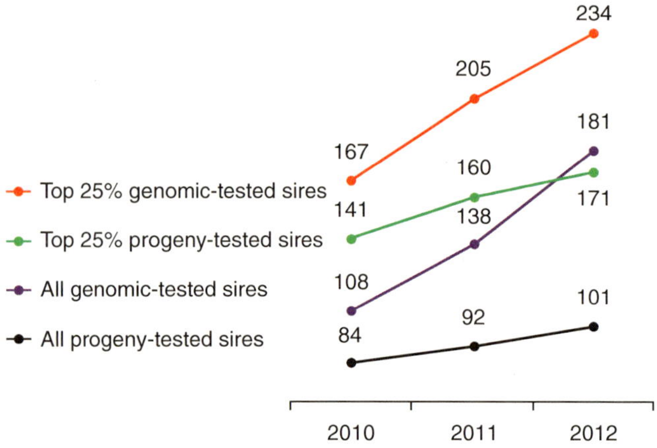

Fig. 6.6 Comparison of genetic merit of Jersey AI sires available for artificial insemination: (1) top 25 % of all genomic-tested sires ($n = 40$); (2) top 25 % of all active progeny-tested AI sires ($n = 34$); (3) all genomic-tested sires ($n = 158$); and (4) all active progeny-tested AI sires ($n = 132$). *Source*: USDA-AIPL data

cause of impaired fertility (Bello et al. 2012), impact of inadequate body condition seems to be greater because it has significant effects on the probability of conception, rate of embryo loss, and proportion of anestrous cows (Weigel 2006). Selection for or against certain traits could lead to improved fertility of dairy cattle.

References

Bauman DE (1978) Bovine somatotropin: review of an emerging animal technology. J Dairy Sci 75:3432–3451

Bello NM, Stevenson JS, Tempelman RJ (2012) Milk production and reproductive performance: modern interdisciplinary insights into an enduring axiom. J Dairy Sci 95(10):5461–5475

Britt JH, Cox NM, Stevenson JS (1981) Advances in reproduction in dairy cattle. J Dairy Sci 64:1378–1402

Bruyère P, Hétreau T, Ponsart C, Gatien J, Buff S, Disenhaus C, Giroud O, Guérin P (2012) Can video cameras replace visual estrus detection in dairy cows? Theriogenology 77:525–530

Caraviello DZ, Weigel KA, Fricke PM, Wiltbank MC, Florent MJ, Cook NB, Nordlund KV, Zwald N, Rawson CL (2006) Survey of management practices on reproductive performance of dairy cattle on large U.S. commercial farms. J Dairy Sci 89:4723–4735

DeJarnette JM, Nebel RL, Marshall CE (2009) Evaluating the success of sex-sorted semen in US dairy herds from on farm records. Theriogenology 71:49–58

Firk R, Stamer E, Junge W, Krieter J (2002) Automation of oestrus detection in dairy cows: a review. Livestock Prod Sci 75:219–232

Foote RH (1996) Dairy cattle reproductive physiology research and management—past progress and future prospects. J Dairy Sci 79:980–990

Fricke PM (2002) Scanning the future—ultrasonography as a reproductive management tool for dairy cattle. J Dairy Sci 85:1918–1926

Fricke PM (2004) Strategies for optimizing reproductive management of dairy heifers. In: Proceeding of Western Canadian Dairy Seminar, Red Deer, AB, Canada, 9–12 Mar 2004, pp 163–176

Lewis GS, Newman SK (1984) Changes throughout estrous cycles of variables that might indicate estrus in dairy cows. J Dairy Sci 67:146–152

Moore K, Thatcher WW (2006) Major advances associated with reproduction in dairy cattle. J Dairy Sci 89:1254–1266

NAHMS (National Animal Health Monitoring Service) (2007) Reproductive practices on U.S. dairy operations, 2007. http://www.aphis.usda.gov/animal_health/nahms/dairy/downloads/dairy07/Dairy07_is_ReprodPrac.pdf. Accessed 15 Aug 2012

NMPF (National Milk Producers Federation) (2012) Dairy producer highlights. NMPF, Arlington, VA

Rodriguez-Martinez H (2012) Assisted reproductive techniques for cattle breeding in developing countries: a critical appraisal of their value and limitations. Reprod Domest Anim 47(suppl 1):21–26

Roelofs J, Lopez-Gatius F, Hunter RHF, van Eerdenburg FJCM, Hanzen C (2010) When is a cow in estrus? Clinical and practical aspects. Theriogenology 74:327–344

Stevenson JS (1999) A review of oestrus behaviour and detection in dairy cows. In: Proceedings of British

Society of Animal Science-sponsored conference, Galway, Ireland, 20–22 Sept 1999
Stevenson JS (2009) Bulls don't do it better. Hoard's Dairyman 154:568
Stevenson JS, Phatak AP (1999) Effective use of heat detection devices. Large Anim Pract 20:28–31
Stevenson JS, Higgins JJ, Jung Y (2009) Pregnancy outcome after insemination of frozen-thawed bovine semen packaged in two straw sizes: A meta-analysis. J Dairy Sci 92:4432–4438
Weigel KA (2006) Prospects for improving reproductive performance through genetic selection. Anim Reprod Sci 96:323–330
Yalow RS (1978) Radioimmunoassay: a probe for the fine structure of biologic systems. Science 200:1236–1245

7 Impact of Swine Reproductive Technologies on Pig and Global Food Production

Robert V. Knox

Abstract

Reproductive technologies have dramatically changed the way pigs are raised for pork production in developed and developing countries. This has involved such areas as pigs produced/sow, more consistent pig flow to market, pig growth rate and feed efficiency, carcass yield and quality, labor efficiency, and pig health. Some reproductive technologies are in widespread use for commercial pork operations [Riesenbeck, Reprod Domest Anim 46:1–3, 2011] while others are in limited use in specific segments of the industry [Knox, Reprod Domest Anim 46:4–6, 2011]. Significant changes in the efficiency of pork production have occurred as a direct result of the use of reproductive technologies that were intended to improve the transfer of genes important for food production [Gerrits et al., Theriogenology 63:283–299, 2005]. While some technologies focused on the efficiency of gene transfer, others addressed fertility and labor issues. Among livestock species, pig reproductive efficiency appears to have achieved exceptionally high rates of performance (PigCHAMP 2011) [Benchmark 2011, Ames, IA, 12–16]. From the maternal side, this includes pigs born per litter, farrowing rate, as well as litters per sow per year. On the male side, boar fertility, sperm production, and sows served per sire have improved as well [Knox et al., Theriogenology, 70:1202–1208, 2008]. These shifts in the efficiency of swine fertility have resulted in the modern pig as one of the most efficient livestock species for global food production. These reproductive changes have predominantly occurred in developed countries, but data suggests transfer and adoption of these in developing countries as well (FAO STAT 2009; FAS 2006) [World pig meat production: food and

R.V. Knox, Ph.D. (✉)
Department of Animal Sciences, University of Illinois, 360 Animal Sciences Laboratory, 1207 West Gregory Drive MC-630, Urbana, IL 61801, USA
e-mail: rknox@illinois.edu

G.C. Lamb and N. DiLorenzo (eds.), *Current and Future Reproductive Technologies and World Food Production*, Advances in Experimental Medicine and Biology 752,
DOI 10.1007/978-1-4614-8887-3_7,

agriculture organization of the United Nations, 2009; FAS, 2006) Worldwide Pork Production, 2006]. Technological advancements in swine reproduction have had profound effects on industry structure, production, efficiency, quality, and profitability. In all cases, the adoption of these technologies has aided in the creation of a sustainable supply of safe and affordable pork for consumers around the world [den Hartog, Adv Pork Prod 15:17–24, 2004].

Keywords
Pork production • Reproductive efficiency • Livestock • Swine • Farrowing

Importance of Reproduction for the Pork Industry for Global Food Production

Before any meaningful discussion of the impact of specific swine reproductive technologies on global food production, it is important to recognize the importance of pork (FAOSTAT 2009; FAS 2006). Around the word, meat is a valuable source of protein that helps to supplement intake of essential nutrients for humans (Taylor and Field 1998). In the dramatically different climates found around the globe, the pig has been shown to be adaptable in temperate and tropical agricultural production systems (Ruvinsky and Rothschild 1998). In fact, there are numerous varieties of swine breeds that have adapted to survive in these diverse environments (Jones 1998). In almost all cases, the pig has been shown to be efficient for conversion of a variety of available food stuffs into high quality protein. In addition, the pig is noted for its high fertility. The efficiency of this large animal is realized from its short interval to maturity, short gestation period, multiple offspring per pregnancy, and quick tendency to rebreed (Taylor and Field 1998; Whittemore 1998). Pork is the leading meat consumed worldwide and in many Asian countries, such as China, pork is consumed in the greatest quantity (Taylor and Field 1998). In developed countries, pork is an important part of the diet, but in addition, many countries are also global leaders in pork production and pork export (FAOSTAT 2009; FAS 2006). There are great contrasts in how swine are raised for food production in developed and developing countries. For example, in many developing countries, the majority of pigs are raised by large numbers of small farmers that maintain only a few animals for supplementing family income or food. However, in developed nations, especially in North America and Europe, the pork industry is quite mature and economically important. In these countries, the farms are larger, more specialized, and may be integrated with crop production systems. Yet there is a growing trend in developing countries for increased herd sizes and modernized production systems. This is a direct result of technology transfer and in almost all cases reproductive technologies are an integral part of these changes. In fact when comparing developed and developing countries based on pork production capability and economic viability, the largest differences follow per capita income and degree of infrastructure. As both of these measures improve, it is evident that rapid swine industry growth can occur in developing countries as a result of improved education and technology transfer.

Impact of Reproduction on Farm Size

Reproductive technologies have had a dramatic effect on increasing farm size throughout the world. While there are significant differences in how modern production systems operate today, the majority of commercial pork is produced in confinement as opposed to outdoor production

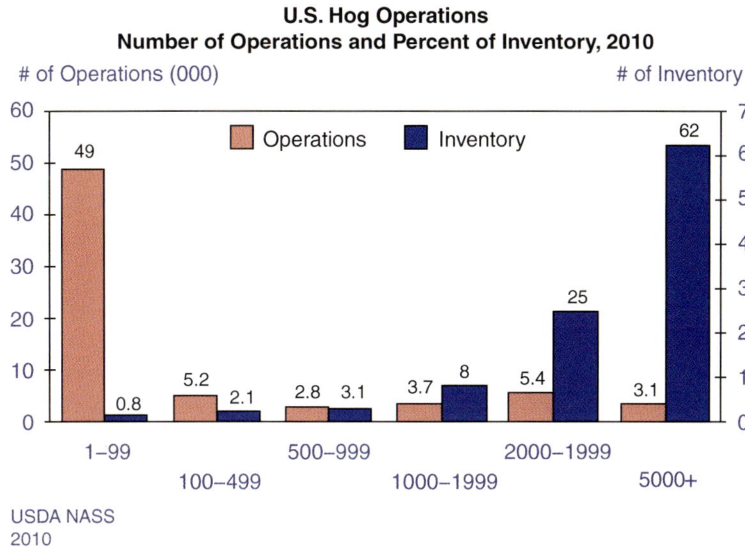

Fig. 7.1 The number of US swine operations based on pig inventory and the percentage of inventory by farm size (USDA-NASS 2010)

systems (USDA-APHIS 2006). Historical pork production systems in North America and Europe were initially outdoor operations with seasonal breeding and farrowing. In these temperate zones with seasonal weather patterns, breeding typically occurred in late fall with farrowing in spring when piglets would have the greatest chance of survival. In a two-litter system, farrowing could occur in fall or spring. However, these outdoor systems could be challenging with loss to predators, parasites, and weather-related deaths (PIH 2007). For practical purposes, confinement systems were developed to control and protect animals and reduce the challenges of animal management in harsh weather conditions. Indoor production noticeably improved feed efficiency, growth, and reproduction. However, the greater production output came at a much higher investment (PIH 2007). With movement of pigs indoors, producers realized the benefits of both group and individual housing for the breeding herd. With indoor systems, automation advancements in air and temperature control, and feeding and watering systems significantly changed the need for labor and allowed herd size to increase. As a result, with advances in swine breeding management, farm size increased over time. In controlled confinement production systems, year round production became possible regardless of weather extremes. With technology transfer from developed countries, these types of systems also arose in more tropical environments for practical reasons related to animal control, feeding, health, and breeding. With the development of the modern confinement systems, labor needs for feeding and watering, waste collection and removal, temperature and air control allowed a redirection of labor to focus on animal production. Subsequently, breeding herds have continued to increase in size and number to take advantage of the efficiencies of larger scale roduction.

The delineation of farms into categorical size is often arbitrary, but the USDA has surveyed farms in the USA (USDA-NASS 2010). From these data (Fig. 7.1) it can be seen that of the 69,000 US swine operations, ~50 % of these farms have <100 pigs in inventory. What is more striking is that the operations with 2,000 head or more hold 87 % of the pig inventory. Further, USDA data suggests that farm size is associated with an increase in pigs per litter (Fig. 7.2). This is likely an association with the adoption of improved technology and management. Other reports suggest that in the USA and Canada, using a data base with over 350 farms, the average sow inventory was more than 1,000 sows/

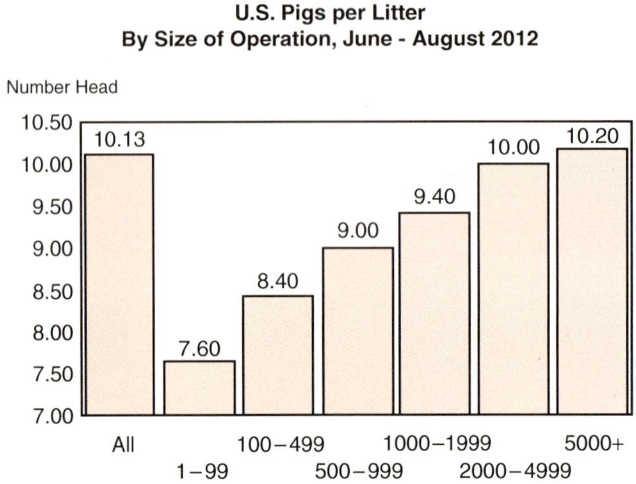

Fig. 7.2 The average number of pigs produced per litter based on size of operation for number of pigs in inventory (USDA-NASS 2012)

farm (PigCHAMP 2011; Knox et al. 2013). On these large farms, farrowing rates average 84 % with 11.9 pigs born/litter and with 10.5 pigs weaned per sow. Data is also available for average number of sows per breeding barn with over 80 % of operations having 500 or more sows/barn (Knox et al. 2013). Although there is great variation in the type of production system, when farms reach the medium to large size, they share common attributes such as higher investment, use of technology, confinement or partial confinement systems, and use of automation where possible. In most modern swine farms, the numbers of sows in a breeding barn can range from 500 to more than 5,000 (Knox et al. 2013). Most of these modern production farms in North America are confinement operations designed for low labor to sow breeding ratios through the use of breeding and gestation stalls, automation or semi-automation in feed and water delivery, and environmental control.

Since most of the commercial production of pork for domestic and export markets originates from the larger operations (Knox et al. 2013), these farms have segregated production phases into multiple sites for the purposes of disease control, economics, and specialization in production efficiency (Knox et al. 2013). In most cases, the medium to large sized farms use pork production as an integral or sole source of business income. Several specialized types of farms important to modern swine breeding operations have been developed and are classified as farrow to finish, breed to wean, gilt development units and boar studs (Taylor and Field 1998; Whittemore 1998; USDA-APHIS 2006; PIH 2007; Faust et al. 1992; Key 2007; Lawrence and Grimes 2006; USDA-NASS 2008). Each of these operations has a unique function and a specialized system for operation. Farrow to finish operations are the most common breeding system worldwide and have a long history of operational success. Although there are many different types, these operations breed, farrowing and grow the offspring to market on the same location. In developed countries, these farms tend to maintain a smaller breeding herd size due to labor. In these farms, labor must be adept and skilled at numerous tasks such as gilt development, breeding, farrowing, lactation management as well as nursery pig growth, and grow-finish management. Breed to wean farms are specialized farms that focus on gilt development, sow breeding and gestation, farrowing and lactation management. However, weaned pigs are transferred off-site for improved health of the pigs. These types of specialized farms can be quite large as herds are managed using as groups of weaned sows (PigCHAMP

2011; Knox et al. 2013). Gilt development units are relatively new, and arose with larger genetic multiplication systems. These farms allowed focus on the development and management of gilts that would eventually need to enter sow breeding groups (Moore 2005). The last specialized farm of mention is the boar stud. This type of farm first appeared in the early 1990s with the development of artificial insemination (AI) and has essentially evolved as the sole repository for genetic improvement from the male side. These farms specialized in boar health, semen production, and sperm fertility (Knox et al. 2008). In many modern production systems, multi-site operations have become standard as a method to reduce the risk of loss of operational capability in the case of a disease outbreak in one of the sites. In addition, multi-site production helps farms conform with local livestock regulations regarding animal production units, and provides proximity of genetic resources for customers in various locations.

The Impact of Reproductive Technologies on the Structure of the Swine Breeding Industry

The successful production of pork depends upon reproductive performance of the pig. In fact, reproductive efficiency is a leading factor related to producer profitability with the most profitable farms producing the greatest numbers of pigs/sow/year (Britt 1986). This measure can be evaluated by an index of farrowing rate, number of pigs produced, and the number litters/sow/year. Newer production measures are being explored for their ability to assess breeding herd performance to include number of piglets alive at day 5 after farrowing, number of pigs weaned, pigs weaned per sow lifetime, and total pounds of pork produced per sow lifetime. These alternative evaluation systems incorporate combinations of some of the most meaningful measures of pig reproduction, including genetic improvement and economic value.

The reproductive performance of the breeding herd sets the flow of animals through the food production chain. The predictable delivery of animals for pork processors, distribution and retail markets is essential for industry competiveness in pricing, quality and providing consumers a dependable and choice for meat purchase. Further, consistent and predictable animal flow allows the industry to meet domestic and export demands. In fact, the price of pigs and pork has been related to the size the export market (NPPC 2012). The sustainability of the pork industry is linked to uniform flow of animals that becomes integral to cost efficiency in energy and labor utilization, transportation, and financial markets.

To ensure a consistent and steady flow of pigs to market, breeding farms must ensure that the proper numbers of healthy pigs enter into the grow-finish operations. While appearing to be a seemingly simple task, realization of this outcome can be a challenge for many breeding farms. Problems occur because most reproductive traits are lowly heritable and therefore reproductive success is mostly influenced by environmental and management factors. This means that for an economically important trait such as litter size, only ~10 % of the response is directly attributable to the genetic makeup of the individual. Despite the low heritability, dramatic improvements have been made over the last 40 years in reproductive performance, especially farrowing rates, litter sizes and pigs/produced/sow/year. For example, previous industry averages for farrowing rates were 73 % in 1998 (PigCHAMP 1998) whereas today, the industry average is 83 % (PigCHAMP 2011). Perhaps even more remarkable, 10 years ago, average total born litter size was 11 pigs, while today, litter size averages 12 pigs born. An examination of litter size over the last 20 years shows increases each year while farrowing rates appeared to remain steady for decades before increasing noticeably in the last 10 years (Fig. 7.3). The dramatic increase in litter size is a result of more sophisticated genetic selection systems, use of records, dramatic improvements in boar and sow fertility, feeding and nutrition, health, reproductive management, and environmental control systems (Whittemore 1998; PIH 2007; PIG 2012; Flowers 1998; Harmon et al. 2001; Merks et al. 2000; PIC 2008).

Fig. 7.3 The change in US average litter size during the years from 2003 to 2012 (USDA 2012)

Reproductive Technologies for Genetic Improvement

Over time, more sophisticated swine breeding systems have been developed with emphasis on rapid distribution and incorporation of traits of economic importance. Multi-trait importance as well as differences in trait heritability pointed toward a need for selection indexes (NSR 2012). In modern genetic production systems, selection of sires and dams can be relatively simple with use of one or more individual phenotypic records to a very complex system that employs multi-trait selection based on weighted phenotypic data, sibling and even relative's records (NSR 2012; Harris et al. 1989; Houska et al. 2004; Konig et al. 2009). In the modern index selection systems that are used today, the same sire can be ranked differently by using a different index that weights more heavily on maternal or terminal traits. These ranking systems can be used to choose a sire based on available records (accuracy), and his expected progeny difference for a particular trait or set of traits. In fact a sire that is ranked highest in a maternal trait such as litter size may not be expected to rank as high in terminal traits such as feed efficiency (NSR 2012). In effect, focus on trait improvement must often be balanced between maternal and terminal traits to achieve optimal results. Modern genetic production systems are most often highly organized and structured for optimal rates of genetic improvement (Gerrits et al. 2005; PIH 2007; Faust et al. 1992; PIG 2012; Houska et al. 2004; Olesen et al. 2000; Spötter and Distl 2006; Clutter 2009).

Some discussion of the economic impact of important traits is essential for an understanding of how reproductive technologies are employed. In a volatile agriculture industry, most notably the swine industry over the past 20 years, efficiency and sustainability have relied on efficient gene transfer and incorporation of these traits into the herd via high reproductive efficiency (Knox 2011; PIH 2007; Faust et al. 1992; Key 2007; PIG 2012; NSR 2012; Harris et al. 1989; Houska et al. 2004; Konig et al. 2009; Olesen et al. 2000; Clutter 2009; Lofgren et al. 1989; Moeller et al. 2004; Schinckel and Bennet 1999). One of the primary drivers has been feed efficiency, as it is the leading cost of pork production. Feed for pork production accounts for >70 % of the cost of production and traits such as feed:gain, average daily gain, and lean deposition of meat are of great financial importance. These traits also happen to be moderately heritable, with 40–60 % of the variation attributed to genetics. As such, selection of sires and dams with improved traits can result in significant advancement in performance, especially when the selection differential is great and reproductive

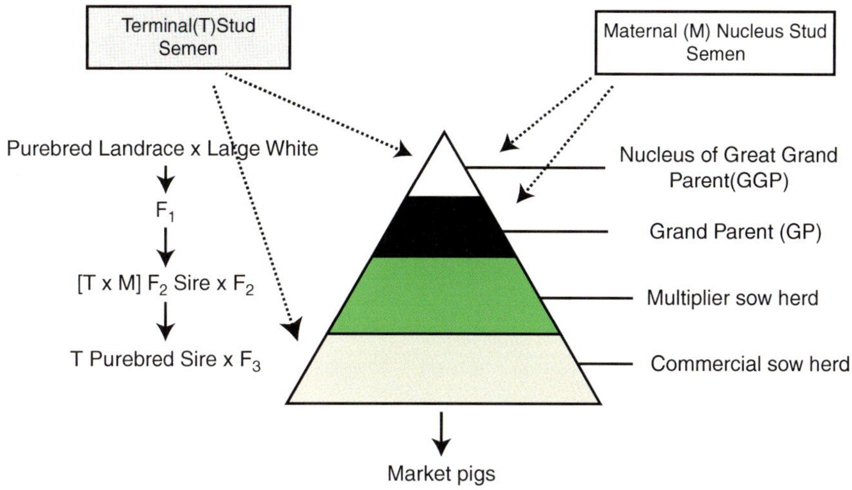

Fig. 7.4 A general structure of the swine genetic pyramid. Semen from nucleus maternal or terminal studs flows to various levels to create purebred or crossbred offspring

fertility optimal. While little economic value is currently placed on pork quality measures, these attributes influence consumer perception and satisfaction. Some pork measures which have a heritable component and impact quality and consumer liking of the product include color, firmness, moisture, and pH (PIH 2007; PIG 2012; von Rohr et al. 1999).

Reproductive traits are also highly important to producer profit. Measures which impact reproductive efficiency will result in increases in pigs produced per sow per year. This measure is a key measure for profit and is a combination of farrowing rate, pigs born alive and litters per sow per year (Britt 1986). Unfortunately, reproductive traits for the most part are lowly heritable and greatly influenced by environment and management. The heritability estimates are ~10 % but may be higher (40 %) for some traits such as age at puberty (Olesen et al. 2000; Spötter and Distl 2006; Schinckel and Bennet 1999; Ehlers et al. 2005; Wuensch et al. 2000). Despite the low heritability, there is clear evidence for the impact of genetic selection on reproductive traits such as litter size. This has been shown in maternal line studies where certain genetic lines excelled in early age at puberty, litter size, farrowing rate, and longevity (Moeller et al. 2004). However, most of these were also somewhat negatively associated with terminal traits.

Today, the swine industry has adopted a genetic supply system that is often modeled on a genetic pyramid (Fig. 7.4). In this pyramid, superior genes flow from a limited number of animals at the top of the selection pyramid and are multiplied by breeding to flow to commercial farms used for market pig production (Whittemore 1998; Ollivier 1998). At the top of the pyramid, the genetic nucleus is selected and maintained to become the foundation for genetic improvement that will flow throughout the production chain. The nucleus or great grandparent (GGP) animals are usually maintained as purebreds. Genetic nucleus farms most often maintain multiple purebred breeds to create offspring that can be used as nucleus replacements and for use in crossbreeding programs that produce F_1 animals for multiplication. The crossbreeding systems are designed to strategically capture the advantages of heterosis for selected maternal traits (Schinckel and Bennet 1999). Depending upon the size of the genetic supplier and the biosecurity system, these farms may maintain boars on the farm site or may have boars in an off-site stud for semen delivery to the nucleus. The nucleus herd is perpetuated by choosing breeding pairs to maintain and advance the traits of purebred lines and families. The offspring from these matings are selected as future sires or dams as replacements based on parental, relative, and individual performance traits. In many of these types of nucleus breeding systems, ongoing challenges involve the ability to maintain genetic diversity, inbreeding, and risk of disease entry from new animals and semen (Ollivier 1998; Bolet et al. 2001). High levels of inbreeding have been shown to suppress genetic advancement in herds that are closed or with limited animal entry for disease prevention. As the commercial swine industry has grown, there has been an even greater demand for genetically superior replacement females. As a result, the genetic supply chain has added daughter nucleus farms, which multiply purebred nucleus gilts in an off-site unit to minimize risk in case of a disease outbreak. These farms collectively cross the purebreds using AI to produce F_1 female multipliers which then supply production farms with commercial production sows. The commercial sows are AI with semen from sires selected to produce high fertility with greater emphasis on terminal pig production traits.

Use of Reproductive Technologies to Control Disease

Swine diseases have been a major limitation to pork production around the world. More recently, swine health has been associated with transfer of diseases to humans as well as other livestock (Straw et al. 1999; Whiting 2003). Certain diseases are highly contagious and indigenous to certain areas while others appear sporadically. The most common disease causing agents of concern are viruses and bacteria (Straw et al. 1999; Allan and Ellis 2000; Guerin and Pozzi 2005). In some cases diseases are controlled by treatment, prevention, testing and elimination to limit entry or spread. Some of the diseases known to be important to swine are spread by animal to animal contact, human or equipment transfer, or even vectors. In instances when infection occurs, options may include elimination of all the animals in the area, depopulation of specific farms, testing and removal of infected animals, or waiting for herd immunity to develop. However, in many areas and herds, some diseases are difficult to manage and control and in these cases, producers must live with the diseases and the associated losses.

Reproductive technology has been used to help control or even eliminate disease. Disease control and gene transfer technology are inherently linked. In nucleus farms where mothers are negative for many diseases, offspring can be delivered from the sterile uterine environment in a surgical facility to a clean site for startup operations. However, this is not a common method and today most disease management is controlled by biosecurity procedures (AASV 2003). Between farms in the genetic pyramid, this can involve directional animal flow, semen sourcing, and people and equipment movement (Harris 2000; Madsen 2005). The highest levels of biosecurity and disease prevention measures are practiced at the top of the pyramid and in the boar studs while less intense biosecurity may often occur at lower levels of the pyramid. In the process of improving herd genetics, new animal entry is recognized as one of the largest risks for disease. However, this cannot be helped as replacement boars and gilts must enter at some time to limit inbreeding. This is true in boar studs and well as nucleus and multiplier farms. Managing risk can be complex and quite variable among farms. In the most controlled farms, herd closure may exist for new animal entry or if allowed, entry is infrequent during the year. Often, new genes can enter in the form of extended or frozen semen. If new animals are allowed to enter, biosecurity procedures can include source herd health verification, and isolation and acclimation before entry into the herd. If semen is the genetic source, studs will routinely test boars and evaluate semen and add antibiotics to the extended semen to limit the risk of disease (Gerrits et al. 2005; Guerin and Pozzi 2005; Madsen 2005; Althouse and Rossow 2011; Maes et al. 2008).

The risk of disease can also come from other sources such as humans. As such, this recognized risk has changed who and how personnel can enter breeding farms. Breeding farms often require that employees have no contact with swine at home or have down time from previous pig exposure before entry into another farm (PIH 2007; PIG 2012; AASV 2003). In some operations this may involve 1–3 days away from pigs for domestic visitors or as long as 7–10 days for international visitors. Risk is managed by assuming that the outside of the farm is considered dirty, and anything inside the building can considered clean. In this approach, people must change clothes and shower, while trucks, cars, and other vehicles may not be allowed access to certain perimeter locations. Equipment and supplies must be cleaned, disinfected, or sterilized before entry (Whittemore 1998; PIH 2007; PIG 2012; Straw et al. 1999; AASV 2003). Many farms use perimeter fencing to limit larger wild animals and use rodent and bird control to limit vector transmission. In light of new information on aerosol transmission of disease spread, some farms have implemented air filtration systems as well.

Replacement gilt entry is an important component for breeding herd management as replacement rates are above 40 % annually for replacing culled sows (PIC 2008; Foxcroft 2001; Knauer et al. 2010; Stalder et al. 2003). Gilts are commonly sourced from high health herds and enter in batches as weaned pigs or more mature females. Batches of gilts may be of similar age or from a wider age range to allow sequential batches of gilts to enter with fewer animal deliveries from outside sources. All animals are required to go through an isolation and acclimation period (Straw et al. 1999). For biosecurity reasons, personnel and equipment at the isolation/acclimation unit are not part of the breeding farm and visits to the isolation unit must be sequenced for biosecurity. In the isolation/acclimation period, animals are observed for signs of disease for 60–90 days (PIH 2007; PIG 2012; PIC 2008; Morrow 1986; Levis 1997a). Acclimation is usually performed in the same unit using culled sows or weaned pigs from the breeding farm to expose the new replacements the microbes present on the breeding farm. This helps the new animals adapt and develop immunity to the new farm microbial environment.

AI among all other reproductive technologies has helped to transform the industry into an efficient multi-site production system to help facilitate gene transfer and control the risk of disease (Riesenbeck 2011; den Hartog 2004; Whiting 2003; Guerin and Pozzi 2005; Harris 2000; Madsen 2005; Maes et al. 2008; Weitze 2000). Entry of new boars and gilts has been recognized as the major risk for spread of disease and use of AI reduced the need to bring in live animals for making genetic improvement. However, one caveat is that while AI is highly effective at gene transfer to many animals and herds great distances apart, this same result can occur if infectious agents are in the semen. For this reason, boar stud health and semen quality is of the highest importance. Boars are routinely tested for a variety of diseases important to meet the requirements of domestic customers or foreign markets. In fact, each country may have its own unique health test requirements before animals or semen can enter based on diseases of local or global importance such as Foot and Mouth Disease, Classical Swine Fever, Porcine Circovirus, Parvovirus, Pseudorabies, and Japanese Encephalitis (Straw et al. 1999; Whiting 2003; Guerin and Pozzi 2005; Madsen 2005; Althouse and Rossow 2011; Maes et al. 2008). In the USA and elsewhere in the world, porcine reproductive and respiratory syndrome (PRRS) has been on the forefront and there have been numerous cases of PRRS virus transfer from studs to breeding farms and from farm to farm by aerosol and other such routes (Nodelijk et al. 2003; Prieto and Castro 2005; Reicks 2008, 2012).

Globally, diseases such as PRRS, which can be transmitted in semen, have changed the way genetic material can be used. PRRS is caused by a unique and problematic virus which can mutate quickly, rendering vaccines only partially effective. The virus survives well at lower temperatures, and it can be spread by air, materials, animals, and semen. Boar studs can be classified by PRRS status such as naïve, negative, vaccinated, or positive. However, most breeding herds require a PRRS negative status, as naïve is difficult to confirm. This has changed the way sow farms source semen especially when considering semen use in a genetic production system. Once a herd is infected, the disease can have serious consequences for the health of the piglets and productivity of the breeding herd. PRRS can cause a variety of short and long-term production losses ranging from abortion to respiratory disease. In boar studs, sires should be randomly or routinely tested at intervals (Reicks 2008) and in the event of a positive test, retesting and confirmation can result in removal of that animal and testing and removal of other sires in the barn (Reicks 2012). In many cases, PRRS infected studs results in entire depopulation of the stud, regardless of their genetic value or potential.

Semen has long been recognized as a potential source of viruses and bacteria. Bacteria are common in ejaculates and while a few are infectious, most are not contagious (Althouse et al. 2000). Bacteria in the inseminate can cause disease and reproductive failure but many vaccines are effective and widely used. Further, antibiotics are added into semen routinely to limit the harmful effects of bacteria on semen fertility and female fertility (AASV 2003; Althouse and Lu 2005; Levis 2000). Since semen is collected in barns, and even though they are clean, the environment is not sterile. Bacteria are often found in the first phase of the ejaculate and as such, the initial phase is not collected. Yet the same media which supports sperm life can also support bacteria as well and is the reason why antibiotics are included in extenders. Antibiotic addition is helpful for preventing growth of bacteria while semen is shipped and stored for extended periods following collection.

Reproductive Technologies and Replacement Gilt Management

Gilt management is an essential area for genetic improvement and herd reproduction and has been recognized as important to sow longevity and lifetime performance. The major limitation to reproduction in gilts is pubertal estrus expression and normal cyclic fertility (PIC 2008; Foxcroft 2001; Levis 1997a; Beltranena et al. 2005). Failure in these areas increases herd entry to service intervals, feed costs, culling, and reduces lifetime pig production. There are limited methods to induce puberty in swine. Acute stressors such as moving and mixing unfamiliar gilts with boar exposure, is the most common method for puberty induction (Koketsu 2007; Engblom et al. 2008). However, the effectiveness and variation in response to these procedures is considerable and may depend upon age, weight, maturity, batches, genetics, season, and health. Gilt selection has gained some attention as offspring of mothers with early age at puberty have offspring that also express early puberty (Foxcroft 2001; Knauer et al. 2010; Levis 1997a; Schukken et al. 1994; Tummaruk et al. 2007). Modern gilt genotypes (Moeller et al. 2004) display fast and efficient lean growth compared to their predecessors, and these traits could change the actual stage of reproductive maturity relative to age and weight, and impact lifetime productivity. Recommendations suggest that modern genotype gilts should be selected for growth rates >600 g per day from weaning until time of puberty induction for early expression of estrus.

Boar exposure is an essential management tool for induction of puberty in gilts. Data reveals that the age of the boar, the frequency of exposure, the type of contact and the ratio of boar:gilts can influence the induction response (Foxcroft 2001; Levis 1997a; Caton et al. 1986; Foxcroft et al. 2001; Knox 2004; Knox and Wilson 2007; Paterson and Lindsay 1981). Young boars <8 months of age are less effective than older age boars at inducing and advancing age of puberty in gilts. Further, high libido boars have a more profound effect on puberty advancement than those with low libido. Frequency of exposure has an effect on gilts presumably through pheromonal and stimulatory cues. Gilts receiving fenceline exposure twice each day show advanced puberty compared to once a day fenceline contact. However when applying physical boar exposure, once daily is just as effective as twice daily physical or fenceline exposure (Zimmerman et al. 1998). Duration of boar contact on puberty induction reveals that 10 min of exposure is optimal with limited benefit from additional time. However, if the gilt to boar ratio exceeds 6:1, then exposure time should be increased to observe the response (Hughes 2001). This may become a more common issue with larger farms and larger numbers of gilts in a pen. In any system, females can be moved to the boar housing area or the boar may be moved adjacent or into the gilt pen. A few new technologies have been developed to improve the effectiveness of fenceline or physical boar contact. One product is a robotic boar that can provide simulated sight, sound, and pheromonal boar stimuli. This novel and unique invention was effective, but not as effective as a real mature boar (Gerritsen et al. 2005). Other advancements in gilt stimulation involve the use of a boar exposure area that has been developed and used successfully over the last decade to intensify boar contact for larger groups of gilts. This system relies on a battery of boars to deliver stimuli to gilts moved into holding and detection areas (Beltranena et al. 2005) with alternating boars facing opposite directions into front and back pens.

Hormonal induction of puberty with pregnant mare serum gonadotropin (PMSG) and human chorionic gonadotropin (hCG) are approved for use in swine in the product known as PG600® (Merck Animal Health, Summit, NJ). Although not a new technology, it is one of only a few hormonal products available for inducing fertility in pigs in the USA (Britt et al. 1989). In other countries and for research purposes, other hormonal products are used such as PMSG with higher dosages, partially purified follicle stimulating hormone (FSH) (Jackson et al. 2006), and synthetic gonadotropin releasing hormone (GnRH). The FSH activity in these products can induce follicle growth but without luteinizing hormone (LH) activity at the right time, many follicles will form cysts or fail to develop to ovulation (Breen and Knox 2012). Therefore the combination of PMSG and hCG together (Huhn et al. 1996), or PMSG followed by hCG, LH (Degenstein et al. 2008) or GnRH (Brussow et al. 1996) can be effective for inducing follicle growth and ovulation. For PG600®, a considerable amount of research has been performed over the last 15 years to improve its effectiveness for puberty induction and fertility (Bartlett et al. 2009; Breen et al. 2005; Knox et al. 2000; Sporke et al. 2005). Overall, approximately 60 ± 10 % of treated animals show estrus within 5 days of treatment provided that the animals meet the age and weight recommendations and are not yet cycling. However, research has shown that several factors such as route of injection, dose, prior boar exposure, boar exposure itself, feeding, and genetics can all impact the response in estrus and ovulation. Interestingly, while 60 % of gilts may show estrus, 70 % actually ovulate. Further, continued boar exposure, following the initial treatment has been shown to be important for subsequent estrus expression and cyclic activity.

Breeding replacement gilts has become an important topic for the industry as lifetime productivity and sow longevity are important factors that must be considered for sustainable breeding systems. It appears that breeding gilts on a PG600® induced estrus at <200 days of age is not advised due to the lower conception rates and litter sizes when compared to breeding at later ages following induction (Kirkwood et al. 2000). Further, for occidental breeds of pigs regardless of age of pubertal estrus expression, data suggests that while the ovaries can respond to exogenous hormones even before 150 days of age, the hypothalamus and pituitary do not function normally until just before 200 days of age (Barb et al. 2010). It may be worthy to note that during gilt development, nutrients are initially used for lean tissue deposition when growth rate is rapid, but as animals mature, fat begins to accumulate and the available nutrient resources tilt toward reproductive function (Whittemore 1998). Puberty induction is performed using boar exposure starting at 150–180 days of age. The available data suggests that age, weight, backfat, and number of cycles before mating are all important for litter size, farrowing rate, and lifetime productivity (Levis 1997a; Schukken et al. 1994; Tummaruk et al. 2001, 2007; Kirkwood et al. 2000). In almost all cases there appears to be an optimal range that must be met for each of these measures. Gilts that are too young or too old, too light or too heavy, too immature or too mature all have reduced fertility compared to the intermediate females for the breed and herd. These data provide strong evidence that within herd, analysis of fertility measures using the categories listed above should be established for setting gilt breeding guidelines.

Detection of Estrus in Sows

For weaned sows, boar exposure following weaning can stimulate follicle development and advance estrus expression (Langendijk et al. 2000a, b; Walton 1986). Boar exposure is an integral component for eliciting the female estrual response and attempts to detect estrus without a boar results in only 50 % accuracy. Sight, sound, and smell, and physical stimulation from the boar aid in eliciting the estrual response in 100 % of fertile females (Hughes et al. 1990). When any one of these components is minimized, accuracy of estrous detection can be reduced by 10–50 %. In the USA, most of sows used for commercial pig production are housed in large breeding barns with 1,000 or more sows in stalls (Knox et al. 2013) and for practical purposes, boars are moved to the alleyway at head of the sows to provide 1–2 min of fenceline exposure. This stimulates sows with sight, sound, and smell of the boar while back pressure is applied by technicians at the rear of the crate. These systems have been developed to optimize detection of large number of sows with few boars and limited labor. Most data suggests that fertility is generally high in weaned sows with 80–90 % of sows expressing estrus within 7 days of weaning (PigCHAMP 2001). However, this measure can vary greatly, depending upon length of lactation, body weight loss of sows in lactation, parity, and season (Stalder 2008; Knox and Rodriguez-Zas 2001; Koketsu and Dial 1997; Love et al. 1993). In these large barns, controlling boar stimulation is challenging but robotic boar movers and leash training of boars have been developed to dramatically improve control of boar movement and exposure (Knox et al. 2013). The use of robotic boars is more common in large operations and are excellent devices that allow controlled duration and focus of boar contact with females. An alternative to use of large occidental breeds of boars which can be aggressive toward handlers and hard to control, has been the development of the F_1 Meishan boar from a Meishan sire and Large White sow. The resulting boars mature earlier, are smaller, and are less aggressive towards handlers and other boars, but display high libido even when used with boars in groups.

Synchronization of Estrus

Almost all breeding herds operate on batch production of pigs which allows for synchronized labor events such as weaning, estrous detection, breeding, pregnancy diagnosis, farrowing, and movement of sows. This system has important implications for managing pig health and disease, and more uniform management for production of market pigs. Although many large farms may farrow during much of the week, batches of sows are weaned to create breeding groups and uniform groups of piglets for growth to market. Batch systems on smaller farms may take advantage of the size of scale by weaning groups of sows at 2 week or at monthly intervals. In any case, production and management of larger groups of pigs allows for more efficient feeding, and marketing of groups of hogs that meet the weight and lean carcass targets for packers. Group weaning is the most successful method for estrus synchronization in adult sows for facilitating production scheduling. However, the occurrence of post-weaning estrus is influenced by many factors, including lactation length, parity, and season (Walton 1986; Knox and Rodriguez-Zas 2001; Koketsu and Dial 1997; Belstra et al. 2004; Langendijk et al. 2000c; Steverink et al. 1999; Weitze et al. 1994). Lactation length has been trending higher in the USA for the last several years. This may reflect the fact that early weaning resulted in a higher incidence of failure of sows to rebreed and if they were bred, lower reproductive performance occurred due to the short time for reproductive system repair following farrowing. However, the trending for increased weaning age in the USA may also reflect recent mandates in the European Union for a minimum 21–28 lactation length before weaning piglets in order to meet animal welfare criteria. Weaning a group of sows results in most of the sows returning to a fertile estrus within 3–7 days. However, extended intervals from weaning to estrus are related to reproductive failure (Knox and Rodriguez-Zas 2001; Steverink et al. 1999) and improving return to estrus within 7 days has implications for fertility. Summary data from industry indicates that 10–20 % of sows fail to return to estrus within 7 days of weaning (Knox and Rodriguez-Zas 2001; Dial et al. 1992; Knox and Probst-Miller 2004; Muirhead 1990; Vargas et al. 2009).

Boar exposure is the most effective tool to advance estrus in weaned sows and data suggests that exposure starting within 1–2 days of weaning improves estrus, with twice daily exposure aiding in detection and accuracy. Exogenous hormones like PMSG and PG600® (Bates et al. 1991, 2000; Bennett-Steward et al. 2008; Knox et al. 2001) improve the percentage of sows that return to estrus within 3–6 days following weaning but only PG600® is approved for use in the USA. These types of products are most commonly used to address the repeated and persistent problems of late summer infertility. Wean to estrus interval affects farrowing rate and litter size with late returning showing reduced performance (Steverink et al. 1999). Further, a single batch of sows weaned on the same day but bred over more days will inevitably show a greater spread in day of farrowing. And when batches of sows are weaned, this results in a wider range in age at weaning for both sows and piglets which each will impact subsequent performance. Sows weaned less than 18 days after farrowing exhibit longer intervals for return to service, may fail to conceive more often and show ovarian cysts (Knox and Rodriguez-Zas 2001; Levis 1997b). These problems can often be compounded in young parity sows especially in late summer and early fall. Many of the early weaning problems were realized when the technology was developed in the 1990s to limit vertical transmission of disease from sows to piglets. Early weaning (<18 days) or medicated early weaning was used to improve piglet health (Whittemore 1998; Straw et al. 1999; Harris 2000). However, this technology worked to some extent on certain diseases but had detrimental effects on sow fertility. It became evident that the problems of delayed returns to estrus, cysts, and poor conception with early weaned sows was in fact due to insufficient time after farrowing for uterine repair, hypothalamic changes, and normal ovarian function. As a result, over the last 10 years, average weaning age has increased from 17 to 20 days (PigCHAMP 2001, 2011). From the available evidence in high performing US herds, it is unlikely that further increases beyond 21 days will occur since many measures of fertility are near optimal and similar to some of the best herds in Europe with longer lactation lengths.

Some recent weaning technologies have been employed to improve sow fertility and piglet growth. Split weaning is a technology used in a small proportion of sows where the heavier half of the litter is weaned after 2 weeks of lactation to allow the smaller pigs in the litter to nurse and increase weaning weight while reducing body weight loss from the sow (Tarocco et al. 2000). Another technology is called bump-weaning and allows transfer of fallout piglets (pigs much lighter than their littermates) or excess piglets from large litters to other sows early or later (nurse sows) in lactation (Reese and Straw 2006). In this system, young parity sows may be given groups of fallout pigs from other litters while older parity sows further along in lactation may receive heavier pigs from young parity sows at risk for excessive body weight loss or for poor lactation. Some European producers also allow younger parity females to lactate up to 30–35 days for use as nurse sows to give them additional time for uterine and hypothalamic repair. There are reproductive cautions with any of these techniques as they can temporarily remove some of the negative feedback from nursing to the hypothalamic pituitary axis (HPX) resulting in a release of hormones that can initiate ovarian function and reduce fertility following weaning.

The synthetic progestagen, allyl-trenbolone, is an approved hormonal feed additive for synchronizing estrus in mature swine. It is known as Matrix® in the USA or Regumate® elsewhere

(Merck Animal Health, Summit, NJ) (Huhn et al. 1996; Sporke et al. 2005). The product allows a producer to take a single female or a group of randomly cycling gilts and synchronize their estrus to occur ~22 days after the start of treatment. Matrix® comes in liquid form in 1 L bottles with enough to treat 10 females. The active ingredient works by suppressing hormone release from the HPX to prevent ovulatory follicle development and estrus. It is delivered as a daily dose of 6.8 mL as a top dress on feed to deliver 15 mg/day for 14 consecutive days. The hormone is cleared within 24 h after the last feeding and the HPX is released from progestagen suppression and initiates hormone release and follicle development on the ovary. As a result, the majority of gilts (>80 %) come into estrus within 5–8 days after the last feeding of Matrix® (Estienne et al. 2001). It is important to utilize daily boar exposure to stimulate the females following treatment to advance estrus and improve synchrony. All measures are normal with duration of estrus 2 days and ovulation between 24 and 48 h after onset of estrus. The synchronized gilts typically have ovulation rates exceeding 16 eggs with some reports indicating increased litter sizes as well. This has great practicality for replacement gilt entry into established breeding groups as most gilts show random cycling over a 21 day period. It is of interest to note that individual feeding is required to apply Matrix® effectively. Group housing of gilts and feeding Matrix® can be difficult for proper delivery so individual feeding may be required. However, some producers use delivery by oral dosing in pens but the effectiveness and consistency must be controlled since the gilts may need to be trained to avoid partial dosing. Also, there has been interest in the use of Matrix® to delay post-weaning estrus in first parity sows (Fernandez et al. 2005; van Leeuwen et al. 2011). Skipping breeding of these young sows increases their conception rate and litter size, but this delay requires an additional 21 days and is cost prohibitive due to the high number of open days and high feed costs. Although not approved in the USA, research has shown that short-term delay with Matrix® feeding for only 5–7 days after weaning can provide the advantages of the extra time in reproductive repair with minimal costs.

Artificial Insemination Technology

Perhaps there is no greater example of a reproductive technology that has had such a dramatic effect on swine production systems than the development of artificial insemination. The technology for manual semen collection from a boar followed by AI was first developed in the mid 1950s and in liquid form, boar sperm survival was short and insemination needed to occur within hours for fertility (Johnson et al. 2000). Early attempts at freezing boar sperm were disappointing as the survival of sperm after thawing was poor and the fertility reduced farrowing rates below 50 % with litter sizes of five pigs or less (Johnson 1985). As a result, while research continued on cryopreservation of boar sperm, new research focused on the preservation and extension of the boar ejaculate in liquid form. As AI technology advanced with liquid semen extenders, breeding strategies on swine farms changed from pen mating where one boar was placed in a pen of females with the intention that all females would be mated, to a hand-mating system where one boar was individually placed with an estrual female and a supervised mating would occur on each day she would stand (PIH 2007). This was important for fertility as sperm numbers and ejaculation frequency could be controlled and boar fertility could be evaluated. This change in breeding procedure aided breeding group establishment, boar fertility and resulted in improved pregnancy rates and litter sizes. During this same period, AI procedures were being developed (Flowers and Esbenshade 1993) and in the early 1990s, training and education programs were in place to help advance the application for on-farm semen collection and semen extension for use with AI. In the early stages of AI development, much of the technology relied on reusable supplies and required producers clean between use these supplies. Further, it was important that producers were competent in several procedures including boar management and training, semen collection, extension, evaluation, and storage. Lastly, producers were required to properly identify estrus and perform AI. Perhaps not surprisingly, this complex array of tasks required a more specialized labor force for success and many farms could not reach the desired level of fertility in all these areas. As a result, many farms designated personnel to specialized tasks such as boar training and semen collection, or semen evaluation and extension, or estrous detection and AI. This specialized approach worked well and with greater proficiency, improvements in fertility resulted in greater success. However, the need for specialized labor was intense and time limitations associated with proper cleaning of reusable supplies, such as semen collection containers and AI rods limited efficiency. The commercial development of disposable supplies such as semen collection bags, AI catheters, as well and longer-term semen extenders, as well as automatic and semiautomatic semen collection and packaging systems helped to revolutionize the swine AI industry resulting in the system in use today with high fertility, efficiency, and control of disease (Levis 2000; Flowers and Esbenshade 1993; Flowers and Alhusen 1992; Singleton 2001).

Adoption of AI in the USA occurred very quickly from the early 1990s to the present day (Riesenbeck 2011; USDA-APHIS 2006; Weitze 2000). Although published figures for use on all swine farms suggests 70 %, in reality, most commercial operations with 500 or more sows, rely exclusively on AI. The technology was shown to help identify the fertility of boars and semen, extend use of valuable sires, reduce labor per sow bred, and allowed dramatic increases in breeding herd size without additional labor (Flowers and Alhusen 1992). AI increased the use of valuable sires from 4 sows hand-mated each week to 10–20 sows when using AI from a single ejaculate. Use of AI also facilitated semen evaluation and helped identify more poor fertility boars (Foxcroft et al. 2008; Rodriguez-Martinez 2003; Ruiz-Sánchez et al. 2006). In addition, the industry began using semen doses from different sires for first and second services and later moved to pooled semen from three to six boars to minimize the impact of low fertility sires used in commercial pig production (Knox et al. 2008). While farms may have their own protocols for breeding herd management, standard AI commonly involves once or twice daily estrous detection with insemination occurring on each day standing (Knox et al. 2008; AASV 2003) up to a maximum of three services per sow (Flowers 1996, 1998; Morrow 1986; Flowers and Esbenshade 1993) given at 24 h intervals while in standing estrus (Knox et al. 2013). AI is performed using three billion viable sperm in 80 mL of extended liquid semen with AI technicians able to inseminate 6–8 females/h (Flowers and Alhusen 1992). Modern semen extenders are classified as short, intermediate, or long term (Levis 2000). Short-term extenders are less expensive and provide sperm shelf-life up to 3 days, while intermediate extenders allow semen use up to 4–6 days, and long-term extenders may extend shelf-life for 7–10 days. On farm semen storage occurs in thermally regulated coolers set at 16–18 °C.

Limitations to achieving high pregnancy rates and large litters from use of AI in swine can involve female fertility, semen fertility, and technician effects. Failure in any of these areas will result in failure of sperm and eggs to meet for fertilization. The greatest limitation to fertilization appears to be the fertile lifespan of the sperm and egg (Soede et al. 1995a). Single inseminations occurring 12–24 h before ovulation result in the greatest pregnancy rate and litter size (Soede et al. 1995a; Nissen et al. 1997). However, inseminations occurring too early or too late relative to ovulation result in lower fertilization, increased embryo loss and higher rates of pregnancy failures. At the present time, the only way to estimate timing for AI is detection of estrus. However, estrus symptoms are highly variable among females during the estrous period. Further, expression of estrus can depend upon factors than can enhance or inhibit expression and observation of symptoms by a technician (Langendijk et al. 2000b, c; Steverink et al. 1999; Knox et al. 2004).

In mature gilts, estrus lasts 1–2 days with ovulation occurring 24–48 h after onset of estrus (Spencer et al. 2010; Almeida et al. 2000;

Bracken et al. 2003; Horsley et al. 2005). Insemination intervals for gilts are very similar to sows with breeding occurring at onset of estrus and 24 h later on each day standing. In weaned sows, estrus varies in relation to interval from weaning to estrus (Weitze et al. 1994) and an important, but imperfect relationship exists with wean to estrus interval and estrus to ovulation interval (Knox and Rodriguez-Zas 2001; Weitze et al. 1994; Knox et al. 2001; Soede et al. 1995a; Nissen et al. 1997; Kemp and Soede 1996). Sows that express estrus soon after weaning show a longer duration of estrus and interval from estrus to ovulation, while those returning later have a shorter duration of estrus and interval from estrus to ovulation. This relationship can affect optimal AI timing, and as a result, most farms perform multiple inseminations. AI timing based on wean to estrus interval has had mixed success (Flowers 2000) and may not be evident due to the use of multiple inseminations or due to differences between farms for wean to estrus interval (Steverink et al. 1999). Another confounding factor is classification of wean to estrus interval based on estrous detection frequency and number of sows that express estrus during days 3–7 after weaning. These issues may limit the identification of relationships between wean to estrus interval and time of ovulation.

Studies evaluating insemination strategies to cover variation in time of ovulation have come to the conclusion that single services most often result in reduced farrowing rate and litter size (Flowers and Esbenshade 1993; Lamberson and Safranski 2000), while others did not report this (Xue et al. 1998). The value of a third service is also controversial since some studies do not report improvements in reproductive performance (Flowers and Esbenshade 1993; Xue et al. 1998) while others (Tilton and Cole 1982) report third services improve the total numbers of pigs born. When using a two insemination system, evaluation of different AI schedules did not alter farrowing rates and litter sizes in females expressing a 2-day estrus (Flowers and Esbenshade 1993). However, many of the previously mentioned studies were aware of reproductive performance differences that were associated with differences in the duration of estrus. Whether or not a second or even a third service had any impact on reproduction may have been related to the interval from insemination to ovulation and to the duration of estrus. This is important since a second AI within 24 h before (Knox et al. 2002) or as late as 3 h after ovulation (Soede et al. 1995b) has been shown to have positive effects, while late inseminations occurring 4 h after ovulation (Rozeboom et al. 1997) have been shown to reduce fertility. Simulated estrous detection frequency and AI schedules for weaned sows observed that overall frequency of estrous detection had minimal impact on fertility but that three and even four inseminations produced higher farrowing rates and litter sizes compared to single inseminations (Lamberson and Safranski 2000). This would support observations that showed detection frequency of once, twice, or three times daily with adjusted AI times for predicted time of ovulation (Knox et al. 2002) had no significant improvement in farrowing rate or litter size although both measures were numerically higher with increased detection frequency. Collectively, these studies suggest a potential for increased reproductive performance when at least one AI occurs before and close to ovulation and inseminations do not occur after ovulation.

The number of sperm in the AI dose has been related to fertility and higher numbers of sperm may help sperm reservoirs remain functional for longer periods of time. More sperm inseminated results in greater numbers of sperm in the uterus following insemination (Pursel et al. 1978). However, the minimal number of sperm cells required in an AI dose to produce optimal farrowing rates and litter sizes appears to be closely linked to the interval from insemination to ovulation and also to the number of inseminations performed. When performing single AI with two billion sperm cells/dose, litter size and farrowing rate were only similar to three billion sperm when AI occurred within 28 h prior to ovulation (Nissen et al. 1997). However, with double insemination at 24 h intervals, fertility with use of two billion sperm was similar to three billion sperm, and each were improved compared to only one billion sperm cells/dose (Watson and

Behan 2002). Collectively, most research studies indicate that two billion sperm cells in an AI dose will not limit reproduction while fewer than two billion cells reduce fertility (Bracken et al. 2003). Further, while research studies show that higher numbers of sperm provide little or no fertility advantage (Steverink et al. 1997), field data with much larger numbers of sows suggests that more sperm cells in the AI dose are related to fertility (Reicks and Levis 2008). These data point to an important observation and reveals that is it is often possible to detect differences under controlled research conditions or fail to detect differences due to limited numbers of observations. However, under commercial conditions, although the same treatments may be used, there is limited ability to control the experimental conditions even when including much larger numbers of sows.

Sperm quality is a component of semen fertility and can include measures of viability, motility, morphology, and other characteristics such as clumped sperm or presence of bacteria. However, the relationship of any measure to fertility has been difficult to establish except with low motility semen, low numbers of viable sperm, and high levels of sperm abnormalities (Flowers 1997). In these cases, fertility is increased by compensation with use of more sperm (Flowers 2002). Fertility differences among boars using similar numbers of sperm have also been demonstrated, but it is often not clear whether this is sperm or seminal plasma related (Caton et al. 1986; Foxcroft et al. 2008; Ruiz-Sánchez et al. 2006; Dyck et al. 2011; Flowers 2008). Methods to improve sperm fertility have included additives to semen to mimic seminal plasma components such prostaglandin, oxytocin, and estrogen (Flowers and Esbenshade 1993; Peña et al. 1998; Willenburg et al. 2003a). In some studies, improvements in fertility suggest physiological effects of the additives on establishing sperm reservoirs and improving fertility.

Insemination volume is important for AI but conclusions from studies involving different volumes are confounded by different numbers of sperm in the inseminate, site of semen deposition, AI timing relative to ovulation, and semen leakage. In conventional AI where semen is deposited into the cervix, 50 mL improved sperm numbers in the uterus, pregnancy rates, and day 25 embryo survival compared to use of only 20 mL. However, although the industry uses 60–80 mL, the actual volume limits are uncertain. It has been reported (Steverink et al. 1998) that with a single AI of one billion sperm, a 5 % leakage reduced fertilization rates. In contrast, when performing AI with only 0.5 billion sperm, a leakage of 44 % was not associated with a change in pregnancy rate or numbers of embryos (Willenburg et al. 2003a). This discrepancy suggests other factors besides the volume must be considered. Although limited information is available, the use of intrauterine insemination (IUI) with lower numbers of sperm and reduced volumes, results in fertility similar to conventional AI (Watson and Behan 2002). Further, surgical and nonsurgical deep uterine insemination (DUI) is successful with even lower sperm numbers and volumes (0.5–7.5 mL).

New AI technologies have been developed for use in the breeding herd for labor efficiency. Some appear to hold true value for users but few research studies and limited information makes these techniques worthy of discussion but hard to place a definitive value on. One of the techniques has been reviewed and involves pre-insemination (Flowers and Esbenshade 1993) which could be performed days or even weeks before AI to improve sperm reservoir formation. The premise for this was that with pre-AI, a cleaning or priming effect on the uterus would help in subsequent fertility. But the outcomes of pre-mating using vasectomized boars, only seminal plasma, or insemination using dead sperm to initiate a uterine immune response, were not clear and proved beneficial in some studies but not in others. The effects of seminal plasma are complex but have been shown to impact reproduction when added to sperm. Yet adding seminal plasma to sperm has drawbacks especially for repeatability and disease control. As a result, synthetic seminal plasma products have been marketed that contain salts, ions, buffers, and antibiotics for pre-insemination technology using 30–100 mL volumes immediately before or at the time of AI.

It is not clear whether this is truly helpful or effective, but clearly seems to not be harmful and could prove helpful.

Other insemination aids are now commonly used and allow insemination of more sows with less physical strain and labor for technicians and in less time. While scientific studies are not available on their efficacy, success of the breeding herds and their continued purchase and use speaks to their effectiveness. Most of the breeding technicians that utilize these systems are quite adept at their use and replace human back pressure and side contact during insemination with devices such as breeding saddles to help simulate the back pressure of the boar and clamp of his front legs when mounted. Hands-free insemination systems are also commonly used and allow simultaneous insemination of three to six females at a time under the supervision of one technician. The insemination occurs by gravity flow and the negative pressure resulting from uterine contractions of the sow. These devices are simple and use a clamp or belt on the sow with an attached rod to hold and elevate the semen container above the female which is then attached to the insemination catheter. Insemination does not occur automatically and uterine contractions must often be elicited through some form of stimulation including boar exposure and back pressure. These systems seem to work quite well and are popular in some large breeding herds. The use of these systems has been reviewed (Levis 2005) and reports suggest these can be valuable but must be used with some caution to prevent excessive leakage.

Boar exposure is performed and required during most inseminations. This proven technique, while effective, requires extra labor to ensure boar contact at the time of insemination. Research reports have reviewed the value of boar exposure at the time of mating and suggest it can be a valued component to AI success (Langendijk et al. 2005; Soede 1993), while others have shown fertility is not affected in the presence or absence of a boar during AI (Willenburg et al. 2003b). The discrepancy may suggest that while the boar can clearly influence some reproductive physiological responses, and induces the standing response in the female, his physical presence alone may have limited effect on the transport of sperm and subsequent fertility. The choice of breeding with or without a boar present may be applicable for advanced reproductive techniques using IUI or timed AI or when considering labor availability and time. Of note however, it is likely that sows in stalls will be much more likely to stand immobile for AI without a boar when compared to gilts.

Boar Studs and Semen Production

Off-site semen production in boar studs is the industry standard for AI. The specialized and isolated unit for boars and semen production was specifically designed for biosecurity, boar training for collection, semen evaluation, and AI dose production. Throughout most developed countries with modern swine production systems, semen is produced in studs. AI use and boars available for AI dose production have been reported for several countries (Riesenbeck 2011). In North America, number of boar studs is estimated at 120 which maintain ~24,000 boars in production (Riesenbeck 2011; Knox et al. 2008; Burke 2000). These sires serve as the genetic resources for the 5.8–7.0 million sows served each year in North America (USDA 2012; Reisenbeck 2011). The number of boars in a stud varies with most farms maintaining 51–500 boars per site. Boars are primarily housed in confinement in stalls with environmental control for high temperatures through the use of evaporative and mechanical cooling systems (Knox et al. 2008) and even air conditioning. The most common age of boars in stud is 1–2 years as annual boar replacement rates average 40 % due to genetic improvement, poor semen quality, and feet and leg problems. Boars are collected manually using a dummy with standardized procedures for minimal contamination and stressors for sperm (AASV 2003). However, new technology was introduced that semi-automates semen collection (Aneas et al. 2008). With this technology, boars are trained to mount a dummy and then the extended boar penis is inserted into a pressured device below the dummy and semen is collected without manual application of pressure. These systems are effective and fertility is similar and has helped reduce long-term physical muscle trauma in the hand of the stud collection technicians. In commercial studs, boars are collected every 5–7 days to maximize sperm collection efficiency. Most mature boars can produce 80–100 billion sperm per ejaculate which is used to produce 27 doses with ~3 billion sperm. In some locations in Europe, doses are now standardized at 1.5 billion, where selected sires producing the same number of sperm results in ~50 doses per ejaculate (Berkvens et al. 2012). In general, ejaculates must pass quality control with >70 % motile, live and normal sperm cells. Pooling of ejaculates is often performed in commercial market pig production and sometimes in multiplier farms, but not in genetic nucleus farms (Clutter 2009; Sonderman and Luebbe 2008).

Semen evaluation is performed on most studs for motility and concentration using microscopes, photometers, or computer assisted semen analysis (CASA). Use of CASA is commonly used in studs with >50 boars, and in the larger production companies in North America (Knox et al. 2008) and Europe (Feitsma et al. 2011). There are several CASA systems available which can perform single or multiple functions. Some perform motility, concentration, morphology, and viability. These machines are fast and can evaluate hundreds of sperm in only seconds. They have been shown to be accurate but require technician training and are subject to variation with improper calibration and use. To date, the CASA systems and most measures can identify subfertile doses, but only a few studies have been able to relate sperm measures to fertility (Feitsma et al. 2011; Didion 2008). Ejaculate discard occurs ~5 % of the time due to poor motility, abnormalities, bacteria, and low concentration of sperm (Sonderman and Luebbe 2008). The majority of discards are also associated with an individual boar or a particular season. Most studs report maintaining quality control samples following extension for evaluation during storage in case of a fertility problem at the sow farms (Knox et al. 2008). There has has also been greater use of third party audits for semen analysis, quality control to help prevent and solve fertility problems for boar studs and sow farms (Althouse 2012).

In the USA, semen doses can range from two to four billion live sperm in volumes of 60–100 mL (Knox et al. 2008; Reicks and Levis 2008). Domestic semen delivery is targeted for next day arrival by designated personnel or commercial courier. Most semen is preserved in the USA using moderate to long-term extenders for use within 4 days of collection (Knox et al. 2008; Levis 2000; Vyt et al. 2004). In Europe, semen is often collected and extended in short-term extenders and delivered the same or next day and used within 3 days of collection. Semen shipping has changed little but there has been more attention on temperature changes in the semen container during shipping using data recorders to identify semen that has been heat or cold shocked in transit.

Ovulation Induction and Fixed Time AI

The higher fertility currently achieved by modern swine breeding herds has allowed a new focus on ovulation induction and fixed time AI to help capture genetic gain, fertility, and labor efficiency. Ovulation induction can reduce the incidence of poor AI timing in response to variation in time of ovulation among a group of weaned sows. Further, use of fixed time AI will facilitate single and low dose AI to increase use of high indexing sires and improve rates of gene transfer. When using ovulation induction and fixed time AI, estrous detection is eliminated. Ovulation induction can begin with a group of sows weaned on the same day or following the last feeding of Matrix® with an ovulation induction hormone administered to all females at the same time 4–6 days later. The result is that a batch of females that would normally express estrus and ovulate over a 5-day period will now ovulate within 24–36 h of each other. This sets up females for AI at an optimal time for fertilization. This has been shown to be effective with different hormones in multiple and single AI systems. The ovulation induction hormones include hCG, porcine LH,

GnRH, and GnRH agonists (Huhn et al. 1996; Degenstein et al. 2008; Brussow et al. 1996, 2009; Nissen et al. 1995, 2000; Webel and Day 1982). While most of the ovulation drugs are injectable and not approved for use in the USA, a new product OvuGel® (JBS United Animal Health, Sheridan, IN) is administered as an intra-vaginal gel and recently received FDA approval for use in swine. The use of this technology could have a dramatic impact on boar stud semen production, AI technology, farrowing labor, genetic improvement, and animal flow (Knox et al. 2011; Stewart et al. 2010; Taibl et al. 2008a, b). It is expected, that as the technology develops, single sire matings will be used leading to rapid identification of boar fertility and genetic value. Hormonal estrus induction protocols followed by ovulation induction protocols have been developed for use in sows in Europe but not for use in the USA. In these combination protocols, females are given 750–1,500 international units (IU) of PMSG or PG600® at weaning followed by hCG, porcine LH, or GnRH 72–96 h later (Huhn et al. 1996). This approach helps limit infertility for fixed time AI following ovulation induction. In most cases, females that are inseminated within 24 h following ovulation induction have acceptable conception rates but it would appear much more work in this area is still required.

Intrauterine and Deep Uterine Insemination

In the last 15 years efforts have been directed towards reducing the number of sperm used for AI to improve the rate of genetic transfer from superior sires. The technologies developed include IUI, DUI, and low dose AI. The IUI approach was reported to be practical for lowering the sperm dose from three to two billion sperm (Watson and Behan 2002), reducing leakage, volume required and time for insemination. This technology is currently being used in a double AI system and is becoming more popular with support from selected genetic suppliers who are able to provide alternative packaging for their customers to access their higher indexing sires. Unfortunately, at the present time the IUI rods have had limited success for passing through the cervix in young sows and gilts but new studies have shown promise for use in parity 1 sows. The DUI technology was developed to aid in use of very low numbers of sperm (millions instead of billions) and relies on the passage of a sterile and disposable, long, flexible catheter deep into one of the uterine horns. For DUI, similar to IUI, passage of a stabilizing rod into the cervix is required and is followed by insertion of an interior catheter deep into a uterine horn for insemination. This type of technique is highly specialized and should be practiced only by trained and experienced personnel. Although it has had limited commercial application due to risks of uterine damage, lowered fertility, and infection in commercial barns, good fertility it has been reported in research studies when using ultra-low doses of sperm in liquid form, with frozen sperm, and with sex-sorted sperm (Bathgate et al. 2008; Martinez et al. 2001, 2002, 2004, 2006; Mozo-Martín et al. 2012; Roca et al. 2003; Vazquez et al. 2005, 2008). This technology appears highly specialized and may not be practical for routine use in commercial breeding barns but might be targeted for nucleus farms.

AI with Frozen Boar Sperm

Widespread use of frozen-thawed boar sperm (FTS) for commercial use has been limited by reduced fertility. Although most species show sperm damage following freezing and thawing, the technology in swine reduces the number of AI doses produced per boar, lowers pregnancy rates and litter sizes compared to use of liquid semen (Johnson 1985). These problems have not prevented the development of the technology, but for the most part, application has been for international shipment, niche markets and for research. However, it is important to note, that improvements in boar fertility (Safranski et al. 2011), freezing procedures, and AI techniques, have resulted in acceptable fertility outcomes and have expanded the potential for commercial application (Knox 2011). Despite the limitations of FTS, there are potential advantages for extended shelf-life from collection to AI, for enhanced testing of semen for disease agents, long-distance shipping, and for greater precision of use in AI with highly valued genetic material. Use of FTS has also been implemented for long-term genetic banking for possible traits of interest in the future, for breed and line genetic diversity, and cryo-banking sperm in the event of catastrophic disease outbreak (Purdy 2008).

Boar sperm in liquid form have a 24 h fertile life following AI, while FTS may survive only up to 12 h or less (Waberski et al. 1994). Most studies using FTS inseminate with two to three billion live sperm/dose with two inseminations occurring during estrus (Spencer et al. 2010; Didion et al. 2012; Eriksson et al. 2002; Hofmo and Grevle 2000; Martin et al. 2000). The fertility limitations are not entirely related to the percentage of live sperm, but may be related to the physiological status of the sperm cell as well. The cryopreservation process has been shown to alter sperm membranes and organelles and cause capacitation-like changes in the cell. Most damage is thought to occur as a result of dehydration and the formation of ice crystals during the cooling stage. The rate at which sperm pass through the critical temperature stage of −5 to −50 °C determines whether the cells remain in equilibrium with the media or become super-cooled and form ice crystals. In most freezing solutions, glycerol is used to increase the osmolarity of the solution which allows water to exit the sperm cell while glycerol enters. During thawing, the process is reversed, and osmotic stress occurs resulting in membrane damage that alters regulation of sperm cell volume, enzyme activity, receptor structure, and cell permeability. Each of these has been related to the capacitation and acrosome reaction processes which are required for fertilization.

The process for freezing semen is not complex, but has time sensitive steps. It involves collection of the sperm rich fraction followed by dilution at 1:1 to 1:4 in standard extender and evaluation for motility >70 %. The solution is slow-cooled to 17 °C for 4 h and then is centrifuged for 10 min at 800×*g* to pellet the sperm and remove the seminal plasma. A protective cooling media containing egg-yolk and other membrane protectants are added to resuspend the sperm at one to two billion sperm/mL which is then cooled to 5 °C for 3 h. Then sperm are diluted to a final concentration of 0.5 to 1 billion cells/mL with the freezing media to obtain a final glycerol concentration of 3–4 %. This solution is placed in the packaging system (straws or packs) and cooled to −5 °C at −6 °C/min, then to −80 °C at 40 °C/min, and then to −150 °C at 70 °C/min. The packages are then placed into liquid nitrogen at −196 °C for storage. Thawing occurs rapidly in 37–70 °C water baths for 5–50 s depending upon temperature. Immediately following thawing, the semen is diluted in extender and held at 20–26 °C for AI within 1–2 h. For evaluation purposes, FTS semen is held at 37 °C for 20 min before evaluation under a microscope to simulate survival in the reproductive tract (Johnson et al. 2000; Spencer et al. 2010; Rath et al. 2009a).

Evaluation for FTS includes multiple measures for motility (>40 %), viability (>50 %), and intact acrosomes (>60 %). The type of packaging continues to evolve and while all are focused on fertility, practicality has gained more attention. Practical packaging for simplicity of use must balance the impact of volume and cooling rate on fertility. Small freezing volumes can improve post-thaw fertility but becomes cumbersome to handle multiple packages to make up a single dose. The 5 mL maxi-straw holds a single dose of five billion sperm, while the smaller straws typically hold 0.25 to 1 billion sperm per straw and require multiple straws for a single dose. Fertility of frozen sperm can be comparable to liquid semen when AI occurs with four billion live sperm within 4 h before ovulation (Waberski et al. 1994). With all of the advances for production and use of FTS, it is not uncommon to find reports of pregnancy rates exceeding 70 % and with litter sizes ≥10 pigs (Spencer et al. 2010; Didion et al. 2012; Eriksson et al. 2002; Martin et al. 2000). It is likely that the advantages for accessing valued genetics, and the benefits associated with those traits will likely drive this technology forward. Recent improvements with low dose insemination using IUI or DUI, fertility enhancers, and ovulation induction with fixed time AI, will open new doors for advancing genetic progress with use of frozen boar sperm.

Sexed Semen

Sexed semen has been of great interest to the swine breeding industry for many years. Depending upon the goals of the breeder, production of an entire litter of boars or gilts may be advantageous. In most litters of pigs, there is large individual variation in the numbers of males to females, but overall the average percentage of boars to gilts is close to 50 %. The only method to date that significantly changes the sex ratio of a litter is when boar sperm are pre-sorted based on the differences in the DNA content in sperm. This method is called the Beltsville Sperm Sexing Technology, and was developed by the USDA in Beltsville, MD. The technology takes advantage of the fact that the male Y chromosome is smaller than the female X chromosome. The sperm cells of males may carry either the X or the Y chromosomes. By sorting sperm cells into Y and X carrying sperm, mature females can be inseminated with sperm containing only the desired sex-determining chromosome. Although the cell-sorting system is highly advanced, the speed at which sperm flow, affects the sort accuracy without excessive damage to the cells. Studies indicate that low flow allows for 2.5 million sperm to be sorted/h, while modified high flow rates can allow 6 million or more/h. Higher flow rates with changes in the system now allow up to 18 million sperm to be sorted/h, but with lower sort accuracy (Johnson 2000). Overall, sorting can require 2–4 days to obtain a conventional AI dose of two billion viable sperm cells of the proper sex (Rath et al. 2009b). The process begins with collection of a boar ejaculate and a small portion of this is treated with stains that bind to DNA and also allows dead sperm to be sorted out. The stained sperm sample is then passed through a cell sorter, which passes an argon laser beam into a stream of the sample to allow the stain in the sperm cells to fluoresce. The equipment is programmed to detect slight differences in the live cells between the lower amount of fluorescence of the Y chromosome bearing sperm cells and the higher fluorescence of the sperm cells carrying the X chromosome. The fluorescence detector is integrated with a nozzle system that directs sperm into separate tubes. This procedure has resulted in 88 % pure populations of the X and Y sorted sperm cells. Litters resulting from inseminations occurring with the sorted sperm have been 97 % of only one sex. Although the prospects are exciting, there are several limitations to the practical use of sex-sorted sperm. One is the need to have semen in frozen form, since liquid extended sorted sperm has fertility limitations of only days. The second limitation is based on the current AI technology in pigs, which uses three billion sperm cells in a double insemination to achieve high farrowing rates and large litter size. There are however, AI methods, such as DUI and ovulation induction, which reduces the requirement for high numbers of sperm and multiple inseminations to allow pregnancy to occur with millions of sorted sperm, instead of billions of sperm (Vazquez et al. 2005).

Real-Time Ultrasound Use for Reproductive Diagnosis

Real-time ultrasound was introduced to the commercial swine industry in the mid 1990s as a method to improve the accuracy of pregnancy diagnosis in swine. The machines that were developed were light, portable and became affordable, and cost effective, especially in larger farms. The technology greatly enhanced the accuracy of diagnosis compared to examination for returns to estrus and helped to reduce nonproductive days (Knox and Flowers 2006; Flowers and Knox 2001). The equipment was quickly adopted and showed great practicality and accuracy for diagnosing larger numbers of animals in breeding group at similar days following breeding in a short period of time. Currently there are numerous machines on the market, and this technology has been widely adopted in most farms with 1,000 or more sows. Most of the larger farms can easily justify the added costs by reducing nonproductive days (days when a sow is not lactating or gestating) and realize the value of identifying nonpregnant and pregnant animals at 4 weeks after breeding. The machines have also been tested for diagnosing pregnancy failures throughout gestation. From a research perspective the use of ultrasound has been important for determining ovarian and uterine status and the time and occurrence of ovulation (Soede et al. 1992; Knox and Althouse 1999; Kauffold et al. 2004). Although not for routine farm use, this technology has helped improve AI timing in weaned sows and gilts and has helped identify the causes of female reproductive failures. It has also facilitated the development of gonadotropin use, ovulation induction, and helped advance fixed time AI.

Farrowing Control

While attention has been focused on achieving high pregnancy rates and large litter sizes, stillborn pigs continues as a persistent problem especially with larger litters. Records from available databases suggest stillborns can range from 5 to 10 % (PigCHAMP 2011). With the steady increase in litter size over years (USDA 2012), there has been a trend for increased stillbirths, and lower birth weight pigs each of which contribute to reduced number of weaned pigs. Intervention is often prescribed if stillbirths are above 5 %. It is also known that attended farrowing can save 0.5–1 pig/sow/year as a result of assisting with delivery, removal of amnion membranes to clear piglet airways, prevention of chilling, and crushing of piglets (PIH 2007; Morrow 1986). However, while many sows start farrowing at night, other start in the daytime, and may proceed into the evening. The end result is that many farrowings or part of the farrowing process are unattended and pigs that could be saved are lost. Technology developed to induce farrowing was intended to ensure that most of the farrowings occurred during the daytime when personnel were present. The procedures were first developed in the 1970s and were advanced and used extensively in some countries in Europe (Huhn et al. 1996). In the USA, while most farms do not induce all sows, many farms induce some sows, especially if they may farrow over a weekend or holiday. Further, there are some farms that induce all sows and use this as a routine procedure. The technique has not had wider adoption due to the variation in the synchrony of farrowing. With the administration of prostaglandin, farrowing may begin within 12–36 h for most sows. The majority of sows actually will farrow during the working day, but as might be expected, some begin early and some start later. Part of the variation in the response may be explained by variation in the actual gestation length of sows which averages 115 days following first service. However, this can often vary as much as ±2 days with delay in farrowing far more common. As a result, improved procedures for administration of prostaglandin and use of oxytocin have been developed (Kaeoket 2006). The reasons for the variation in induction are not clear but could be related to actual day of conception instead of service. There are currently two approaches for farrowing control with the first involving administration of the luteolysin in single or split injections in either muscle or in the vulva. The use of the natural form of prostaglandin or its analogues has each been effective. The luteolysin causes regression of the corpus luteum and clearance of progesterone from circulation. Once this occurs, uterine contractions will be initiated (Guthrie 1985). The current recommendation is for sows to be induced on day 114 and not before as early induction reduces piglet survival. Many farms also include administration of oxytocin the next day for sows that are not yet farrowing. An alternative procedure that is not approved in the USA is the delay of early farrowing with short-term feeding of a synthetic progestagen starting on day 111 and continuing until day 113 to avoid weekend farrowings. This has been tested for short-term delay and to synchronize farrowing in batches of sows (Huhn et al. 1996; Guthrie 1985; Gaggini et al. 2013). Longer delays appear harmful to pig survival. The farms that employ synchronized farrowing use it to improve the synchrony of weaned pigs, facilitate early cross-fostering, avoid weekend farrowing, improve supervised farrowing, and improve piglets born alive.

Conclusions

The increase in the value of multiple genetic traits for efficient pig production has been a significant driver for the adoption of new reproductive technologies around the world. Technologies such as AI have revolutionized how breeding herds are managed and has facilitated increases in herd size to help producers capture the economies of scale. While reproductive advancements such as the implementation of boar studs, semen analysis, AI, and IUI were initiated in developed countries, adoption and use of some of these has occurred through collaborative efforts with scientists and producers in developing countries. Collectively, these technologies have helped producers worldwide improve the efficiency of production through genetics, management, disease control, and economic success. The end result is that greater amounts of quality pork can be produced when using selected reproductive technologies for production of pigs and pork.

References

AASV (2003) Boar stud guidelines: health, hygiene, and sanitation guidelines for boar studs providing semen to the domestic market. J Swine Health Prod 11:204–206

Allan GM, Ellis JA (2000) Porcine circoviruses: a review. J Vet Diagn Invest 12(1):3–14

Almeida FR, Novak S, Foxcroft GR (2000) The time of ovulation in relation to estrus duration in gilts. Theriogenology 53(7):1389–1396

Althouse GC (2012) How have third-party quality assurance programs been applied to boar studs? In: Safranski T (ed) Midwest boar stud manager's conference IV, St. Louis, MO

Althouse GC, Lu KG (2005) Bacteriospermia in extended porcine semen. Theriogenology 63(2):573–584

Althouse GC, Rossow K (2011) The potential risk of infectious disease dissemination via artificial insemination in swine. Reprod Domest Anim 46:64–67

Althouse GC, Kuster CE, Clark SG, Weisiger RM (2000) Field investigations of bacterial contaminants and their effects on extended porcine semen. Theriogenology 53(5):1167–1176

Aneas SB, Gary BG, Bouvier BP (2008) Collectis® automated boar collection technology. Theriogenology 70(8):1368–1373

Barb CR, Hausman GJ, Kraeling RR (2010) Luteinizing hormone secretion as influenced by age and estradiol in the prepubertal gilt. Anim Reprod Sci 122(3–4):324–327

Bartlett A, Pain SJ, Hughes PE, Stott P, van Wettere WHEJ (2009) The effects of PG600 and boar exposure on oestrus detection and potential litter size following mating at either the induced (pubertal) or second oestrus. Anim Reprod Sci 114(1–3):219–227

Bates RO, Day BN, Britt JH, Clark LK, Brauer MA (1991) Reproductive performance of sows treated with a combination of pregnant mare's serum gonadotropin and human chorionic gonadotropin at weaning in the summer. J Anim Sci 69(3):894–898

Bates RO, Kelpinski J, Hines B, Ricker D (2000) Hormonal therapy for sows weaned during fall and winter. J Anim Sci 78(8):2068–2071

Bathgate R, Eriksson BM, Thomson PC, Maxwell WM, Evans G (2008) Field fertility of frozen-thawed boar sperm at low doses using non-surgical, deep uterine insemination. Anim Reprod Sci 103(3–4):323–335

Belstra BA, Flowers WL, See MT (2004) Factors affecting temporal relationships between estrus and ovulation in commercial sow farms. Anim Reprod Sci 84(3–4):377

Beltranena E, Patterson JL, Foxcroft GR (2005) Designing effective boar stimulation systems as a critical feature of the gilt development unit. In: Allen D. Leman swine conference pre-conference workshop, College of Veterinary Medicine, University of Minnesota, St. Paul, MN

Bennett-Steward K, Aramini J, Pelland C, Friendship R (2008) Equine chorionic gonadotrophin and porcine luteinizing hormone to shorten and synchronize the wean-to-breed interval among parity-one and parity-two sows. J Swine Health Prod 16:182–187

Berkvens P, Feitsma H, Eggert J (2012) Low concentration, high fertility; every day Dutch AI practice. In: Safranski T (ed) Miwest boar stud manager's conference IV, St. Louis, MO

Bolet G, Bidanel JP, Ollivier L (2001) Selection for litter size in pigs. II. Efficiency of closed and open selection lines. Genet Sel Evol 33(5):515–528

Bracken CJ, Safranski TJ, Cantley TC, Lucy MC, Lamberson WR (2003) Effect of time of ovulation and sperm concentration on fertilization rate in gilts. Theriogenology 60(4):669–676

Breen SM, Knox RV (2012) The impact of dose of FSH (folltropin) containing LH (lutropin) on follicular development, estrus and ovulation responses in prepubertal gilts. Anim Reprod Sci 132(3–4):193–200

Breen SM, Farris KL, Rodriguez-Zas SL, Knox RV (2005) Effect of age and physical or fence-line boar exposure on estrus and ovulation response in prepubertal gilts administered PG600. J Anim Sci 83:460–465

Britt JH (1986) Improving sow productivity through management during gestation, lactation and after weaning. J Anim Sci 63(4):1288–1296

Britt JH, Day BN, Webel SK, Brauer MA (1989) Induction of a fertile estrus in prepuberal gilts by treatment with a combination of pregnant mare's serum gonadotropin and human chorionic gonadotropin. J Anim Sci 67:1148–1153

Brussow KP, Jochle W, Huhn U (1996) Control of ovulation with a GnRH analog in gilts and sows. Theriogenology 46(6):925

Brussow KP, Schneider F, Kanitz W, Ratky J, Kauffold J, Wahner M (2009) Studies on fixed-time ovulation induction in the pig. Soc Reprod Fertil Suppl 66:187–195

Burke P (2000) Productivity assessment of liquid boar semen usage. In: Johnson LA, Guthrie HD (eds) Boar semen preservation IV. Allen Press, Lawrence, KS, p 149

Caton JS, Jesse GW, Day BN, Ellersieck MR (1986) The effect of duration of boar exposure on the frequency of gilts reaching first estrus. J Anim Sci 62:1210–1214

Clutter A (2009) Genetic selection for lifetime reproductive performance. Soc Reprod Fertil Suppl 66:293–302

Degenstein KL, O'Donoghue R, Patterson JL, Beltranena E, Ambrose DJ, Foxcroft GR, Dyck MK (2008) Synchronization of ovulation in cyclic gilts with porcine luteinizing hormone (pLH) and its effects on reproductive function. Theriogenology 70(7):1075–1085

den Hartog L (2004) Developments in global pig production. Adv Pork Prod 15:17–24

Dial GD, Marsh WE, Polson DD, Vaillancourt JP (1992) Reproductive failure: differential diagnosis. In: Leman AD et al (eds) Diseases of swine, 7th edn. Iowa State University Press, Ames, IA, pp 88–137

Didion BA (2008) Computer-assisted semen analysis and its utility for profiling boar semen samples. Theriogenology 70(8):1374–1376

Didion BA, Braun G, Duggan M (2012) Field fertility of frozen boar semen: a retrospective report comprising over 2700 AI services spanning a four year period. In: Safranski T (ed) Midwest boar stud manager's conference IV, St. Louis, MO

Dyck MK, Foxcroft GR, Novak S, Ruiz-Sanchez A, Patterson J, Dixon WT (2011) Biological markers of boar fertility. Reprod Domest Anim 46:55–58

Ehlers MJ, Mabry JW, Bertrand JK, Stalder KJ (2005) Variance components and heritabilities for sow productivity traits estimated from purebred versus crossbred sows. J Anim Breed Genet 122(5):318–324

Engblom L, Lundeheim N, Strandberg E, Schneider P, Dalin AM, Andersson K (2008) Factors affecting length of productive life in Swedish commercial sows. J Anim Sci 86(2):432–441

Eriksson BM, Petersson H, Rodriguez-Martinez H (2002) Field fertility with exported boar semen frozen in the new flatpack container. Theriogenology 58:1065–1079

Estienne MJ, Harper AF, Horsley BR, Estienne CE, Knight JW (2001) Effects of P.G. 600 on the onset of estrus and ovulation rate in gilts treated with regumate. J Anim Sci 79(11):2757–2761

FAOSTAT (2009) World pig meat production: food and agriculture organization of the United Nations. http://faostat.fao.org/site/569/DesktopDefault.aspx?PageID=569. Accessed 28 March 2013

FAS (2006) Worldwide pork production. USDA Foreign Agricultural Service. http://www.fas.usda.gov/

Faust MA, Tess MW, Robison OW (1992) A bioeconomic simulation model for a hierarchical swine breeding structure. J Anim Sci 70(6):1760–1774

Feitsma H, Broekhuijse M, Gadella BM (2011) Do CASA systems satisfy consumers demands? A critical analysis. Reprod Domest Anim 46:49–51

Fernandez L, Diez C, Ordonez J, Carbajo M (2005) Reproductive performance in primiparous sows after postweaning treatment with a progestagen. J Swine Health Prod 13:28–30

Flowers WL (1996) Performance expectations of different mating systems. Proc Allen D Leman Swine Conf 25:63–66

Flowers WL (1997) Management of boars for efficient semen production. J Reprod Fertil Suppl 52:67–78

Flowers WL (1998) Insemination programs for swine to increase fertility. J Anim Sci 76(Suppl 3):39–46

Flowers WL (2000) Influence of adjusting timing and frequency of mating on the anticipated duration of estrus on reproductive performance of sows. Department of Animal Sciences Report No. 248. http://www.ncsu.edu/project/swine_extension/swinereports/2000/flowers2.htm. Accessed 26 Jan 2012

Flowers WL (2002) Increasing fertilization rate of boars: influence of number and quality of spermatozoa inseminated. J Anim Sci 80(Suppl 1):E47–E53

Flowers WL (2008) Genetic and phenotypic variation in reproductive traits of AI boars. Theriogenology 70(8):1297–1303

Flowers WL, Alhusen HD (1992) Reproductive performance and estimates of labor requirements associated with combinations of artificial insemination and natural service in swine. J Anim Sci 70(3):615–621

Flowers WL, Esbenshade KL (1993) Optimizing management of natural and artificial matings in swine. J Reprod Fertil 48:217–228

Flowers W, Knox R (2001) Pregnancy diagnosis in swine. Pork industry handbook. N.C. Cooperative Extension Service, Raleigh, NC

Foxcroft GR (2001) Gilt management for the new millennium-research to reality. In: Manitoba swine seminar, Manitoba

Foxcroft GR, Almeida F, Patterson JL, Willis HJ. (2001) Age and weight at puberty in relation to lifetime performance. In: Proceedings of the 6th international conference on pig reproduction, Columbia

Foxcroft GR, Dyck MK, Ruiz-Sanchez A, Novak S, Dixon WT (2008) Identifying useable semen. Theriogenology 70(8):1324–1336

Gaggini TS, Perin J, Arend LS, Bernardi ML, Wentz I, Bortolozzo FP (2013) Altrenogest treatment associated with a farrowing induction protocol to avoid early parturition in sows. Reprod Domest Anim 48:390–395

Gerrits RJ, Lunney JK, Johnson LA, Pursel VG, Kraeling RR, Rohrer GA, Dobrinsky JR (2005) Perspectives for artificial insemination and genomics to improve global swine populations. Theriogenology 63(2):283–299
Gerritsen R, Langendijk P, Soede NM, Kemp B (2005) Effects of (artificial) boar stimuli on uterine activity in estrous sows. Theriogenology 64(7):1518–1525
Guerin B, Pozzi N (2005) Viruses in boar semen: detection and clinical as well as epidemiological consequences regarding disease transmission by artificial insemination. Theriogenology 63(2):556–572
Guthrie HD (1985) Control of time of parturition in pigs. J Reprod Fertil Suppl 33:229–244
Harmon J, Levis D, Zulovich J, Hoff S, Bodman G (2001) Swine breeding and gestation facilities handbook. MidWest plan service, vol 43. Iowa State University, Ames, IA
Harris DL (2000) Multi-site pig production. Iowa State University Press, Ames, IA
Harris DL, Lofgren DL, Stewart TS, Schinckel AP (1989) Adapting best linear unbiased prediction (BLUP) for timely genetic evaluation: I. Progeny traits in a single contemporary group for each sex. J Anim Sci 67(12):3209–3222
Hofmo PO, Grevle IS (2000) Development and commercial use of frozen use of frozen boar semen in Norway. In: Johnson LA, Guthrie HD (eds) Boar semen preservation IV. Allen Press, Lawrence, KS, pp 71–86
Horsley BR, Estienne MJ, Harper AF, Purcell SH, Baitis HK, Beal WE, Knight JW (2005) Effect of P.G. 600 on the timing of ovulation in gilts treated with altrenogest. J Anim Sci 83(7):1690–1695
Houska L, Wolfová M, Fiedler J (2004) Economic weights for production and reproduction traits of pigs in the Czech Republic. Livest Prod Sci 85(2–3):209–221
Hughes PE (2001) Factors affecting gilt age and liveweight at puberty. In: Proceedings of the 6th international conference on pig reproduction, Columbia
Hughes PE, Pearce GP, Paterson AM (1990) Mechanisms mediating the stimulatory effects of the boar on gilt reproduction. J Reprod Fertil Suppl 40:323–341
Huhn U, Jochle W, Brussow KP (1996) Techniques developed for the control of estrus, ovulation and parturition, in the East German pig industry: a review. Theriogenology 46:911–924
Jackson AL, Breen SM, Rodriguez-Zas SL, Knox RV (2006) Evaluation of methodology for administration of porcine FSH for use in estrus induction and for increasing ovulation rate in prepubertal gilts. Theriogenology 66(4):1042–1047
Johnson LA (1985) Fertility results using frozen boar spermatozoa; 1970-1985. In: Johnson LA Larsson K (eds) Proceedings of the first international conference on deep freezing of boar semen, Uppsala, pp 199–222
Johnson LA (2000) Sexing mammalian sperm for production of offspring: the state-of-the-art. Anim Reprod Sci 60–61:93–107
Johnson LA, Weitze KF, Fiser P, Maxwell WMC (2000) Storage of boar semen. Anim Reprod Sci 62:143–172
Jones GF (1998) Genetic aspects of domestication, common breeds and their origin. In: Rothschild MF, Ruvinsky A (eds) The genetics of the pig. C.A.B. International, Wallingford
Kaeoket K (2006) The effect of dose and route of administration of R-cloprostenol on the parturient response of sows. Reprod Domest Anim 41(5):472–476
Kauffold J, Rautenberg T, Richter A, Waehner M, Sobiraj A (2004) Ultrasonographic characterization of the ovaries and the uterus in prepubertal and pubertal gilts. Theriogenology 61(9):1635–1648
Kemp B, Soede NM (1996) Relationship of weaning-to-estrus interval to timing of ovulation and fertilization in sows. J Anim Sci 74(5):944–949
Key N (2007) The changing economics of U.S. hog production. [Economic research report no. 52, economic research service]. United States Department of Agriculture, Washington, DC
Kirkwood RN, Aherne FX, Monaghan PG, Misutka SC (2000) Breeding gilts at natural or a hormone-induced estrus: effects on performance over four parities. J Swine Health Prod 8:177–179
Knauer M, Stalder KJ, Serenius T, Baas TJ, Berger PJ, Karriker L, Goodwin RN, Johnson RK, Mabry JW, Miller RK, Robison OW, Tokach MD (2010) Factors associated with sow stayability in 6 genotypes. J Anim Sci 88(11):3486–3492
Knox RV (2004) The real impact of boars in breeding programs. Adv Pork Prod 15:307–314
Knox RV (2011) The current value of frozen-thawed boar semen for commercial companies. Reprod Domest Anim 46:4–6
Knox RV, Althouse GC (1999) Visualizing the reproductive tract of the female pig using real-time ultrasound. J Swine Health Prod 7:207–215
Knox R, Flowers W (2006). Real time ultrasound for pregnancy diagnosis in swine. Pork Information Gateway. http://www.porkgateway.org
Knox R, Probst-Miller S (2004) Evaluation of transrectal real-time ultrasound for use in identifying sources of reproductive failure in weaned sows. J Swine Health Prod 12:71–74
Knox RV, Rodriguez-Zas SL (2001) Factors influencing estrus and ovulation in weaned sows as determined by transrectal ultrasound. J Anim Sci 79(12): 2957–2963
Knox RV, Wilson WD (2007) Induction of estrus and control of the estrous cycle in swine. In: Youngquist RS, Threlfall WR (eds) Current therapy in large animal theriogenology, 2nd edn. Saunders Elsevier, St. Louis, MO, pp 757–764
Knox RV, Rodriguez-Zas SL, Tudor KW, Robb JA (2000) Effect of subcutaneous or intramuscular administration of PG600 on estrus and ovulatory responses in prepubertal gilts. J Anim Sci 78:1732–1737
Knox RV, Rodriguez-Zas SL, Miller GM, Willenburg KL, Robb JA (2001) Administration of P.G. 600 to sows at weaning and the time of ovulation as determined by transrectal ultrasound. J Anim Sci 79(4):796–802

Knox RV, Miller G, Willenburg KL, Rodriguez-Zas SL (2002) Effect of frequency of boar exposure and adjusted mating times on measures of reproductive performance in weaned sows. J Anim Sci 80:892–899

Knox RV, Breen SM, Willenburg KL, Roth S, Miller GM, Ruggiero KM, Rodriguez-Zas S (2004) Effect of housing system and boar exposure on estrus expression in weaned sows. J Anim Sci 82:3088–3093

Knox R, Levis D, Safranski T, Singleton W (2008) An update on North American boar stud practices. Theriogenology 70(8):1202–1208

Knox RV, Willenburg KL, Rodriguez-Zas SL, Greger DL, Hafs HD, Swanson ME (2011) Synchronization of ovulation and fertility in weaned sows treated with intravaginal triptorelin is influenced by timing of administration and follicle size. Theriogenology 75(2):308–319

Knox RV, Rodriguez Zas SL, Sloter NL, McNamara KA, Gall TJ, Levis DG, Safranski TJ, Singleton WL (2013) An analysis of survey data by size of the breeding herd for the reproductive management practices of North American sow farms. J Anim Sci 91(1):433–445

Koketsu Y (2007) Longevity and efficiency associated with age structures of female pigs and herd management in commercial breeding herds. J Anim Sci 85(4):1086–1091

Koketsu Y, Dial GD (1997) Factors influencing the postweaning reproductive performance of sows on commercial farms. Theriogenology 47:1445–1461

Konig S, Simianer H, Willam A (2009) Economic evaluation of genomic breeding programs. J Dairy Sci 92(1):382–391

Lamberson WR, Safranski TJ (2000) A model for economic comparison of swine insemination programs. Theriogenology 54(5):799–807

Langendijk P, van den Brand H, Soede NM, Kemp B (2000a) Effect of boar contact on follicular development and on estrus expression after weaning in primiparous sows. Theriogenology 54(8):1295–1303

Langendijk P, Soede NM, Bouwman EG, Kemp B (2000b) Responsiveness to boar stimuli and change in vulvar reddening in relation to ovulation in weaned sows. J Anim Sci 78(12):3019–3026

Langendijk P, Soede NM, Kemp B (2000c) Effects of boar contact and housing conditions on estrus expression in weaned sows. J Anim Sci 78(4):871–878

Langendijk P, Soede NM, Kemp B (2005) Uterine activity, sperm transport, and the role of boar stimuli around insemination in sows. Theriogenology 63(2): 500–513

Lawrence JD, Grimes G (2006) Production and marketing characteristics of U.S. pork producers, 2006. Farm marketing-AgEBB. Iowa State University Department of Economics, Ames, IA, pp 1–22

Levis DL (1997a) Management of replacement gilts for efficient reproduction. In: Cooperative extension, University of Nebraska, Lincoln, NE, pp 1–11

Levis D (1997b) EC97-275 Effect of lactation length on sow reproductive performance. University of Nebraska Extension, Lincoln, NE, pp 1–13

Levis D (2000) Liquid boar semen production: current extender technology and where do we go from here? In: Johnson LA, Guthrie HD (eds) Boar semen preservation IV. Allen Press, Lawrence, KS, pp 121–128

Levis DG (2005) Applying new technologies to optimize reproduction. Seminar: optimizing reproductive efficiency. In: Proceedings 36th American association of swine veterinarians annual meeting, Toronto

Lofgren DL, Harris DL, Stewart TS, Schinckel AP (1989) Adapting best linear unbiased prediction (BLUP) for timely genetic evaluation: II. Progeny traits in multiple contemporary groups within a herd. J Anim Sci 67(12):3223–3242

Love RJ, Evans G, Klupiec C (1993) Season effects on fertility in gilts and sows. J Reprod Fertil Suppl 48:191–206

Madsen KS (2005) Management of disease control and epidemics in AI in Denmark. Theriogenology 63(2):585–594

Maes D, Nauwynck H, Rijsselaere T, Mateusen B, Vyt P, de Kruif A, Van Soom A (2008) Diseases in swine transmitted by artificial insemination: an overview. Theriogenology 70(8):1337–1345

Martin M, Edgerton S, Wiseman B (2000) Frozen semen: a breeding protocol that results in high fecundity. J Swine Health Prod 8(6):275–277

Martinez EA, Vazquez JM, Roca J, Lucas X, Gil MA, Parrilla I, Vazquez JL, Day BN (2001) Successful nonsurgical deep intrauterine insemination with small numbers of spermatozoa in sows. Reproduction 122(2):289–296

Martinez E, Vazquez J, Roca J, Lucas X, Gil M, Parrilla I, Vazquez J, Day B (2002) Minimum number of spermatozoa required for normal fertility after deep intrauterine insemination in non-sedated sows. Reproduction 123(1):163–170

Martinez EA, Caamaño JN, Gil MA, Rieke A, McCauley TC, Cantley TC, Vazquez JM, Roca J, Vazquez JL, Didion BA, Murphy CN, Prather RS, Day BN (2004) Successful nonsurgical deep uterine embryo transfer in pigs. Theriogenology 61(1):137–146

Martinez EA, Vazquez JM, Parrilla I, Cuello C, Gil MA, Rodriguez-Martinez H, Roca J, Vazquez JL (2006) Incidence of unilateral fertilizations after Low dose deep intrauterine insemination in spontaneously ovulating sows under field conditions. Reprod Domest Anim 41(1):41–47

Merks JWM, Ducro-Steverink DWB, Feitsma H (2000) Management and genetic factors affecting fertility in sows. Reprod Domest Anim 35:261–266

Moeller SJ, Goodwin RN, Johnson RK, Mabry JW, Bass TJ, Robison OW (2004) The national pork producers council maternal line national genetic evaluation program: a comparison of six maternal genetic lines for female productivity measures over four parities. J Anim Sci 82:41–53

Moore C (2005) Parity segregation. In: Proceedings of the London swine conference. http://www.londonswineconference.ca/proceedings/…/LSC2005_CMoore.p

Morrow DA (1986) Current therapy in theriogenology, 2nd edn. Saunders, Philadelphia, PA

Mozo-Martín R, Gil L, Gómez-Rincón CF, Dahmani Y, García-Tomás M, Úbeda JL, Grandía J (2012) Use of a novel double uterine deposition artificial insemination technique using low concentrations of sperm in pigs. Vet J 193(1):251–256
Muirhead MR (1990) Reproductive failure in the sow. Vet Annu 30:92–102
Nissen AK, Lehn-Jensen H, Hyttel P, Greve T (1995) Follicular development and ovulation in sows: effect of hCG and GnRH treatment. Acta Vet Scand 36:123–133
Nissen AK, Soede NM, Hyttel P, Schmidt M, D'Hoore L (1997) The influence of time of insemination relative to time of ovulation on farrowing frequency and litter size in sows, as investigated by ultrasonography. Theriogenology 47:1571–1582
Nissen AK, Schmidt M, Hyttel P, Greve T (2000) Ovulation and embryonic developmental rate following hCG-stimulation in sows. Acta Vet Scand 41:321–328
Nodelijk G, Nielen M, De Jong MCM, Verheijden JHM (2003) A review of porcine reproductive and respiratory syndrome virus in Dutch breeding herds: population dynamics and clinical relevance. Prev Vet Med 60(1):37–52
NPPC (2012) Benefits of expanding U.S. pork exports. http://www.nppc.org/issues/international-trade/
NSR (2012) Swine testing and evaluation system. National Swine Registry. http://www.nationalswine.com/Genetic_Tech_pages/STAGES.html
Olesen I, Groen AF, Gjerde B (2000) Definition of animal breeding goals for sustainable production systems. J Anim Sci 78:570–582
Ollivier L (1998) Gentic improvement of the pig. In: Rothschild MF, Ruvinsky A (eds) The genetics of the pig. C.A.B. International, Wallingford
Paterson AM, Lindsay DR (1981) Induction of puberty in gilts. Anim Prod 32:51–54
Peña FJ, Domínguez JC, Carbajo M, Anel L, Alegre B (1998) Treatment of swine summer infertility syndrome by means of oxytocin under field conditions. Theriogenology 49(4):829–836
PIC (2008) Fundamentals of gilt and sow management, pp 1–32. http://www.pic.com
PIG Factsheets: Statistics (2012) Pork information gateway. http://www.porkgateway.org/PigLibrary/LT/Factsheets.aspx
PigCHAMP (1998) Breeding herd summary—USA. PigCHAMP Data Share
PigCHAMP (2001) http://www.pigchampinc.com/2001Datashare.htm. Accessed 6 Jan 2005
PigCHAMP (2011). Data summary. Benchmark.Farms. Com. http://www.pigchamp.com/LinkClick.aspx?fileticket=NMdM5F73gKE%3d&tabid=115
PIH (2007) Pork Industry Handbook TPI Purdue extension. Purdue University, West Lafayette, IN
Prieto C, Castro JM (2005) Porcine reproductive and respiratory syndrome virus infection in the boar: a review. Theriogenology 63(1):1–16
Purdy PH (2008) Swine gene banking: a quality control perspective on collection, and analysis of samples for a national repository. Theriogenology 70(8):1304–1309
Pursel VG, Schulman LL, Johnson LA (1978) Distribution and morphology of fresh and frozen-thawed sperm in the reproductive tract of gilts after artificial insemination. Biol Reprod 19(1):69–76
Rath D, Bathgate R, Rodriguez-Martinez H, Roca J, Strzezek J, Waberski D (2009a) Recent advances in boar semen cryopreservation. Soc Reprod Fertil Suppl 66:51–66
Rath D, Moench-Tegeder G, Taylor U, Johnson LA (2009b) Improved quality of sex-sorted sperm: a prerequisite for wider commercial application. Theriogenology 71(1):22–29
Reese D, Straw BE (2006) Paper 212: the case against evening-up litters until weaning. In: Nebraska swine report, University of Nebraska Extension, Lincoln, NE, pp 7–10
Reicks DL (2008) PRRS monitoring techniques and success stories. In: Proceedings of the midwest boar stud manager's conference II, St. Louis, MO
Reicks DL (2012) Interpretation of PRRS tests and Implications for stud management. In: Midwest boar stud manager's conference IV, St. Louis, MO
Reicks DL, Levis DG (2008) Fertility of semen used in commercial production and the impact of sperm numbers and bacterial counts. Theriogenology 70(8):1377–1379
Reisenbeck A (2011) Review on international trade with boar semen. Reprod Domest Anim 46(suppl 2):1–3
Roca J, Carvajal G, Lucas X, Vazquez JM, Martinez EA (2003) Fertility of weaned sows after deep intrauterine insemination with a reduced number of frozen-thawed spermatozoa. Theriogenology 60(1):77–87
Rodriguez-Martinez H (2003) Laboratory semen assessment and prediction of fertility: still utopia? Reprod Domest Anim 38(4):312–318
Rozeboom KJ, Troedsson MH, Shurson GC, Hawton JD, Crabo BG (1997) Late estrus or metestrus insemination after estrual inseminations decreases farrowing rate and litter size in swine. J Anim Sci 75(9):2323–2327
Ruiz-Sánchez AL, O'Donoghue R, Novak S, Dyck MK, Cosgrove JR, Dixon WT, Foxcroft GR (2006) The predictive value of routine semen evaluation and IVF technology for determining relative boar fertility. Theriogenology 66(4):736–748
Ruvinsky A, Rothschild MF (1998) Systematics and evolution of the pig. In: Rothschild MF, Ruvinsky A (eds) The genetics of the pig. C.A.B. International, Wallingford
Safranski TJ, Ford JJ, Rohrer GA, Guthrie HD (2011) Plenary contribution to international conference on boar semen preservation 2011. Genetic selection for freezability and its controversy with selection for performance. Reprod Domest Anim 46:31–34
Schinckel AP, Bennet G (1999) The economic impact of genetic improvement. Natl Swine Improv Fed 1(9/2006):1–8
Schukken YH, Buurman J, Huirne RB, Willemse AH, Vernooy JC, van den Broek J, Verheijden JH (1994)

Evaluation of optimal age at first conception in gilts from data collected in commercial swine herds. J Anim Sci 72(6):1387–1392
Singleton WL (2001) State of the art in artificial insemination of pigs in the United States. Theriogenology 56(8):1305–1310
Soede NM (1993) Boar stimuli around insemination affect reproductive processes in pigs: a review. Anim Reprod Sci 32:107–125
Soede NM, Noordhuizen JPTM, Kemp B (1992) The duration of ovulation in pigs, studied by transrectal ultrasonography, is not related to early embryonic diversity. Theriogenology 38:653–666
Soede N, Wetzels C, Zondag W, de Koning M, Kemp B (1995a) Effects of time of insemination relative to ovulation, as determined by ultrasonography, on fertilization rate and accessory sperm count in sows. J Reprod Fertil 104(1):99–106
Soede NM, Wetzels CC, Zondag W, Hazeleger W, Kemp B (1995b) Effects of a second insemination after ovulation on fertilization rate and accessory sperm count in sows. J Reprod Fertil 105(1):135–140
Sonderman JP, Luebbe JJ (2008) Semen production and fertility issues related to differences in genetic lines of boars. Theriogenology 70:1380–1383
Spencer KW, Purdy PH, Blackburn HD, Spiller SF, Stewart TS, Knox RV (2010) Effect of number of motile, frozen-thawed boar sperm and number of fixed-time inseminations on fertility in estrous-synchronized gilts. Anim Reprod Sci 121:259–266
Sporke J, Patterson JL, Beltranena E, Foxcroft GR (2005) Gilt development unit management using Matrix and PG600 in a commercial swine operation. In: Allen D. Leman swine conference, University of Minnesota, St. Paul, MN
Spötter A, Distl O (2006) Genetic approaches to the improvement of fertility traits in the pig. Vet J 172(2):234–247
Stalder K (2008) Parity's impact on productivity, Feb 15 genetics-reproduction. Natl Hog Farmer 53:28
Stalder KJ, Lacy RC, Cross TL, Conatser GE (2003) Financial impact of average parity of culled females in a breed-to-wean swine operation using replacement gilt net present value analysis. J Swine Health Prod 11:69–74
Steverink DW, Soede NM, Bouwman EG, Kemp B (1997) Influence of insemination-ovulation interval and sperm cell dose on fertilization in sows. J Reprod Fertil Suppl 111:165–171
Steverink DW, Soede NM, Bouwman EG, Kemp B (1998) Semen backflow after insemination and its effect on fertilisation results in sows. Anim Reprod Sci 54(2):109–119
Steverink DW, Soede NM, Groenland GJ, van Schie FW, Noordhuizen JP, Kemp B (1999) Duration of estrus in relation to reproduction results in pigs on commercial farms. J Anim Sci 77(4):801–809
Stewart KR, Flowers WL, Rampacek GB, Greger DL, Swanson ME, Hafs HD (2010) Endocrine, ovulatory and reproductive characteristics of sows treated with an intravaginal GnRH agonist. Anim Reprod Sci 120(1–4):112–119
Straw BE, D'Allaire WLMS, Taylor DJ (eds) (1999) Diseases of swine, 8th edn. Iowa State University Press, Ames, IA
Taibl JN, Breen SM, Webel SK, Knox RV (2008a) Induction of ovulation using a GnRH agonist for use with fixed time AI in weaned sows. Theriogenology 70:1400 (abstract)
Taibl JN, Breen SM, Webel SK, Swanson ME, Knox RV (2008) Effect of synchronizing ovulation in weaned sows using Ovugel with single fixed time AI on pregnancy rate and litter size. In: VIII International conference on pig reproduction, Banff
Tarocco C, De Rensis F, Kirkwood RN, Yang R (2000) Effect of split weaning interval on return to estrus and sow fertility. J Swine Health Prod 8:221–223
Taylor RE, Field TG (1998) Scientific farm animal production, 6th edn. Prentice Hall, Upper Saddle River, NJ
Tilton JE, Cole DJA (1982) Effect of triple versus double mating on sow productivity. Anim Sci 34(03):279–282
Tummaruk P, Lundeheim N, Einarsson S, Dalin AM (2001) Effect of birth litter size, birth parity number, growth rate, backfat thickness and age at first mating of gilts on their reproductive performance as sows. Anim Reprod Sci 66(3–4):225–237
Tummaruk P, Tantasuparuk W, Techakumphu M, Kunavongkrit A (2007) Age, body weight and backfat thickness at first observed oestrus in crossbred landrace×Yorkshire gilts, seasonal variations and their influence on subsequence reproductive performance. Anim Reprod Sci 99(1–2):167–181
USDA-NASS (2010) National agricultural statistics service: hogs and pigs. http://www.nass.usda.gov/Statistics_by_Subject/index.php?sector=ANIMALS%20&%20PRODUCTS. Accessed 28 March 2013
USDA-NASS (2012) National agricultural statistics service: hogs and pigs. http://www.nass.usda.gov/Statistics_by_Subject/result.php?CA117921-CBC5-3549-833E-4A86059B1834&sector=ANIMALS%20%26%20PRODUCTS&group=LIVESTOCK&comm=HOGS. Accessed 28 March 2013
USDA-APHIS (2006) National animal health monitoring system: swine. http://www.aphis.usda.gov/animal_health/nahms/swine/index.shtml. Accessed 28 March 2013
USDA-NASS (2008) Hogs and pigs. National Agricultural Statistics Service, Washington, DC. http://www.nass.usda.gov/QuickStats/index2.jsp
van Leeuwen JJJ, Martens MRTM, Jourquin J, Driancourt MA, Kemp B, Soede NM (2011) Effects of altrenogest treatments before and after weaning on follicular development, farrowing rate, and litter size in sows. J Anim Sci 89(8):2397–2406
Vargas AJ, Bernardi ML, Bortolozzo FP, Mellagi APG, Wentz I (2009) Factors associated with return to estrus in first service swine females. Prev Vet Med 89(1–2):75–80

Vazquez JM, Martinez EA, Roca J, Gil MA, Parrilla I, Cuello C, Carvajal G, Lucas X, Vazquez JL (2005) Improving the efficiency of sperm technologies in pigs: the value of deep intrauterine insemination. Theriogenology 63(2):536–547

Vazquez JM, Roca J, Gil MA, Cuello C, Parrilla I, Caballero I, Vazquez JL, Martinez EA (2008) Low-dose insemination in pigs: problems and possibilities. Reprod Domest Anim 43(Suppl 2):347–354

von Rohr P, Hofer A, Kunzi N (1999) Economic values for meat quality traits in pigs. J Anim Sci 77(10):2633–2640

Vyt P, Maes D, Dejonckheere E, Castryck F, Van Soom A (2004) Comparative study on five different commercial extenders for boar semen. Reprod Domest Anim 39(1):8–12

Waberski D, Weitze KF, Gleumes T, Schwartz M, Willmen T, Petzoldt R (1994) Effect of time of insemination relative to ovulation on fertility with liquid and frozen boar semen. Theriogenology 42:831–840

Walton JS (1986) Effect of boar presence before and after weaning on estrus and ovulation in sows. J Anim Sci 62:9–15

Watson PF, Behan JR (2002) Intrauterine insemination of sows with reduced sperm numbers: results of a commercially based field trial. Theriogenology 57:1683–1693

Webel SK, Day BN (1982) The control of ovulation. In: Cole DJA, Foxcroft GR (eds) Control of pig reproduction. Butterworths, London

Weitze KF (2000) Update on the worldwide application of swine AI. In: Johnson LA, Guthrie HD (eds) Boar semen preservation IV. Allen Press, Lawrence, KS, pp 141–146

Weitze KF, Wagner-Reitschel H, Waberski D, Richter L, Krieter J (1994) The onset of heat after weaning, heat duration, and ovulation as major factors in AI timing in sows. Reprod Domest Anim 29:433–443

Whiting TL (2003) Foreign animal disease outbreaks, the animal welfare implications for Canada: risks apparent from international experience. Can Vet J 44(10): 805–815

Whittemore C (1998) The science and practice of pig production, 2nd edn. Blackwell, Oxford

Willenburg KL, Miller GM, Rodriguez-Zas SL, Knox RV (2003a) Influence of hormone supplementation to extended semen on artificial insemination, uterine contractions, establishment of a sperm reservoir, and fertility in swine. J Anim Sci 81(4):821–829

Willenburg KL, Miller GM, Rodriguez-Zas SL, Knox RV (2003b) Effect of boar exposure at time of insemination on factors influencing fertility in gilts. J Anim Sci 81:9–15

Wuensch U, Nitter G, Bergfeld U, Schueler L (2000) Genetic and economic evaluation of genetic improvement schemes in pigs. II. Comparison of selection strategies in a three-way crossbreeding scheme. Arch Tierz 43(2):139–149

Xue J, Dial GD, Trigg T, Davies P, King VL (1998) Influence of mating frequency on sow reproductive performance. J Anim Sci 76:2962–2966

Zimmerman DR, McGargill T, Rohda N (1998) Efficacy of once (1x) vs twice (2x) daily physical or fence-line contact with boars for stimulating earlier puberty in gilts. In: Nebraska swine report, Lincoln, NE, pp 3–4

Impacts of Reproductive Technologies on Beef Production in South America

8

José Luiz Moraes Vasconcelos,
Ocilon Gomes de Sá Filho, and Reinaldo F. Cooke

Abstract

The majority of beef cow herds in South America are constituted by *Bos indicus* females, which have particular reproductive features that contribute to reduced reproductive efficiency compared with that of *B. taurus* cohorts. Hence, several alternatives to enhance reproductive efficiency of *B. indicus* heifers and cows have been developed to address their inherent reproductive shortcomings. These research-based technologies are being described in detail within this chapter and have already made an impact on South American *B. indicus*-based production systems. These include the following: (a) hormonal protocols to induce puberty in nulliparous heifers or estrous cyclicity in postpartum cows to maximize their reproductive performance during the subsequent breeding season, (b) hormonal protocols to synchronize estrus and/or ovulation in *B. indicus* females to exploit their reproductive responses to artificial insemination, and (c) genetic and environmental factors that influence reproductive success in beef herds, including reproductive diseases and excitable temperament of *B. indicus* females, that have been investigated to support/promote the development of appropriate mitigation technologies.

Keywords

Artificial insemination • *Bos indicus* • Beef females • Estrus synchronization • Reproduction • South America

J.L.M. Vasconcelos (✉) • O.G. de Sá Filho
Departamento de Produção Animal, Faculdade de Medicina Veterinária e Zootecnia–Universidade Estadual Paulista, Botucatu, São Paulo 18618-000, Brazil
e-mail: vasconcelos@fmvz.unesp.br

R.F. Cooke
Eastern Oregon Agricultural Research Center, Oregon State University, Burns, OR 97720, USA

G.C. Lamb and N. DiLorenzo (eds.), *Current and Future Reproductive Technologies and World Food Production*, Advances in Experimental Medicine and Biology 752,
DOI 10.1007/978-1-4614-8887-3_8,

Introduction

In South America, the majority of beef cattle herds are constituted by *B. indicus* breeds due to their capacity in coping with severe environmental challenges, such as hot weather and local diseases. However, *B. indicus* females have particular reproductive characteristics that contribute to reduced reproductive efficiency compared with that of *B. taurus* cohorts raised in subtropical and temperament environments. As an example, the use of artificial insemination (AI) after estrus detection is often limited in tropical areas because *B. indicus* cattle have shorter estrus, as well as higher incidence of estrus occurring during the night, compared with *B. taurus* breeds (Pinheiro et al. 1998). Moreover, *B. indicus* cows have longer gestation and postpartum anestrous compared with *B. taurus* cohorts (Abeygunawardena and Dematawewa 2004), which often hinders the attainment of 365-day calving intervals. These challenges are even more critical in primiparous cows maintained on low-quality tropical pastures (Meneghetti and Vasconcelos 2008; Vasconcelos et al. 2009a), given that nutrient intake during the postpartum period is not sufficient to meet their con-current requirements for growth and lactation. Consequently, many producers maintain nonpregnant primiparous cows in their herds for an entire production year to be inseminated during the subsequent breeding season (BS), which contributes to reduce the profitability of cow-calf enterprises.

Besides the differences regarding estrus behavior and anestrous length in postpartum cows, *B. indicus* heifers also reach puberty at an older age and higher percentage of body weight relative to mature body weight compared with *B. taurus* heifers (Dobson and Kamonpatana 1986), further decreasing the proportion of pubertal females and consequent pregnancy rates during their first BS (Patterson et al. 1991; Restle et al. 1999). Moreover, there are several other divergences in reproductive characteristics between *B. indicus* and *B. taurus* females that will not be discussed herein, including size of ovarian follicular/corpus luteum (Pinheiro et al. 1998; Figueiredo et al. 1997), amount of follicles before divergence (Sergerson et al. 1984; Buratini et al. 2000), follicular diameter at divergence (Sartori et al. 2001; Sartorelli et al. 2005), and size of dominant follicle at acquisition of ovulatory capacity (Sartori et al. 2001; Gimenes et al. 2008). Based on all the reproductive challenges inherent to *B. indicus* females, the objective of this chapter is to discuss novel pharmacological and management technologies to improve reproductive performance of cow-calf systems from tropical areas in South America.

Reproductive Challenges in *B. indicus* Females

Anestrus and Estrus Detection

The underlying physiology for anovulation appears to be similar between postpartum cows and prepubertal heifers (Wiltbank et al. 2002). In both instances, follicular waves develop and progress through to follicle deviation, preceded by an increase in peripheral follicle-stimulating hormone (FSH) concentrations. However, in anovular beef females, the hypothalamus is sensitive to the negative feedback of estradiol on luteinizing hormone (LH) secretion. Therefore, low peripheral estradiol concentrations originated from the ovary limits growth and adequate estradiol production by the dominant follicle, preventing the LH surge required for ovulation (Wiltbank et al. 2002). Further, the hypothalamic sensitivity to estradiol is associated with the presence of estradiol receptors in the anterior and medial basal hypothalamus, which declines gradually after parturition or as puberty approaches (Wiltbank et al. 2002) and permits adequate follicle growth, LH surge, and subsequent ovulation.

Nutrition is one of the primary factors affecting estrous resumption in beef females, given that inadequate nutrient intake may intensify the negative feedback of estradiol on LH secretion (Randel 1990). Several studies demonstrated that *B. indicus* cows with inadequate or losing body condition score (BCS) during the BS, indicating inadequate nutritional status, are less likely to show estrus behavior and to conceive (Vasconcelos et al. 2009a; Meneghetti et al. 2009; Sá Filho et al. 2009a, b, 2010). Nutritional strategies to reduce the length of postpartum anestrus will not be discussed in the present book chapter, but cattle producers should always ensure that beef females are receiving adequate nutrition to fully express their reproductive potential. However, even when adequate nutrition is provided, offspring–dam interactions such as physical, auditory, visual, or olfactory, decrease LH secretion and delay estrous resumption in suckled females (Stevenson et al. 1994). Interestingly, a short-term temporary weaning (TW; 48 h) increases LH pulse frequency to adequate levels (Williams et al. 1983; Edwards 1985), improving estrous detection rates (Vasconcelos et al. 2009a) and ovulation in response to exogenous gonadotropin-releasing hormone (GnRH) (Vasconcelos et al. 2009b) in postpartum anestrous cows. Hence, strategies to increase the frequency of LH pulses, such as TW, are effective in inducing ovulation in non-cycling *B. indicus* females.

Another approach that has been successfully used for induction of cyclicity in anovular *B. indicus* females is treatment with intravaginal inserts that release exogenous progesterone (P4; Vasconcelos et al. 2009a; Meneghetti et al. 2009; Sá Filho et al. 2009a, b, 2010). In non-cycling females, treatment with progestogens increased LH secretion (Williams et al. 1983; Garcia-Winder et al. 1987) and reduced the occurrence of premature luteolysis after first postpartum ovulation (Vasconcelos et al. 2009a), promoting estrus behavior (Claro Júnior et al. 2010) and pregnancy establishment (Vasconcelos et al. 2009a). Given that P4 inhibits estradiol receptors in several tissues (Hseuh et al. 1976; Pavlik and Coulson 1976), it is likely that hypothalamic estradiol receptors are suppressed when cattle are exposed to exogenous P4, thereby reducing the estradiol negative feedback on LH secretion (Day and Anderson 1998). Further, another major factor contributing to inadequate reproductive performance of beef cows is the short luteal phase that often occurs after first postpartum ovulation, whereas treatment with exogenous P4 prevents premature regression of the corpus luteum (Sá Filho et al. 2009c). Based on the aforementioned rationale that strategies to increase the frequency of LH pulses improve fertility in *B. indicus* females, Vasconcelos et al. (2009a) evaluated the effects of treatments with an intravaginal insert containing 1.9 g of P4 (CIDR, Pfizer Animal Health, São Paulo, SP, Brazil) and/or 48 h TW on reproductive performance of postpartum and anestrous Nelore (*B. indicus*) × Angus (*B. taurus*) cows ($n=286$) throughout the BS. Experimental groups consisted in: (1) no treatment; (2) 48 h TW beginning on day 0 of BS; (3) CIDR device between day −7 and 0 of BS; and (4) CIDR device between day −7 and 0 of BS followed by 48 h TW. Artificial insemination following estrus detection was performed between day 0 and 25 of the BS, and cows were exposed to natural service between day 26 and 80 of the BS. Results are shown in Table 8.1. Estrus detection during the first 3 or 25 days of the BS were greater in cows receiving TW, whereas conception rate in cows inseminated within the first 3 days of BS was greater in cows receiving exogenous P4. Further, combining P4 with TW increased pregnancy rate within the first 3 and 25 days of the BS, but pregnancy rate at end of BS did not differ among treatment groups. Therefore, combining exogenous P4 with TW improved the percentage of cows becoming pregnant at the beginning of the BS by inducing cyclicity enhancing estrus behavior and increasing proportion of cows that conceived early into the BS. However, these treatments have the inconvenience of being dependent of estrus detection, which is a limiting factor in *B. indicus* cattle as previously discussed.

Hormonal treatments developed since the last decade have precisely controlled follicular and luteal development, as well as ovulation, allowing artificial insemination (AI) without the need for estrus detection (i.e., fixed-time AI; TAI) and regardless of cycling status. Based on the aforementioned reproductive challenges of *B. indicus* females, these hormonal protocols have largely benefited reproductive efficiency of *B. indicus* herds reared in South American tropical environments. Three steps are necessary for the success

Table 8.1 Percentage of cows inseminated (first 3 and 25 days of breeding season), conception (first 3 and 25 days of breeding season), and pregnancy rates (first 3 days, 25 days, and in 80 days of breeding season) in anestrous Nelore (*B. indicus*) × Angus (*B. taurus*) cows receiving or not an intravaginal insert containing 1.9 g of progesterone (CIDR) between day −7 and 0 of breeding season and/or 48 h temporary weaning (TW) between day 0 and 2 of breeding season

	Without CIDR		With CIDR	
Reproductive variable	Without TW	With TW	Without TW	With TW
% Cows inseminated (3 days)[ab]	2.7 % (2/73)	48.6 % (34/70)	17.8 % (13/73)	41.4 % (29/70)
Conception (3 days)[bc]	50.0 % (1/2)	8.8 % (3/34)	46.1 % (6/13)	58.6 % (17/29)
Pregnancy rate (3 days)[ac]	1.4 % (1/73)	4.3 % (3/70)	8.2 % (6/73)	24.3 % (17/70)
Conception (4–25 days)	50.0 % (7/14)	16.7 % (2/12)	50.0 % (4/8)	60.0 % (3/5)
% Cows inseminated (25 days)[a]	21.9 % (16/73)	60.0 % (42/70)	28.8 % (21/73)	48.6 % (34/70)
Pregnancy rate (25 days)[bc]	10.9 % (8/73)	7.1 % (5/70)	13.7 % (10/73)	28.6 % (20/70)
Pregnancy rate (80 days)	58.9 % (43/73)	65.7 % (46/70)	61.6 % (45/73)	58.6 % (41/70)

Adapted from Vasconcelos et al. (2009a, b)
[a]CIDR effect ($P<0.05$)
[b]TW effect ($P<0.05$)
[c]CIDR × TW interaction ($P<0.05$)

of a protocol designed to synchronize ovulation. First, a new follicular wave has to be initiated to recruit young follicles with a similar and known stage of development. Currently, there are two main pharmacological approaches to synchronize the emergence of a new follicular wave during ovulation synchronization protocols: (a) inducing ovulation of the dominant follicle by administration of GnRH (e.g., Ovsynch protocol; Pursley et al. 1995) or (b) inducing follicular atresia by administration of estrogens and P4 (Bó et al. 1994). Protocols using P4 + estradiol are less expensive and more efficient in initiating a new follicular wave in postpartum *B. indicus* cows compared with GnRH-based protocols (Baruselli et al. 2004), whereas the specific reasons for this outcome will be discussed later in this chapter.

The second step in an ovulation synchronization protocol is decreasing circulating concentrations of P4 to allow the beginning of proestrus and development of the ovulatory follicle, which can be performed by the withdrawal of P4 inserts and induction of luteolysis with prostaglandin F2α (PGF2α)-analogues, such as dinoprost or cloprostenol. When a cow is in anestrus, treatments with PGF2α-analogues are not necessary, since a corpus luteum (CL) is not present. However, evaluation of CL presence is not practical in most field/commercial scenarios; hence luteolytic drugs are often administered to all females assigned to the protocol. One important characteristic to be considered is that PGF2α treatment is ineffective in inducing luteolysis during the first 5 days of estrous cycle (Henricks et al. 1974); therefore, treatment with P4 inserts for at least 6 days prior to administration of PGF2α ensures that all cows have a CL with luteolytic capacity. Moreover, a 50 % reduction in the recommended dose of dinoprost tromethamine (12.5 mg) administered intramuscularly is as effective in causing luteolysis as the standard recommended dose (25 mg; Table 8.2) in *B. indicus* cows with a CL older than 7 days.

The third step within an ovulation synchronization protocol is the induction of ovulation, which can be performed with hormones that act directly on preovulatory follicles, such as human chorionic gonadotropin (hCG) or LH, or hormones that induce a preovulatory LH surge such as GnRH or esters of estradiol. However, because intervals between injection and ovulation vary for each hormone, caution should be adopted when scheduling treatments and AI based on the hormone(s) being used. Previous studies indicated that replacing GnRH by esters of estradiol in TAI protocols results in similar pregnancy rates. For example, pregnancy rates in *B. indicus* cows did not differ when the ovulatory stimulus after luteolysis was GnRH or estradiol benzoate

Table 8.2 Ovulation, conception, and pregnancy rates of suckled *B. indicus* cows submitted to ovulation synchronization protocols with different dosages of PGF2α (Study 1), ovulatory stimuli (Study 2), or previous uses of the intravaginal progesterone insert (Study 3 and 4)

Item	Ovulation rate	Conception rate	Pregnancy rate
Study 1—PGF2α dosage[a]			
Half-dose	88.7 % (110/124)	69.2 % (76/110)	61.3 % (76/124)
Full-dose	94.4 % (119/126)	68.2 % (81/119)	64.3 % (81/126)
Study 2—ovulatory stimulus[b]			
ECP	89.5 % (282/315)	56.7 % (160/282)	50.8 % (160/315)
EB	90.9 % (289/318)	57.1 % (165/289)	51.9 % (165/318)
GnRH	91.0 % (284/312)	59.5 % (169/284)	54.2 % (169/312)
Study 3—CIDR® uses[c]			
CIDRn	91.7 % (496/541)	56.0 % (278/496)	51.4 % (278/541)
CIDR9d	90.6 % (482/532)	59.3 % (286/482)	53.8 % (286/532)
CIDR18d	90.1 % (373/414)	55.8 % (208/373)	50.2 % (208/414)
Study 4—CIDR® uses[c]			
CIDRn	79.7 % (55/69)	54.5 % (30/55)	43.5 % (30/69)
CIDR9d	80.0 % (144/180)	61.1 % (88/114)	48.9 % (88/180)
CIDR18d	83.9 % (188/224)	60.1 % (113/188)	50.4 % (113/224)
CIDR27d	85.7 % (90/105)	57.8 % (52/90)	49.5 % (52/105)

Adapted from Meneghetti et al. (2009)

[a]Cows were treated with 12.5 mg (half-dose) or 25 mg (full-dose) of dinoprost tromethamine on day 7 of a protocol using progesterone associated with benzoate estradiol. Progesterone devices were maintained between day 0 and 9 and fixed-time AI (FTAI) were performed on day 11. Ovulation was induced with estradiol benzoate administered on day 10. Dependent variables were not affected by PGF2α dosage ($P>0.10$)

[b]Cows were treated with estradiol cypionate (ECP) on day 9, estradiol benzoate (EB) on day 10, or gonadotropin releasing (GnRH) hormone on day 11 relative to initiation of a protocol using progesterone associated with benzoate estradiol. Progesterone devices were maintained between day 0 and 9, and FTAI were performed on day 11. Dependent variables were not affected by ovulatory stimulus ($P>0.10$)

[c]Cows received (day 0–9 of the protocol) a non-previously used CIDR® (CIDRn), a CIDR® used previously for 9 days (CIDR9d), a CIDR® used previously for 18 days (CIDR18d), or a CIDR® used previously for 27 days (CIDR27d) at initiation of the protocol. Dependent variables were not affected by CIDR® type ($P>0.10$)

(EB; Barros et al. 2000). In another study with *B. indicus* cows receiving P4-including protocols, similar synchronization, conception, and pregnancy rates were detected among cows receiving EB, estradiol cypionate (ECP), or GnRH to induce ovulation (Table 8.2). Because esters of estradiol are less expensive than GnRH, whereas ECP requires less labor than EB treatment (EB treatment has to be administered 24 h after P4 insert withdrawal and 30 h before TAI, requiring cows to be handled once more than in ECP and GnRH treatments), ECP treatment administered concurrently with P4 insert withdrawal and 48 h before TAI is being recommended as the ovulatory stimulus to be used in TAI protocols for *B. indicus* cows. It is important to note that the efficiency of the second and third steps is dependent on the success of the first step within an ovulation synchronization protocol. Therefore, different approaches for synchronizing the follicular wave within TAI protocols have been extensively investigated for *B. indicus* females and will also be discussed in more detail later in this chapter.

Synchronization Protocols Using Progesterone in Association with Estradiol

Treatment with exogenous P4 (such as CIDR) and estradiol (EB or ECP) suppresses FSH and follicular growth, synchronizing emergence of the follicular wave in bovine females regardless of the stage of follicular development at the time of treatment (Bó et al. 1994). The interval between treatment application and beginning of a new follicular wave ranges from 3 to 6 days (Martinez and Adams 2000; Kim et al. 2005). Still, this treatment yields satisfactory results in anestrous and cycling beef cows (Rhodes et al. 2002) and was more efficient in synchronizing follicular wave emergence compared with GnRH in postpartum cows under tropical climates (Baruselli et al. 2004). Furthermore, these protocols provide exogenous P4 to prevent premature luteolysis after the first postpartum ovulation. Meneghetti et al. (2009) developed an ovulation synchronization protocol for postpartum *B. indicus* cows aiming to: (a) synchronize follicular wave emergence with P4+estradiol; (b) increase dominant follicle development with an intravaginal P4 insert+TW; (c) prevent premature luteolysis after a synchronized ovulation by providing exogenous P4 to anestrous cows before the synchronized ovulation; and (d) reduce the cost of the synchronization protocol by decreasing the dosage of PGF2α, replacing GnRH with estradiol as an ovulatory stimulus and using the same P4 insert more than once. The protocol consisted in CIDR insertion+2 mg of EB on day 0, CIDR withdrawal+12.5 mg of dinoprost tromethamine+0.5 mg of ECP+TW on day 9, TAI+calf return on day 11, and resulted in synchronization and pregnancy rates of approximately 90 % and 50 %, respectively (Tables 8.2 and 8.3). Moreover, with this protocol, the CIDR may be used as many as four times with no detrimental effects on pregnancy rates (Table 8.2).

In a complementary study (Sá Filho et al. 2009a), the factors affecting pregnancy rates of heifers and cows receiving the synchronization protocol developed by Meneghetti et al. (2009) were evaluated. Lactating and non-lactating beef cows and heifers from various locations in Brazil during the 2006–2007 (n=27,195) and 2007–2008 (n=36,838) BS were submitted to the aforementioned ovulation synchronization protocol (Meneghetti et al. 2009). During both breeding seasons, cattle were originated from 71 pasture-based ranches located in seven Brazilian states with BS averaging 120 days. Pregnancy was diagnosed by transrectal ultrasonography 28–35 days after TAI. Pregnancy rate did not differ between BS. Across both BS (2006–2007 and 2007–2008), overall pregnancy rate at TAI was 49.6 %. Pregnancy rate did not differ among state, but varied among ranches (results ranging from 26.8 to 68.0 %), as well as cow group within ranch. Pregnancy rate was influenced by breed (*B. indicus*=48.3 %, *B. taurus*=61.7 % and crossbred *B. indicus*×*B. taurus*=50.7 %), category (nulliparous=39.6 %, suckled primiparous=45.2 %, suckled multiparous=51.8 %, and non-suckled multiparous=46.1 %), BCS (on a 1–9 scale; Wagner et al. 1988) at TAI (≤4=43.0 %, 5=49.6 %, ≥6=52.7 %). Days postpartum at beginning of protocol did not affect pregnancy rate (30–60 days=47.6 %, 61–90 days=51.7 %, and 91–150 days=50.8 %). Pregnancy rate was also consistently affected by sire (results ranging from 7.2 to 77.3 %) and AI technician (results ranging from 15.1 to 81.8 %). Therefore, a 50 % pregnancy rate to TAI in *B. indicus* cattle was shown to be achievable with the protocol developed by Meneghetti et al. (2009) and differences in breed, BCS, category, sire, and AI technician may account for some of the variation in reproductive performance that occurs between and within ranches.

In another study (Sá Filho et al. 2009a) evaluating additional strategies to improve reproductive performance of *B. indicus* cows submitted to the treatment proposed by Meneghetti et al. (2009), replacing TW with treatment with a one-time treatment of equine chorionic gonadotropin (eCG; 400 IU), but not with FSH, was beneficial to pregnancy rates in postpartum *B. indicus* cows (Table 8.3). The length of the half-life of eCG is relatively long (Carruthers 1986), and it has the capacity to bind to both follicular LH and FSH receptors (Murphy and Martinuk 1991), which may stimulate follicle development and estradiol synthesis (Kuran et al. 1996), as well as P4 secretion by the corpus luteum (Baruselli et al. 2004). The eCG treatment may also provide a more adequate endocrine environment during proestrus (greater circulating estradiol concentrations) and diestrus (greater circulating P4 concentrations), which is beneficial to fertility. Interestingly, treatments with either 200 or 400 IU of eCG did not

Table 8.3 Ovulation, conception, and pregnancy rates of suckled *B. indicus* cows submitted to an ovulation synchronization protocol using progesterone associated with estradiol but with different hormonal strategies to improve fertility

Item	Ovulation rate	Conception rate	Pregnancy rate
Study 1—eCG to replace TW[a]			
Control	81.9 % (181/221)	50.8 % (92/181)	41.6 % (92/221)
eCG300	80.8 % (168/208)	57.7 % (97/168)	46.6 % (97/208)
eCG400	83.9 % (187/223)	64.7 % (121/187)	54.3 % (121/223)
TW	86.6 % (207/239)	59.4 % (123/207)	51.5 % (123/239)
Study 2—FSH to replace eCG[b]			
TW	–	–	41.9 % (83/198)
eCG400	–	–	43.3 % (81/187)
Folltropin	–	–	34.3 % (74/216)
Pluset	–	–	32.1 % (36/112)
Study 3—eCG associated with TW[c]			
TW	88.9 % (264/297)	158/264 (59.8)	53.2 % (158/297)
TW + eCG200	93.0 % (278/299)	155/278 (55.7)	51.8 % (155/299)
TW + eCG400	90.0 % (261/291)	152/261 (58.2)	52.2 % (152/291)

Adapted from Sá Filho et al. (2009a)

[a]Cows received no treatment (control), 300 IU of eCG (eCG300), 400 IU of eCG (eCG400), or temporary weaning (48 h; TW) on day 9 relative to initiation of an ovulation synchronization protocol. Pregnancy rates were greater in eCG400 and TW treatments compared with control ($P<0.05$)

[b]Cows received temporary weaning (48 h; TW), 400 IU of eCG (eCG400), 20 mg of Folltropin® (Folltropin—FSH), or 20 mg of Pluset® (Pluset—FSH) on day 9 relative to initiation of an ovulation synchronization protocol. Pregnancy rates were greater in eCG400 and TW treatments compared with Folltropin and Pluset ($P<0.05$)

[c]Cows received temporary weaning (48 h; TW), temporary weaning + 200 IU of eCG (TW + eCG200), or temporary weaning + 400 IU of eCG (TW + eCG400) on day 9 relative to initiation of an ovulation synchronization protocol. Dependent variables were not affected by treatments ($P>0.10$)

improve conception or pregnancy rates in cows that were temporarily weaned (Table 8.3), indicating that TW likely stimulated adequate secretion of LH in a manner that additional gonadotropin support was not required for final follicular development (Sá Filho et al. 2009a). Thus, the TAI protocol using P4 + estradiol should be associated with TW or eCG, but not with both, to maximize reproductive performance of postpartum *B. indicus* cows.

Synchronization Protocols Using GnRH

Protocols for TAI-based GnRH and PGF2α yield acceptable results in *B. taurus* females (pregnancy rate >50 %; Lamb et al. 2001; Larson et al. 2006), but reduced pregnancy rates have been reported in *B. indicus* cows (Barros et al. 2000; Saldarriaga et al. 2007; Vasconcelos et al. 2009b). A commercial source of estradiol is not available for estrous cycle manipulation in many countries; therefore other hormones (e.g., GnRH, LH, or hCG) have been used to induce ovulation of dominant follicles and follicular wave synchronization. The mechanism by which GnRH induces new wave emergence is based on inducing ovulation of the dominant follicle, and its success is dependent on the presence of a dominant follicle with ovulatory capacity at the time of treatment, which is acquired when ovarian follicles are >8.5 mm in diameter in *B. indicus* cattle (Gimenes et al. 2008). In cows ovulating to the GnRH treatment, the emergence of a new follicular wave was described to occur within 2 days (Bó et al. 2002). The induction of ovulation in a high percentage of treated cows at the start of treatment is critical for obtaining satisfactory results, as shown previously in dairy cattle (Vasconcelos et al. 1999). However, the likelihood of ovulation to a GnRH treatment is reduced in anestrous cows compared with cycling cohorts (Fernandes et al. 2001; Vasconcelos et al. 2009b), because follicles rapidly become atresic after the acquisition of ovulatory capacity in anovular females (Wiltbank et al. 2002). Furthermore, anestrous cows that do not ovulate to the first GnRH treatments, but do ovulate upon the second GnRH, are likely to experience premature luteolysis because ovulation occurs with no preexposure to P4 (Vasconcelos et al. 2009a, b; Sá Filho et al. 2009c). Although exogenous P4 can be provided between first and second ovulatory stimuli to avoid premature luteolysis in GnRH-based protocols, cows that do not ovulate to the first ovulatory stimulus do not have a synchronized follicular wave, reducing the probability of success of the protocol (Vasconcelos et al. 1999, 2009b). These are some of the reasons why GnRH-based protocols are often less efficient in postpartum cows compared with those based on synchronization of follicular wave emergence with P4+estradiol (Baruselli et al. 2004) and highlight the importance of the ovulation to the first ovulatory stimulus for the success of GnRH-based protocols.

In a series of two studies in suckled *B. indicus* cows (experiment 1: $n=139$; experiment 2: $n=376$) submitted to a GnRH-based protocol for synchronization of ovulation, the effects of TW (48 h) prior to each ovulatory stimulus were evaluated (Vasconcelos et al. 2009b). Treatments consisted in GnRH on day 0 followed by PGF2α on day 7 and EB on day 8. Cows were assigned to TW or not before GnRH (TW1) and/or after PGF2α (TW2). Anestrous cows receiving TW1 had greater follicular diameter on day 0 (10.6 vs. 9.9 mm) and greater ovulation rate to GnRH treatment (85.4 % vs. 49.0 %) than control cows. However, the same advantages of TW1 were not detected in cycling cows. Regardless of cyclicity, cows receiving TW2 had larger follicular diameter on day 9 (10.8 vs. 10.4 mm) and greater ovulation rate to EB treatment (79.1 % vs. 58.3 %) than control cows. Pregnancy rates at TAI, which was performed 30 h after EB treatment, were greater in cows receiving both TW1 and TW2 compared with control cows (29.8 % vs. 10.6 %). Thus, the inclusion of TW in strategic times of GnRH-based protocols may improve fertility by causing an increase on follicular development and ovulation rates.

When circulating concentrations of P4 are maintained between 1 and 2 ng/mL, the period of dominant follicle growth is prolonged due to an increase on the frequency of LH pulses (Stock and Fortune 1993). Based on this rationale, it is plausible to speculate that prolonged treatments with P4 in anestrous cows may have beneficial effects on follicular development and, consequently, on the likelihood of ovulation to a GnRH treatment. Sá Filho et al. (2009d) evaluated treatments with P4 and TW to increase the follicular size prior to the first ovulatory stimulus of a GnRH-based protocol in *B. indicus* cows ($n=283$) and recommended the following protocol based on their results: CIDR insertion on day 0 (morning), CIDR withdrawal + TW on day 7 (morning), GnRH + calf return on day 9 (morning), PGF2α + TW on day 15 (afternoon), and TAI + GnRH + calf return on day 17 (afternoon). Independently of cycling status, this protocol resulted in 94.4 % ovulation rate to the first GnRH treatment, 95.8 % ovulation rate to the second GnRH treatment, as well as conception and pregnancy rates of 55.1 % and 52.8 %, respectively. Sá Filho et al. (2009d) also compared the effects of the inclusion of a CIDR between the first GnRH and the PGF2α treatments of a synchronization protocol consisting of TW on day 0, GnRH + calf return on day 2, PGF2α + TW on day 8, and TAI + GnRH + calf return on day 10 ($n=301$). However, pregnancy rates were similar between cows receiving (28.5 %) or not the CIDR (32.7 %). Given that treatments with exogenous P4 associated with GnRH-based synchronization protocols were described to improve fertility in anestrous *B. taurus* females (Lamb et al. 2001; Larson et al. 2006), and 51 % of cows in the study of Sá Filho et al. (2009d) were cycling, further studies are still required to better evaluate this hormonal treatment in anestrous *B. indicus* cattle.

Nulliparous and Primiparous Cows

Prepubertal Heifers

In cow-calf enterprises using *B. indicus* breeds, it is relatively common to find prepubertal heifers with adequate age and body weight for establishment of puberty. The key factor responsible for establishment of puberty is an increase in LH

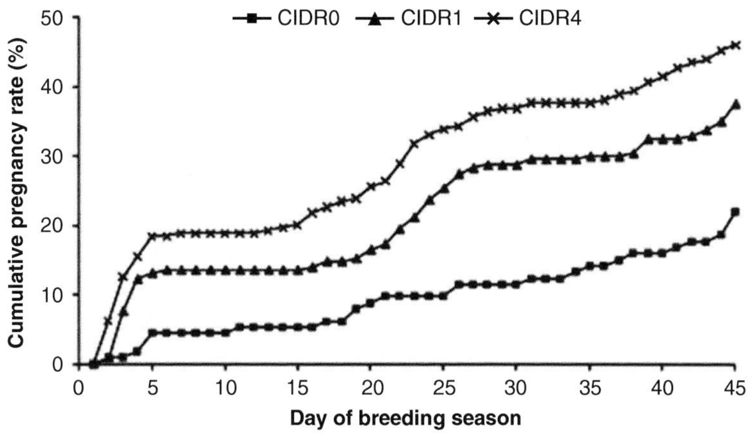

Fig. 8.1 Cumulative percentage of pregnancy in prepubertal *B. indicus* heifers throughout a 45-day AI breeding season. Heifers were assigned to receive, between day −12 and 0, no treatment (CIDR0), a new intravaginal insert (CIDR1), or a similar intravaginal insert that had previously been used three times (CIDR4). Heifers were artificially inseminated 10–14 h after detection of estrus; those that returned in estrus after the first service were re-inseminated. Adapted from Claro Júnior et al. (2010)

release from the anterior pituitary in response to GnRH from the hypothalamus (Day et al. 1987). Treatment for 12 days with a new CIDR or a CIDR that had previously been used three times, with each use occurring for 9 days, hastened puberty and improved estrus detection and pregnancy rates at the beginning of the BS in prepubertal *B. indicus* heifers (Fig. 8.1; Claro Júnior et al. 2010). However, the treatment with a previously used CIDR resulted in greater follicular diameter at the end of treatments, as well as greater conception rate during the first week and overall pregnancy rate within 45 days of breeding compared with the treatment with a new CIDR (Table 8.4). Thus, a previously used CIDR is preferable to a new CIDR to improve reproductive performance during the BS in prepubertal *B. indicus* heifers, likely by providing a lower but more appropriate amount of supplemental P4. Furthermore, in prepubertal *B. indicus* heifers with adequate age and puberty, treatment with P4 for 12 days may be used prior to the synchronization of ovulation, with the goal of increasing the percentage of cycling animals at the beginning of the protocol.

In a series experiments, Rodrigues et al. (2011) reported that administration of eCG and ECP at CIDR removal further hastened puberty attainment and enhanced pregnancy rates in prepubertal *B. indicus* heifers receiving an used CIDR prior to breeding, as in Claro Júnior et al. (2010). More specifically, Rodrigues et al. (2011) only evaluated prepubertal heifers receiving an used CIDR for 12 days prior to BS ($n=896$) and reported that administration of 200 IU of eCG at CIDR removal increased estrus detection and ovulation rates within the first 7 days of BS, as well as pregnancy rates to AI upon estrus detection within 28 of BS (34.3 % vs. 27.5 % for estrus detection, 71.9 % vs. 53.3 % for ovulation rate, and 22.4 % vs. 17.3 % for pregnancy rates in heifers receiving or not 200 IU of eCG at CIDR removal, respectively). Rodrigues et al. (2011) also reported ($n=401$) that administration of 200 IU of eCG+0.5 mg of ECP at CIDR removal increased, compared with heifers receiving eCG or no treatment at CIDR removal, estrus detection and ovulation rates (56.1 %, 34.8 %, and 21.5 % for estrus detection rate and 88.3 %, 75.0 %, and 45.6 % for ovulation rate, respectively), as well as pregnancy rates within the first 7 days of the BS in heifers bred to AI (16.7 %, 10.6 %, and 8.9 %, respectively) or natural service (17.6 %, 13.2 %, and 9.4 %, respectively). However, no advantages on these reproductive variables were detected when heifers only received a 0.5 mg of ECP at CIDR removal instead of 200 IU of eCG or 200 IU of eCG+0.5 mg of ECP.

In another series of experiments, Rodrigues et al. (2012) evaluated the aforementioned protocols based on the used CIDR for 12 days prior to BS+eCG and/or ECP treatment at CIDR removal

Table 8.4 Reproductive performance of prepubertal *B. indicus* heifers that received, between day −12 and 0, no treatment (CIDR0), a new intravaginal insert containing 1.9 g of progesterone (CIDR1), or a similar intravaginal insert that had previously been used three times (CIDR4)

	Treatment		
Reproductive variable[1]	CIDR0	CIDR1	CIDR4
Progesterone on day 0 (ng/mL)	0.37 ± 0.16[a]	2.31 ± 0.11[a]	1.20 ± 0.11[a]
Follicle diameter on day 0 (mm)	9.45 ± 0.24[a]	9.72 ± 0.17[a]	11.42 ± 0.16[a]
Uterine score on day 0	1.49 ± 0.06[a]	1.88 ± 0.04[a]	2.24 ± 0.04[a]
Interval to estrus (day)	4.4 ± 0.28[a]	3.48 ± 0.13[a]	3.24 ± 0.14[a]
Estrus detection in 7 days, % (*n*)	19.5[a] (22/113)	42.6[a] (101/237)	39.3[a] (94/239)
Conception rate in 7 days, % (*n*)	27.3[a] (6/22)	33.7[a] (34/101)	46.8[a] (44/94)
Pregnancy rate in 7 days, % (*n*)	5.3[a] (6/113)	14.3[a] (34/237)	18.4[a] (44/239)
Estrus detection in 45 days, % (*n*)	52.2[a] (59/113)	72.1[a] (171/237)	75.3[a] (180/239)
Pregnancy rate in 45 days, % (*n*)	27.4[a] (31/113)	39.2[a] (93/237)	47.7[a] (114/239)
Pregnancy rate in 90 days, % (*n*)	72.6[a] (82/113)	83.5[a] (198/237)	83.7[a] (200/239)

Adapted from Claro Júnior et al. (2010)
The breeding season started on day 1 and consisted of AI after estrus detection between day 1 and 45 and exposure to bulls between day 46 and 90
[a]Within a row, means without a common superscript differed ($P < 0.05$)

to induce puberty, in addition to subsequent TAI protocol for prepubertal *B. indicus* heifers ($n = 3{,}022$). Authors again reported that administration of 200 IU of eCG + 0.5 mg of ECP at CIDR removal was the most efficient alternative to induce puberty, whereas a 12-day interval from CIDR removal to the beginning of a TAI protocol (described by Meneghetti et al. 2009, without TW) resulted in greater conception and pregnancy rates to TAI compared with 10- or 14-day intervals (51 %, 39 %, and 46 % for conception rates and 45 %, 33 %, and 40 % for pregnancy rates, respectively). Rodrigues et al. (2012) also assigned pubertal or prepubertal heifers ($n = 2{,}288$) to an induction protocol (used CIDR for 12 days + 0.5 mg ECP treatment at CIDR removal), followed by a 12-day interval before initiating a TAI protocol (Meneghetti et al. 2009, without TW). Authors reported that prepubertal and pubertal heifers effectively synchronized by the induction and TAI protocols had similar pregnancy rates to TAI (46 % vs. 49 %, respectively). Based on these outcomes, Rodrigues et al. (2012) concluded that it is possible to breed prepubertal heifers to the first ovulation following puberty induction and yield similar pregnancy rates compared with that of cycling heifers, differing from previous research reporting that fertility is greater during the third ovulation compared with the pubertal ovulation (Byerley et al. 1987).

Pubertal Nulliparous Heifers and Non-lactating Cycling Cows

Cycling cattle treated with exogenous P4 may have circulating concentrations of this hormone, a combination of exogenous and CL-originated P4, elevated to levels above the optimal. For instance, TAI protocols containing exogenous P4 to *B. indicus* pubertal heifers and non-lactating cows resulted in lower pregnancy rates compared with anestrous cows (Dias et al. 2009; Sá Filho et al. 2009a). Carvalho et al. (2008) and Meneghetti et al. (2009) reported negative effects of elevated circulating P4 concentrations during TAI protocols on fertility in cattle, likely because elevated P4 is known to reduce LH pulse frequency (Roberson et al. 1989) and subsequent development of the dominant follicle (Stock and Fortune 1993). Peres et al. (2009) evaluated strategies to increase fertility of *B. indicus* pubertal heifers and non-lactating cows submitted to a TAI protocol consisting of a CIDR insertion + EB on day 0, CIDR withdrawal + ECP on day 9, and TAI on day 11. In the first study, pubertal heifers ($n = 1{,}153$) received on day 0 a new or a CIDR previously used for 18 days, as well PGF2α in

Table 8.5 Ovulation, conception, and pregnancy rates of *Bos indicus* pubertal heifers (Study 1) and non-lactating cows (Study 2) submitted to a synchronization of ovulation protocol based on progesterone and estradiol with different alternatives to improve fertility

Item	Ovulation rate	Conception rate	Pregnancy rate
Study 1—CIDR® uses and eCG[a]			
CIDR® 1st use/0 IU eCG	84.3 % (161/191)	49.1 % (79/161)	41.4 % (79/191)
CIDR® 1st use/200 IU eCG	89.9 % (169/188)	51.5 % (87/169)	46.3 % (87/188)
CIDR® 1st use/300 IU eCG	94.6 % (177/187)	52.0 % (92/177)	49.2 % (92/187)
CIDR® 3rd use/0 IU eCG	83.4 % (176/211)	49.4 % (87/176)	41.2 % (87/211)
CIDR® 3rd use/200 IU eCG	87.2 % (170/195)	54.7 % (93/170)	47.7 % (93/195)
CIDR® 3rd use/300 IU eCG	93.9 % (170/185)	47.1 % (80/170)	44.2 % (80/181)
Study 2—Time of PGF2α treatment and eCG[b]			
PGF2α day 7/0 IU eCG	80.3 % (143/178)	59.4 % (85/143)	47.7 % (85/178)
PGF2α day 7/300 IU eCG	90.7 % (156/172)	62.1 % (97/156)	56.4 % (97/172)
PGF2α day 9/0 IU eCG	64.4 % (112/174)	42.0 % (47/112)	27.0 % (47/174)
PGF2α day 9/300 IU eCG	89.3 % (159/178)	50.9 % (81/159)	45.5 % (81/178)

Adapted from Peres et al. (2009)

[a]Pubertal heifers received a non-previously used CIDR® (1st use) or a CIDR® used previously for 18 days (3rd use) at initiation of a TAI protocol and 0, 200, or 300 IU of eCG on day 9 relative to CIDR® insertion, in a 2×3 factorial design. There were effects of eCG treatment on ovulation, conception, and pregnancy rates ($P<0.05$)

[b]Non-lactating cows were treated with 12.5 mg of dinoprost tromethamine on day 7 (48 h before CIDR® withdrawal; PGF2α day 7) or on day 9 (immediately after CIDR® withdrawal; PGF2α day 9) and 0 IU of eCG or 300 IU of eCG on day 9 of an ovulation synchronization protocol. There were effects of time of PGF2α treatment on ovulation, conception, and pregnancy rates ($P<0.05$); eCG affected ovulation and pregnancy rates ($P<0.05$)

addition to 0, 200, or 300 IU eCG on day 9 of the protocol. Heifers treated with a new CIDR had greater serum concentration of P4 on day 9 (3.06±0.1 vs. 2.53±0.1 ng/mL) and smaller follicle diameter at TAI (11.61±0.11 vs. 12.05±0.12 mm) compared with cohorts receiving the used CIDR. However, no differences were detected in ovulation, conception, and pregnancy rates between heifers receiving new or previously used CIDR (Table 8.5). Treatment with eCG, independently of CIDR usage, improved follicle diameter at TAI (11.50±0.10, 11.90±0.11, and 12.00±0.10 mm for 0, 100, and 200 IU, respectively), serum P4 concentration on day 18 (2.77±0.11, 3.81±0.11, and 4.87±0.11 ng/mL), and ovulation and pregnancy rates (Table 8.5). In a companion study, non-lactating *B. indicus* cows ($n=702$) received PGF2α treatment on day 7 or 9 and, as well as 0 or 300 IU eCG on day 9. Cows receiving PGF2α on day 7 had reduced serum P4 concentrations on day 9 (3.05±0.21 vs. 4.58±0.21 ng/mL), a larger follicle at TAI (11.54±0.21 vs. 10.84±0.21 mm), and greater ovulation, conception, and pregnancy rates (Table 8.5). Treatment with eCG increased serum P4 concentration on day 18 (3.24±0.14 vs. 4.55±0.14 ng/mL) as well as ovulation and pregnancy rates (Table 8.5). Therefore, administration of PGF2α earlier in the protocol in addition to eCG treatment may improve fertility of non-lactating cycling cattle submitted to TAI protocols using P4 associated with estradiol.

Primiparous Cows

As previously discussed, nutrient intake during the postpartum period is often not sufficient to meet requirements for growth and lactation of primiparous cows, which further contributes to postpartum anestrus and inadequate pregnancy rates in their second BS. Hence, several beef producers maintain non-pregnant primiparous cows in the herd for an entire production year to be inseminated during their third BS. Before the 1990s, when treatments for induction of cyclicity were yet to be developed, many producers in Brazil adopted the strategy of exposing nulliparous heifers to BS before primiparous/multiparous cows, in manner to allow heifers to calve earlier and have more days postpartum during their second BS.

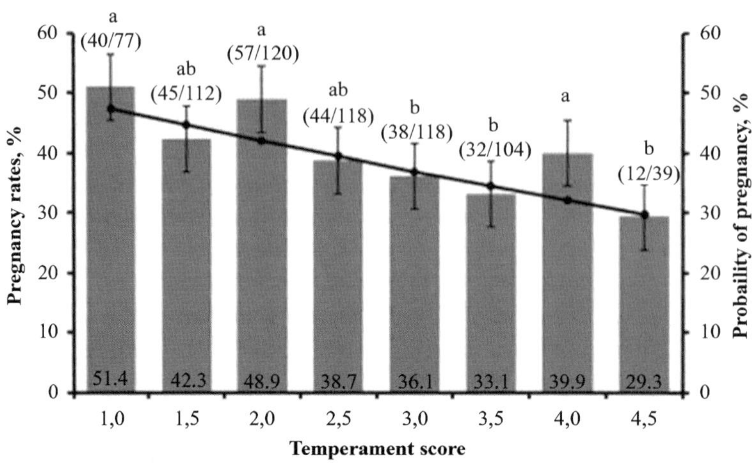

Fig. 8.2 Pregnancy rates (*bars*) and probability of pregnancy (*line*) to FTAI (TAI) in Nelore (*B. indicus*) beef cows according to temperament score, which was calculated by averaging cow chute score and exit score at the time TAI. Exit score was calculated by dividing exit velocity results into quintiles and assigning cows with a score from 1 to 5 (exit score: 1 = slowest cows; 5 = fastest cow). Pregnancy rates tended ($P = 0.08$) to be negatively affected whereas probability of pregnancy was negatively associated (linear effect, $P < 0.01$) with temperament score. Values within *bars* correspond to means. Values in *parenthesis* correspond to pregnant cows divided by total cows assigned to the TAI protocol. Means with different *superscripts* (*a* vs. *b*) differ at $P < 0.05$. Adapted from Cooke et al. (2011)

the speed of the cow exiting the squeeze chute by measuring rate of travel over a 1.9-m distance. Further, within each ranch group, cows were divided in quintiles according to their exit velocity, and assigned a score from 1 to 5 (exit score; 1 = cows within the slowest quintile; 5 = cows within the fastest quintile). Individual temperament scores were calculated by averaging cow chute score and exit score. Cows were also classified according to the final temperament score (temperament type) as adequate temperament (temperament score ≤3) or excitable temperament (temperament score >3).

Cooke et al. (2011) reported that pregnancy rates to TAI tended to be negatively affected by temperament score (Fig. 8.2). The probability of cows becoming pregnant to FTAI was negatively associated with temperament score (linear effect, Fig. 8.2). Accordingly, pregnancy rates were reduced in cows with excitable temperament compared to cows with adequate temperament (35.3 % vs. 42.8 % of pregnant cows/total cows, respectively). These results demonstrated that excitable temperament was detrimental to pregnancy rates to TAI in *B. indicus* beef cows, likely by stimulating, during handling for TAI, neuroendocrine stress responses that directly impair physiological mechanisms required for fertility in females. In fact, pregnancy rates were reduced by 17 % when comparing cows with excitable temperament and cows with adequate temperament or by 43 % when comparing cows with the highest temperament score with those with the lowest temperament score. Based on these outcomes, authors concluded that temperament has a significant impact on reproductive performance of *B. indicus* cows exposed to TAI protocols, whereas strategies to ameliorate temperament or the cowherd (such as selecting cattle for adequate temperament or encouraging interaction between young cattle and humans) are necessary to optimize reproductive and overall efficiency of Brazilian cow-calf operations.

Conclusions

Currently, there are several hormonal treatments allowing South American *B. indicus* producers to reach the benchmark of 50 % pregnancy rate to TAI, as well as enhance the percentage of cows becoming pregnant to AI at the beginning of the BS. This outcome is expected to optimize the use of labor and increase calf weaning age and weight via genetic improvement and concentration of births at the beginning of the calving season. In *B. indicus* females, these protocols must provide a source of exogenous P4 and stimulate the final development of the dominant follicle. In postpartum females, this stimulus may be provided by TW or eCG treatment, whereas non-lactating cycling females should be treated with eCG or advancing the time of PGF2α treatment. Administration of bST may be used as another alternative to enhance pregnancy rates by stimulating early embryo development. Hormonal protocols to induce puberty in replacement heifers, based on exogenous P4, eCG, and ECP, are also available and allow *B. indicus* heifers to have adequate pregnancy rates during their first ovulation. However, the particularities within *B. indicus* females must be considered when choosing for a protocol to maximize pregnancy rates. Reproductive diseases and excitable temperament may prevent optimal reproductive outcomes to TAI protocols. Differences in BCS, category, sire, and AI technician also accounts for some of the variation in reproductive performance that occurs between and within farms and should always be taken into consideration to anticipate and improve reproductive results.

References

Abeygunawardena H, Dematawewa CMB (2004) Prepubertal and postpartum anestrus in tropical zebu cattle. Anim Reprod Sci 82–83:373–387

Albuquerque JP, Dias HP, Bueno IC, Rodrigues ADP, Pereira MHC, Carvalho ER et al (2012) Use of bovine somatotropin (bST) associated with protocols of timed artificial insemination (TAI) in beef cows. Anim Reprod 9:500 (Abstr.)

Aono FH, Cooke RF, Alfieri AA, Vasconcelos JLM (2013) Effects of vaccination against reproductive diseases on reproductive 1 performance of beef cows submitted to fixed-timed AI. Theriogenology 79: 242–248

Armstrong JD, Harvey RW, Poore MA, Simpson RB, Miller DC, Gregory GM et al (1995) Recombinant bovine somatotropin increases milk yield and calf gain in diverse breeds of beef cattle: associated changes in hormones and indices of metabolism. J Anim Sci 73:3051–3061

Barros CM, Moreira MBP, Figueiredo RA, Teixeira AB, Trinca LA (2000) Synchronization of ovulation in beef cows (B. indicus) using GnRH, PGF2α, and estradiol benzoate. Theriogenology 53: 1121–1134

Baruselli PS, Reis EL, Marques MO, Nasser LF, Bó GA (2004) The use of hormonal treatments to improve reproductive performance of anestrous beef cattle in tropical climates. Anim Reprod Sci 82–83: 479–486

Berg DK (2010) Embryo loss in cattle between days 7 and 16 of pregnancy. Theriogenology 73:250–260

Bilby TR, Guzeloglu A, Kamimura S, Pancarci SM, Michel F, Head HH et al (2004) Pregnancy and bovine somatotropin in nonlactating dairy cows: I. Ovarian, conceptus, and insulin-like growth factor system responses. J Dairy Sci 87:3256–3267

Bó GA, Adams GP, Pierson RA, Tribulo HE, Caccia M, Mapletoft RJ (1994) Follicular wave dynamics after estradiol-17β treatment of heifers with or without a progestogen implant. Theriogenology 41:1555–1569

Bó GA, Baruselli PS, Moreno D, Cutaia L, Caccia M, Tribulo R et al (2002) The control of follicular wave development for self-appointed embryo transfer programs in cattle. Theriogenology 57:53–72

Buratini J Jr, Price CA, Visintin JA, Bó GA (2000) Effects of dominant follicle aspiration and treatment with recombinant bovine somatotropin (BST) on ovarian follicular development in Nelore (*B. indicus*) heifers. Theriogenology 54:421–431

Byerley DJ, Staigmiller RB, Berardinelli JB, Short RE (1987) Pregnancy rates of beef heifers bred either on puberal or third estrus. J Anim Sci 65:645–650

Carruthers TD (1986) Principles of hormone therapy in theriogenology. In: Morrow DA (ed) Current therapy in theriogenology, vol 2. Diagnosis, treatment and prevention of reproductive diseases in small and large animals. WB Saunders, Philadelphia, PA, p 4

Carvalho JBP, Carvalho NAT, Reis EL, Nichi M, Souza AH, Baruselli PS (2008) Effect of early luteolysis in progesterone-based timed AI protocols in B. indicus, B. indicus x B. taurus and B. taurus heifers. Theriogenology 69:167–175

Claro Júnior I, Sá Filho OG, Peres RFG, Aono FHS, Day ML, Vasconcelos JLM (2010) Reproductive performance of prepubertal B. indicus heifers after progesterone-based treatments. Theriogenology 74:903–911

Cooke RF, Arthington JD, Austin BR, Yelich JV (2009a) Effects of acclimation to handling on performance, reproductive, and physiological responses of Brahman-crossbred heifers. J Anim Sci 87:3403–3412

Cooke RF, Arthington JD, Araujo DB, Lamb GC (2009b) Effects of acclimation to human interaction on performance, temperament, physiological responses, and pregnancy rates of Brahman-crossbred cows. J Anim Sci 87:4125–4132

Cooke RF, Bohnert DW, Meneghetti M, Losi TC, Vasconcelos JLM (2011) Effects of temperament on pregnancy rates to fixed-timed AI in *Bos indicus* beef cows. Livest Sci 142:108–113

Cooke RF, Bohnert DW, Cappellozza BI, Mueller CJ, DelCurto T (2012) Effects of temperament and

acclimation to handling on reproductive performance of *Bos taurus* beef females. J Anim Sci 90: 3547–3555
Day ML, Anderson LH (1998) Current concepts on the control of puberty in cattle. J Anim Sci 76:1–15
Day ML, Imakawa K, Wolfe PL, Kittok RJ, Kinder JE (1987) Endocrine mechanisms of puberty in heifers. Role of hypothalamo-pituitary estradiol receptors in the negative feedback of estradiol on luteinizing hormone secretion. Biol Reprod 37:1054–1065
Dias CC, Wechsler FS, Day ML, Vasconcelos JLM (2009) Progesterone concentrations, exogenous eCG and timing of prostaglandin F2α treatment affect fertility in postpuberal Nelore heifers. Theriogenology 72:378–385
Dobson H, Kamonpatana M (1986) A review of female cattle reproduction with special reference to a comparison between buffaloes, cows and zebu. J Reprod Fertil 77:1–36
Dunne LD, Diskin MG, Sreenan JM (2000) Embryo and foetal loss in beef heifers between day 14 of gestation and full term. Anim Reprod Sci 58:39–44
Edwards S (1985) The effects of short term calf removal on pulsatile LH secretion in the postpartum beef cow. Theriogenology 23:777–785
Fernandes P, Teixeria AB, Crocci AJ, Barros CM (2001) Timed artificial insemination in beef cattle using GnRH agonist, PGF2alpha and estradiol benzoate. Theriogenology 55:1521–1532
Figueiredo RA, Barros CM, Pinheiro OL, Soler JMP (1997) Ovarian follicular dynamics in Nelore breed (*B. indicus*). Theriogenology 47:1489–1505
Flores EF, Weiblen R, Vogel FSF, Roehe PM, Alfieri AA, Pituco EM (2005) Infection with bovine viral diarrhea virus (BVDV) in Brazil, history, current situation and prospects. Pesqui Vet Bras 25:125–134
Fordyce GE, Dodt RM, Wythes JR (1988) Cattle temperaments in extensive beef herds in northern Queensland. 1. Factors affecting temperament. Aust J Exp Agric 28:683–687
Garcia-Winder M, Lewis PE, Townsend EC, Inskeep EK (1987) Effects of norgestomet on follicular development in postpartum beef cows. J Anim Sci 64:1099–1109
Gimenes LU, Sá Filho MF, Carvalho NAT, Torres-Júnior JRS, Souza AH, Madureira EH et al (2008) Follicle deviation and ovulatory capacity in B. indicus heifers. Theriogenology 69:852–858
Grooms DL (2004) Reproductive consequences of infection with bovine viral diarrhea virus. Vet Clin North Am Food Anim Pract 20:5–19
Henricks DM, Long JT, Hill JR (1974) The various effects of prostaglandin $F_2\alpha$ during various stages of estrous cycle of beef heifers. J Reprod Fertil Suppl 41: 113–120
Hseuh AJW, Peck EJ Jr, Clark JH (1976) Control of uterine estrogen receptor levels by progesterone. Endocrinology 98:438–444
Junqueira JRC, Freitas JC, Alfieri AF, Alfieri AA (2006) Reproductive performance evaluation of a beef cattle herd naturally infected with the BoHV-1, BVDV and Leptospira hardjo. Semina Cien Agr 57:471–480
Kahrs RF (1977) Infectious bovine rhinotracheitis a review and update. J Am Vet Med Assoc 171: 1055–1064
Kim U, Suh G, Nam H, Kang H, Kim I (2005) Follicular wave emergence, luteal function and synchrony of ovulation following GnRH or estradiol benzoate in a CIDR-treated, lactating Holstein cows. Theriogenology 63:260–268
Kuran M, Hutchinson JSM, Broadbent PJ (1996) The response of bovine granulosa cells to different gonadotrophins in culture. Anim Reprod Sci 45:1–12
Lamb GC, Stevenson JS, Kesler DJ, Garverick HA, Brown DR, Salfen BE (2001) Inclusion of an intravaginal progesterone insert plus GnRH and prostaglandin F2α for ovulation control in postpartum suckled beef cows. J Anim Sci 79:2253–2259
Larson JE, Lamb GC, Stevenson JS, Johnson SK, Day ML, Geary TW et al (2006) Synchronization of estrus in suckled beef cows for detected estrus and artificial insemination and timed artificial insemination using gonadotropin-releasing hormone, prostaglandin F2α and progesterone. J Anim Sci 84:332–342
Leite RC (1999) Control of bovine virus diarrhea (BVD) and infectious bovine rhinotracheitis (IBR). Rev Bras Reprod Anim 23:531–535
Martinez MF, Adams GP, Kastelic JP, Bergfelt D, Mapletoft RJ (2000) Induction of follicular wave emergence for estrus synchronization and artificial insemination in heifers. Theriogenology 54:757–769
Mainar-Jaime RC, Berzal-Herranz B, Arias P, Rojo-Vázquez FA (2001) Epidemiological pattern and risk factors associated with BVDV infection in a non-vaccinated dairy-cattle population from the Asturias region of Spain. Prev Vet Med 52:63–73
McEwan B, Carman S (2005) Animal health laboratory reports—cattle. Bovine abortion update, 1998–2004. Can Vet J 46:46
Meneghetti M, Vasconcelos JLM (2008) Calving date, body condition score, and response to a timed artificial insemination protocol in first-calving beef cows. Arq Bras Med Vet Zootec 60:786–793
Meneghetti M, Sá Filho OG, Peres RFG, Lamb GC, Vasconcelos JLM (2009) Fixed-time artificial insemination with estradiol and progesterone for B. indicus cows I: basis for development of protocols. Theriogenology 72:179–189
Miller JM, Van Der Maaten MJ (1987) Experimentally induced infectious bovine rhinotracheitis virus infection during early pregnancy: effect on the bovine corpus luteum and conceptus. Am J Vet Res 47:223–228
Mineiro ALBB, Bezerra EEA, Vasconcellos AS, Costa FAL, Macedo NA (2007) Leptospiral infection in bovine and its association with reproductive failure and climatic conditions. Arq Bras Med Vet Zootec 59:1103–1109
Moreira F, Orlandi C, Risco CA, Mattos R, Lopes F, Thatcher WW (2001) Effects of presynchronization

and bovine somatotropin on pregnancy rates to a timed artificial insemination protocol in lactating dairy cows. J Dairy Sci 84:1646–1659

Murphy BD, Martinuk SD (1991) Equine chorionic gonadotropin. Endocr Rev 12:1305–1319

Patterson DJ, Corah LR, Brethour JR, Spire MF, Higgins JJ, Kiracofe GH et al (1991) Evaluation of reproductive traits in B. taurus and B. indicus crossbred heifers: effects of postweaning energy manipulation. J Anim Sci 69:2349–2361

Pavlik EJ, Coulson PB (1976) Modulation of estrogen receptors in four different target tissue: differential effects of estrogen vs. progesterone. J Steroid Biochem 7:369–376

Peres RFG, Claro Júnior I, Sá Filho OG, Nogueira GP, Vasconcelos JLM (2009) Strategies to improve fertility in B. indicus postpubertal heifers and non-lactating cows submitted to fixed-time artificial insemination. Theriogenology 72:681–689

Pinheiro OL, Barros CM, Figueiredo RA, Valle ER, Encarnação RO, Padovani CR (1998) Estrus behavior and the estrus to ovulation interval in Nelore cattle (*B. indicus*) with natural estrus or estrus induced with prostaglandin F2 alpha or norgestomet and estradiol valerate. Theriogenology 49:667–681

Pursley JR, Mee MO, Wiltbank MC (1995) Synchronization of ovulation in dairy cows using PGF2α and GnRH. Theriogenology 44:915–923

Randel RD (1990) Nutrition and postpartum rebreeding in cattle. J Anim Sci 68:853–862

Restle J, Polli VA, De Senna DB (1999) Efeito de grupo genético e heterose sobre a idade e peso à puberdade e sobre o desempenho reprodutivo de novilhas de corte. Pesqui Agropecu Bras 34:701–707

Rhodes FM, Burke CR, Clark BA, Day ML, Macmillan KL (2002) Effect of treatment with progesterone and oestradiol benzoate on ovarian follicular turnover in postpartum anoestrous cows and cows which have resumed oestrous cycles. Anim Reprod Sci 69:139–150

Roberson MS, Wolfe MW, Stumpf TT, Kittok RJ, Kinder JE (1989) Luteinizing hormone secretion and corpus luteum function in cows receiving two levels of progesterone. Biol Reprod 41:997–1003

Rodrigues A, Peres R, Lemes A, Martins T, Aono F, Pereira M et al (2011) Puberty induction in Nelore heifers receiving eCG and/or estradiol cypionate at the end of the estrus synchronization protocol. J Anim Sci 89:593–594 (Abstr.)

Rodrigues ADP, Pereira MHC, Carvalho ER, Lemes AP, Martins T, Peres RFG et al (2012) Association of puberty induction protocol and timed-AI protocol in Nelore heifers. Anim Reprod 9:480 (Abstr.)

Sá Filho OG, Meneghetti M, Peres RFG, Lamb GC, Vasconcelos JLM (2009a) Fixed-time artificial insemination with estradiol and progesterone for B. indicus cows II: strategies and factors affecting fertility. Theriogenology 72:210–218

Sá Filho OG, Patterson DJ, Vasconcelos JLM (2009b) Development of estrous synchronization protocols using melengestrol acetate in B. indicus cattle. J Anim Sci 87:1981–1990

Sá Filho OG, Thatcher WW, Vasconcelos JLM (2009c) Effect of progesterone and/or estradiol treatments prior to induction of ovulation on subsequent luteal lifespan in anestrous Nelore cows. Anim Reprod Sci 112:95–106

Sá Filho OG, Vilela ER, Geary TW, Vasconcelos JLM (2009d) Strategies to improve fertility in postpartum multiparous B. indicus cows submitted to a fixed-time insemination protocol with gonadotropin-releasing hormone and prostaglandin F2{alpha}. J Anim Sci 87:2806–2814

Sá Filho OG, Dias CC, Lamb GC, Vasconcelos JLM (2010) Progesterone-based estrus synchronization protocols in non-suckled and suckled primiparous B. indicus beef cows. Anim Reprod Sci 119:9–16

Saldarriaga JP, Cooper DA, Cartmill JA, Zuluaga JF, Stanko RL, Williams GL (2007) Ovarian, hormonal, and reproductive events associated with synchronization of ovulation and timed appointment breeding of B. indicus-influenced cattle using intravaginal progesterone, gonadotropin-releasing hormone, and prostaglandin $F_2\alpha$. J Anim Sci 85: 151–162

Sartorelli ES, Carvalho LM, Bergfelt DR, Ginther OJ, Barros CM (2005) Morphological characterization of follicle deviation in Nelore (*B. indicus*) heifers and cows. Theriogenology 63:2382–2394

Sartori R, Fricke PM, Ferreira JCP, Ginther OJ, Wiltbank MC (2001) Follicular deviation and acquisition of ovulatory capacity in bovine follicles. Biol Reprod 65:1403–1409

Sergerson EC, Hansen TR, Libby DW (1984) Ovarian and uterine morphology and function in Angus Brahman cows. J Anim Sci 59:1026–1049

Stevenson JS, Knoppel EL, Minton JE, Salfen BE, Garverick HA (1994) Estrus, ovulation, luteinizing hormone and suckling-induced hormones in mastectomized in cows with and without unrestricted presence of the calf. J Anim Sci 72:690–699

Stock AE, Fortune JE (1993) Ovarian follicular dominance in cattle: relationship between prolonged growth of the ovulatory follicle and endocrine parameters. Endocrinology 132:1108–1114

Takiuchi E, Alfieri AF, Alfieri AA (2001) Bovine herpesvirus type 1: infection and diagnosis methods. Semina Cien Agr 22:203–209

Vasconcelos JLM, Silcox RW, Rosa GLM, Pursley JR, Wiltbank MC (1999) Synchronization rate, size of the ovulatory follicle and pregnancy rate after synchronization of ovulation beginning on different days of the estrous cycle in lactating dairy cows. Theriogenology 52:1067–1078

Vasconcelos JLM, Sá Filho OG, Perez GC, Silva ATN (2009a) Intravaginal progesterone device and/or temporary weaning on reproductive performance of anestrous crossbred Angus x Nelore cows. Anim Reprod Sci 111:302–311

each of the "clean," "green," and "ethical" dimensions of a modern ruminant industry (Durmic and Blache 2012).

CGE Management: A Staged Approach to Adoption

Nutrition is the core environmental factor in reproduction and is therefore a core management tool (Fig. 9.3). We use the operational term "focus feeding" for optimized management of nutrition, a concept that has much in common with the "nutrient synchrony" described by Hersom (2008) in the context of intensively fed animals.

The framework in Fig. 9.3 is best viewed as an ideal because full implementation still requires research and development. In some areas, it requires major changes in management. However, it is important to note that the suggested interventions are generally simple, "low tech" and applicable on a large scale. Moreover, some of the tools will be familiar to many farmers because they have been known, perhaps even part of normal practice, for many years. A possible exception is ultrasound for which the investment in materiel and skills would be beyond most farmers. A service industry is therefore required and has been developing rapidly in Australia and New Zealand over the past 20 years.

Given this situation, it is simply not feasible to expect producers to adopt the whole CGE management package in one step. Some of the tools require a new level of precision in management and, for others, we would need to convince farmers to abandon "traditional" practices (e.g., delayed first mating of pubertal females). To do so would require the use of "Fourth Generation" adoption practices because the traditional top-down approaches used by governmental agricultural agencies have failed for animal industries that are not vertically aligned (for an example, see Sneddon 2009).

In the next section of this chapter, we will expand on the 4-stage approach to the adoption of the CGE package, with progressive levels of difficulty. For individual managers, the optimum solution could be anywhere in this continuum, and perhaps with the management tools in combinations that differ to the suggestions below, because the overall plan would depend on the economic, geographical, and societal environment.

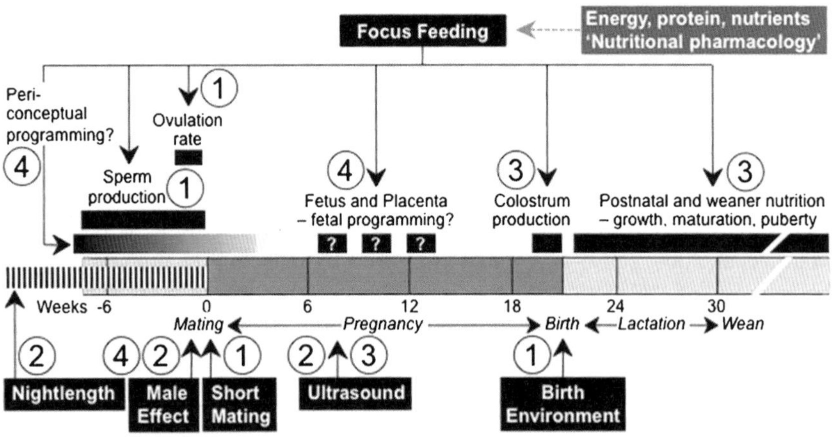

Fig. 9.3 A "Clean-Green-Ethical Package" for managing reproduction in sheep: periods of focus feeding to supply energy and protein are used to control the reproductive process and improve reproductive success. With the advent of "nutritional pharmacology," forages can also be used to supply bioactive secondary compounds that improve health and reduce emissions. To accurately synchronize the periods of feeding with specific reproductive events, mating must be controlled and brief, or ultrasound must be used to classify the mothers based on the age of their fetuses. Finally, the survival of the newborn must be maximized by a combination of good genetics and good management. The numbers in *circles* indicate a staged approach for maximizing reproductive output from *Basic* (*1*) to *Intermediate* (*2*), *Advanced* (*3*), and *The Future* (*4*). Modified after Martin and Kadokawa (2006)

Second Major Factor: Predictable Timing of Reproductive Events

As can be seen from Fig. 9.3, an essential goal is the accurate coordination of nutritional supplements with events in the reproductive process because it would ensure that costly supplements are kept to a minimum while appropriate metabolic signals are used to manage the reproductive outcome. At the simplest level, farmers control timing by managing the presence of the male and thus the time of fertile mating. More refined options allow better control and open up opportunities to use aspects of the reproductive biology of males other than gamete supply.

Stage 1: Basic

(a) *Restrict the duration of births in the herd or flock*

In extensive production systems, as well as smallholder systems, it is common for the males to be with the females for a very long period (sometimes throughout the year) so that the females have a "maximum opportunity to conceive." However, this is short-sighted because it prevents any precision in the management of birth, postnatal development, and marketing. In fact, the apparently self-evident principle of maximizing opportunity to conceive is misguided because the actual gains are much smaller than most appreciate (Fig. 9.4). It is very difficult to see how any gains from mating beyond two cycles can outweigh the longer-term benefits listed below.

The advantages of a restricted mating period:

1. Pressure on the genetics of female fertility—a 2-cycle mating period gives the females only two chances to conceive (Fig. 9.4) and uncovers any underlying infertility—failure of this test should lead to culling.
2. Reduced neonatal mortality—with a short, concentrated period of births, intensive management of the birth environment to reduce neonatal mortality becomes cost-effective. In addition, as we shall see below, concentrated births are also compatible with focus feeding to improve fecundity and increase the production of colostrum. In combination with ultrasound scanning, short mating becomes even more cost-effective.
3. Avoiding the "tail"—a cohort of offspring of an even age and size is easier to manage and to market.
4. Avoiding the effects of the "autumn feed gap"—small ruminants mate naturally in autumn and, in many environments, this is the worst season of the year for pasture quality and quantity. With extended mating periods at this time, the females that fail to conceive early will inevitably be attempting to do so in the second or third cycle, by which time they will have lost significant body mass, leading to lower conception rates and fecundity as the season progresses. The only way to avoid this problem is an expensive program of feed supplementation.
5. Increased value from ultrasound scanning—if conceptions are limited to one or two cycles, ultrasound scanning (see Stages 2, 3) also becomes more cost-effective because it will allow accurate segregation of females into groups carrying zero, one, or two fetuses. The nonpregnant females can be culled for profit and to improve flock fertility. The segregation of mothers carrying single and multiple fetuses is also the first step towards precision management of the birth environment, as mentioned above, and towards "focus feeding" for colostrum production (see Stage 3).

To overcome the insecurity of a decision to restrict the duration of mating, data can be collected on flock fertility. The simplest way is the use of a harness with marking crayons on the males, and then changing the color to determine whether the mated (marked) females come back into estrus. This works well for sheep. Alternatively, as we will discuss

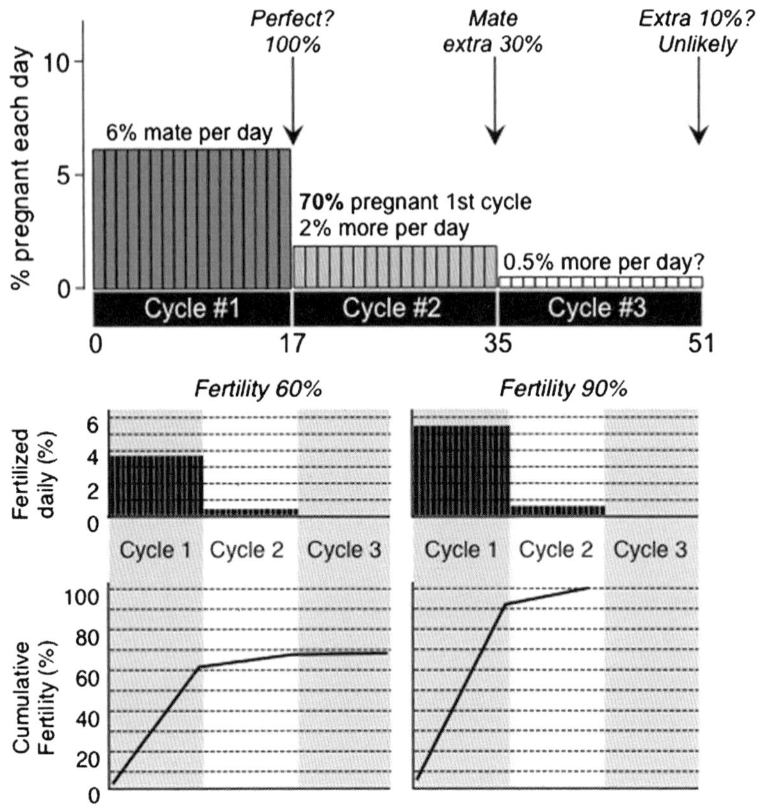

Fig. 9.4 Theoretical aspects of mating outcomes over a series of three estrous cycles for a sheep flock. *Top*: even assuming a low value for conception rate at 70 %, it is clear that that gains for a third cycle, in particular, are negligible. *Bottom*: cumulative flock conceptions comparing 60 % ewe fertility (*left*) with 90 % ewe fertility (*right*)

below, ultrasound can be used to identify pregnant and nonpregnant females.

(b) *Feed males for fertility*

Restricting the duration of mating immediately places pressure on the males, so they need to be managed correctly—adequate numbers, maximum mass of testis, anatomically sound, healthy, and fit. To maximize testis mass and therefore sperm production, males need to be fed a supplement for 8 weeks before mating (review: Martin et al. 2010). An important issue here is the concept of "fit but not fat"—males that are overweight and do not get exercise can perform poorly, even when they have maximum testicular mass. This is especially important in many genotypes of sheep and goats if the normal practice is to mate before midsummer because estrus in the females will be synchronized by the "male effect" (see below). Instead of the males having to serve 5–6 % of the females every day, as happens during the breeding season, they might encounter as many as 30 % of the females in estrus on some days. In Australia and New Zealand, the number of males might have to be increased from 1.5 to 4.0 per 100 females.

(c) *Flush for fecundity—maximize potential litter size (ovulation rate)*

Ovulation rate determines the upper limit of prolificacy, and thus productivity. Ovulation rate is under primary genetic control so it can be improved through selection, but the expression of that genetic potential is greatly influenced by the nutritional regime before mating (review: Scaramuzzi et al. 2011). This is evident from the correlations between body condition and litter size but, more importantly in the context of focus feeding, there is also an acute effect—in sheep, supplementation for only 4–6 days in the final stages of the estrous cycle can increase the frequency of twin ovulations by 20–30 % (Viñoles et al. 2005, 2009). In the absence of very precise control over the time of ovulation, supplementation for only 4–6 days is not realistic and it is necessary to offer the supplement for an entire cycle to ensure that all females have an opportunity to respond. If supplements are not feasible, high quality pasture can be used (Viñoles et al. 2009). Importantly, we were originally concerned that nutritional supplements immediately after ovulation would increase embryo mortality in early pregnancy (up to Day 14 after mating) because high levels of nutrition can increase progesterone clearance and thus reduce the circulating concentrations of progesterone (Parr 1992; Parr et al. 1993). This would have nullified any advantage in ovulation rate that might be gained from supplementation. However, the initial conclusions by Parr et al. (1993) seem to depend on comparison of quite extreme nutritional treatments and, in any case, the plasma concentrations of progesterone recover within 13 h after feeding a diet that provides double the requirements for maintenance so any decrease caused by enhanced clearance is transitory and not likely to have a deleterious effect on the uterine environment. Recent studies have shown that, although early embryo mortality is a major limitation to reproductive efficiency, there is little risk of it being induced by supplements under farm conditions (Viñoles et al. 2012).

(d) *Manage the birth environment—maximize neonatal survival*

High rates of neonatal mortality have obvious consequences for profitability and also carry a risk of market failure because the problem raises questions about the ethical credentials of the industry. Neonatal mortality therefore goes hand-in-hand with any plan to improve fecundity through genetic selection or focus feeding. In addition, if the neonates are lost because of mismanagement, then any investment in genetics has been wasted. The simplest approach is to better manage birth—provide a calm environment, and shelter, feed, and water close to the birth site. This will increase the amount of time the mother spends at the birth site and therefore improve the development of the mother–young bond (Nowak 1996).

Stage 2: Intermediate

The aim in Stage 2 is to continue to work towards short mating with the ultimate target of one cycle or less. To achieve a mating period shorter than one cycle, some form of synchronizing treatment is needed. Remarkably, a potential solution can be found in three fundamental aspects of the reproductive process that impose limitations on the management decision for the optimum time of the year for mating—puberty, seasonal breeding, and postpartum anestrus. In all three situations, the lack of ovulation is due to lack of GnRH output by the reproductive centers of the brain (Fig. 9.1). Science has been working for decades on ways to override this block with exogenous hormones, and great progress has been made, but these treatments now raise issues for human health ("clean") and pose risks with liberation of sex steroids into environment ("green") with, for example, the disposal of used intravaginal devices. In addition, for extensive management systems, the treatments are too costly. For these reasons, we need efficient, non-pharmacological methods for precise control of the timing of reproductive events. Here, we describe two possibilities (control of night-length and the "male effect") for managing the season of breeding, one of which also offers quite precise control over the prediction of reproductive events, and then explore the use of ultrasound as an alternative for when the first two are not feasible.

(a) *Changing the night-length*

Controlling the night-length for about 6 weeks can be readily used to "switch on" the reproductive centers of the brain in goats and sheep. This is attractive for farmers with flocks or herds of seasonal genotypes, but is probably limited to the males for practical reasons (they are fewer in number). The treatment can be used to ensure that their reproductive axis is working at maximum efficiency when the males are used for short mating, as described above, or for the "male effect" (Delgadillo et al. 2002; Zarazaga et al. 2011).

(b) *The "male effect"*
In many genotypes of sheep and goats, the sudden introduction of novel males can induce ovulation in females that are reproductively quiescent because they are prepubertal, out of season, or lactating (reviews: Martin et al. 1986; Rosa and Bryant 2002; Ungerfeld 2007). Most importantly, the ovulations induced by the "male effect" are usually sufficiently synchronized among a herd or flock of females to allow the use of strategies such focus feeding and even artificial insemination (Martin and Scaramuzzi 1983). However, the male effect is not universally effective—it works best with moderately seasonal genotypes and, currently, it can only be applied when the females are not ovulating spontaneously. Moreover, even with genotypes and environments where the male effect is reliable, the initially tight synchrony of ovulations in the flock or herd broadens quickly so there is little choice but to mate for about the equivalent of one cycle. Therefore, focus feeding for fecundity still cannot be refined to the theoretical minimum of 4–6 days—in sheep, at least 11 days is still needed, but there is nevertheless a significant saving in feed costs over the requirements for mating over a full cycle. Despite these limitations, the degree of synchrony that is offered by the male effect presents opportunities for far more precise management of the production system, particularly for the application of ultrasound, focus feeding during pregnancy, and the management of birth.

(c) *Ultrasound for pregnancy diagnosis*
Ultrasound pregnancy diagnosis is now a routine procedure in small ruminants, even for the large flocks and herds that are typical in Australia and New Zealand where numerous commercial operators offer reasonable rates and accurate results. There is no reason for animal managers to avoid it. At the simplest level, scanning will identify pregnant and nonpregnant females—even this information is very valuable because it offers: (1) an opportunity for re-mating if there has been a disaster; (2) culling for improvement of fertility; and (3) planning of conditions for birth.

Stage 3: Advanced

The aim with Stage 3 is precision management in extensive animal production systems. We are unlikely to achieve the levels of precision seen in intensive industries, such as pigs and dairy cattle, but great improvements are well within reach.

(a) *Ultrasound for number and age of fetuses*
The current best commercial practice in ultrasound provides number and approximate age of the fetuses, giving a date for birth within about 17 days and reliable options for focus feeding during pregnancy and for management of births. Ultimately, we might achieve a broad level of expertise and equipment that will allow us to determine the age of the fetus to within 5 days, but this is currently the regime of researchers (e.g., González de Bulnes et al. 1998).

If conceptions are limited to two cycles, ultrasound scanning becomes more cost-effective because it will provide two extra levels of information. Thus, skilled operators with modern instruments can:

1. Identify single-bearing and multiple-bearing females—pregnant females can be segregated into groups carrying single or multiple fetuses—the data can be used for selection for fecundity and those carrying multiple fetuses can be offered better management of multiple births, where there is inherent high rate of neonatal mortality.

2. Estimate fetus age (to within a cycle length)—segregation of fetuses into two broad age groups will increase the precision of focus feeding for colostrum production.

(b) *Colostrum production and survival of the newborn*

In addition to the nutritional and immunological benefits, colostrum in the gut improves the ability of the newborn to recognize its mother, thus contributing to the establishment of the mother–young bond (Goursaud and Nowak 1999). With the implementation of short mating and ultrasound assessment of pregnancy, focus feeding can be used in the last week of gestation to more than double the amount of colostrum available at birth (review: Banchero et al. 2006) and improve neonatal survival (Goodwin and Norton 2004). The value of ultrasound is again obvious—for animals in good body condition, feeding for colostrum can be limited to mothers predicted to have multiple births, thus avoiding the risk of dystocia caused by excessively large fetuses.

(c) *Early mating of young females (management of puberty)*

The traditional practice in the extensive industries is to delay the mating of young females until they are 18–24 months old (for sheep, these females are labelled as "maidens" in Australia and "hoggets" in New Zealand). The broader consequences for the industry are high greenhouse gas emissions intensity, delays in genetic progress, and reduction of the effective size of the breeding flock. This is a complex problem, with a mix of sociological and biological causes and solutions:

1. Sociological: farmer skepticism is a major aspect of the decision to delay first mating in young females, with the widely held belief that mating young animals impairs their performance in subsequent years, even though research indicates that this is not that case (Kenyon et al. 2004).
2. Biological: in young females, conception is limited by the factors that determine whether they have reached puberty at attempted mating (live weight, condition score, photoperiod). It is assumed that young females mated at 18–24 months have a higher weaning rate than those mated at 12 months simply because they are heavier at mating but this is a very low-level interpretation of the problem and should not preclude new management strategies. It is important to focus on ensuring ovulation and conception and then the survival of the newborn (a challenge with young experienced females). The male effect (see Kenyon et al. 2006), ultrasound, and birth-site management are obvious targets, as outlined above for adult females. Clearly, with body mass being such a large factor in the physiological triggering of puberty, "focus feeding" needs to be considered, as well as delaying the first breeding season by a few weeks to allow extra weight gain. In recent studies with young sheep, it has been shown that genetic selection can be used to increase growth rate and thus advance the onset of puberty, even if the extra growth is primarily muscle (Rosales Nieto et al. 2013); this finding plus the general industry need to mate females early should be driving new research in this domain. Finally, there is also scope for the use of alternative forages, from both nutritional and "bioactive" perspectives, to improve performance—young animals often suffer a period of "ill-thrift" after weaning and two major contributing factors are poor forage quality and internal parasitism.

If young females do not fall pregnant at a reasonably early opportunity, they should be culled since they may well perform poorly throughout their life. However, underlying this general principle is the assumption that they are not being culled because they have been handicapped by poor management.

Stage 4: The Future

The ultimate versions of the CGE management package promise much greater precision for the coordination of focus feeding with the stages of the reproductive process. Complexity seems inevitable but it is a threat that needs to be avoided for the extensive industries because low labor inputs are essential to keep input costs down. The next generation of management tools, proposed below, will not be realized without considerable fundamental research followed by industrial development. In general, we still do not understand the biological principles that underpin the phenomena.

(a) *The ultimate male effect*

The male effect offers great promise as a way of controlling the timing of the reproductive process and thus permitting the ultimate in focus feeding (e.g., flushing for only 4–6 days). The major attraction is the sheer simplicity and practicality of the idea—putting males in the field with the females and then knowing with confidence when the females will ovulate, conceive, and give birth. However, even in situations where the male effect is currently reliable, there can be wide variation among females in delay to the induced ovulation, in the incidence of short cycles following the first ovulation, and in the outcomes of mating (reviews: Rosa and Bryant 2002; Ungerfeld 2007). The loss of tight synchrony of ovulations in the flock or herd means that we cannot pursue the ultimate versions of focus feeding for fecundity and for colostrum, or manage birth conditions in the most cost-effective way. Moreover, the ultimate version of focus feeding would also take advantage of the biological phenomenon of "programming" of gametes, zygotes, and fetuses. This will not be feasible without tight coordination of metabolic inputs with reproductive events.

Thus, for the male effect, there are three major limitations that need to be resolved through research and development:

1. The season problem: on the distant horizon, there is hope for application to cycling females but further research is required. Originally, a male effect during the breeding season was considered impossible because it was thought that luteal progesterone would block the GnRH response to the sociosexual signals—this has now been refuted (Hawken et al. 2007; 2009) and, in sheep, there is evidence of cycle synchronization among females in the flock (Hawken et al. 2008).
2. The genotype problem: the sociosexual factors responsible for the male effect converge at sites in the preoptic-hypothalamic continuum that control the output of GnRH (Fig. 9.1). It is highly likely that these pathways are the same in all genotypes. For example, even in the very seasonal Suffolk sheep, sociosexual signals can elicit an endocrine response (Blache et al. 2003). We therefore need to unravel the control processes that currently allow photoperiod to override the sociosexual factors. Inevitably, we will need to harness the power of genetics as we seek a solution.
3. The short cycle problem: following the first ovulation induced by the males, the corpus luteum that is formed either persists until normal luteolysis at the end of the luteal phase of the cycle, or it regresses prematurely 5–6 days after ovulation. Premature regression is followed immediately by another ovulation and this corpus luteum persists normally. The consequence of this phenomenon is a degradation of the initial highly synchronized ovulation in the flock as animals allocate themselves into two separate populations. The synchrony that remains is still very useful (as outlined above) but the prevention of the short cycle is a high priority. In fact, we do have a solution—if the females are all treated with a single injection of progesterone at the time of male induction, the short cycles are prevented, although our ultimate goal is to avoid such interventions. The cause of the short cycles has been investigated in several laboratories (review: Chemineau et al.

2006) but the issue does not appear to be fully resolved at this stage.

(b) "*Programming*" *the future productivity of the offspring*
Feeding during pregnancy for fetal programming (sperm production, ovulation rate, milk production) depends on precise knowledge over the timing of the process of fetal development. This concept originally arose in the context of human health, and so is now known as "developmental origins of human disease," but we see the biological principle as having much broader applications than disease prevention. First, from the beginning, the phenomenon was seen as an issue only during fetal development, probably associated with tissue differentiation, but there is now gathering evidence that the early conceptus, and even the oocyte, can be "programmed" (Bloomfield et al. 2003; Thompson 2006; Hernandez et al. 2010; Fleming et al. 2012). Second, several non-disease aspects of sheep production are known to be affected by the nutrition of the mother during fetal life (review: Martin et al. 2004): (1) initiation and development of secondary follicles in the skin; (2) inhibitory effects of undernutrition during gestation on muscle fiber formation; (3) development of the reproductive axis. Recently, tolerance to a high-salt diet has been added to the list (Digby et al. 2010, 2011).

Before these effects can be incorporated into management, we need to quantify the consequences for productivity and add far more precision to the nature of the metabolic signals and the timing during development when those signals will alter the outcome for the offspring.

The Genetic Frontiers

To date, much of our research has targeted the physiological, behavioral, and managerial aspects of CGE management. Clearly, we also need to consider the role of genetics because, as can be seen from much of the above discussion, genotypic factors are a major restraint to the full implementation of the CGE package. Martin and Greeff (2011) reviewed the situation for Merino sheep under extensive grazing and selected several areas for discussion based on evidence of genetic variation (known breed differences or within-breed variation) and of heritability of the trait under consideration (Fig. 9.5).

In brief,

(a) *Enhancing the power of the male effect*
The scope for genetics-driven research on the male effect is clear: (1) it is highly likely that all breeds have the anatomy and physiology,

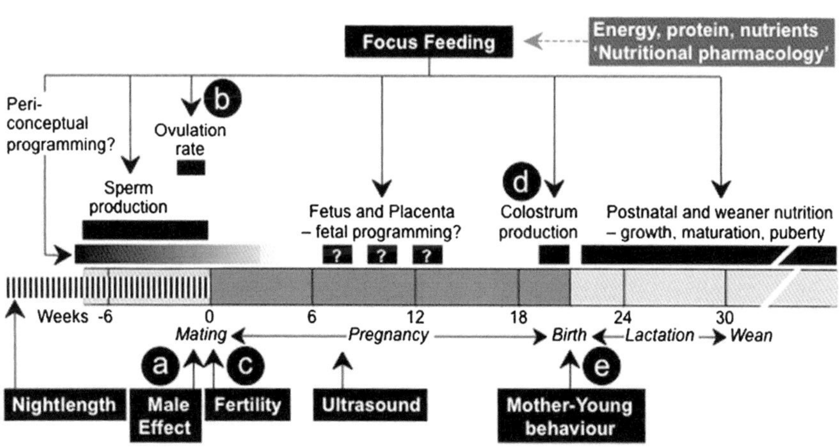

Fig. 9.5 Aspects of the "CGE Management Package" for sheep showing five areas (*a–e*) where there is scope for genetic selection. Redrawn after Martin and Greeff (2011)

and thus the genes, that underpin the phenomenon; (2) there are profound differences among genotypes in responsiveness to the male effect; (3) there is considerable variation among genotypes, and among individuals within a genotype, in the way they express their breeding season, probably due to differences in the strength of the photoperiod-driven "filter" (Fig. 9.1). The outcome is genetic variation in the strength of the sociosexual signals from the males and in the responsiveness of the females to those signals.

(b) *Increasing fecundity*
Basically, our aim should be the identification of animals that have the genetic potential to produce a maximum of two ovulations, perhaps with the final outcome of single or twin births being decided by the breeder using focus feeding.

(c) *Increasing fertility in the face of short period of mating*
In extensively managed flocks and herds, the practice of long periods of mating has been placed little pressure on female fertility, so there should be great scope for improvement.

(d) *Increasing colostrum production*
Two areas could be exploited: (1) variation between genotypes (e.g., milk breeds, meat, and fiber breeds) in the quantity produced; (2) variation among individuals in the timing of production, leading to variation in the synchrony of parturition and colostrum supply.

(e) *Enhancing mother–young bonding*
Variation among genotypes in neonatal survival can be dissected into two behavioral traits: time taken by the mother to recognize its newborn, and time taken by the newborn to recognize its mother.

Genetics-driven, long-term solutions to some of the major limitations in the CGE framework are highly likely in the near future because we are on the verge of a technology-led revolution in the generation of genetic data: electronic identification, DNA pedigrees, and the automatic recording of body weights and number of lambs born. This combination of technologies will make it possible to assess large numbers of animals for a wide variety of production traits under extensive production systems. The outcome will be new directions in research, with genetics as well as physiology and behavior, leading to an expansion of the scope of the CGE concept. Importantly with respect to our primary premise, genetic improvement is a simple, low-cost management tool and the benefits are cumulative. This is perfect for extensive grazing industries.

Application of CGE Management in Other Socioeconomic Environments

The CGE concept was developed in response to consumer demand in high-profit export markets for Australian ruminant-derived products. Implementation of CGE management into our major production systems, involving large numbers of animals on large areas of land with minimal inputs, constrains management options to those that are simple and inexpensive. Does this mean that the whole concept is irrelevant to smallholders in developing economies? On the contrary—the smallholders also need simple and inexpensive management options. Further, we would hypothesize that CGE production would also enable smallholder producers to target consumers with discretionary spending power, perhaps allowing them to get higher prices and emerge from a subsistence economy.

Consider the example of sheep in Tunisia: use of the "male effect" to control the timing of reproduction (Lassoued et al. 2004a); use of "focus feeding" to improve reproductive outcomes (Lassoued et al. 2004b); use of novel local forages to replace expensive imported supplements (Rekik et al. 2010, 2012); better management of prepubertal development (Ben Salem et al. 2009). Thus we see the application of selected tools from Stages 1 and 2. It is not difficult to see that the next steps could include dietary components for controlling gastrointestinal nematodes and methane emissions, and the provision of practical ultrasound services. Ultimately, they will need to harness the power of genetics—this will require the identification of individual animals, a difficult hurdle in traditional smallholder industries, but animal identification is also the first step towards traceability, a prerequisite for high-profit export markets.

Conclusions

Understanding the reproductive responses of animals to exteroceptive factors such as photoperiod, nutrition and sociosexual signals is an obvious step towards the replacement of exogenous hormones and drugs to control and improve the productivity of small ruminants. In addition, we now have forages that can be used to reduce emissions and improve health, while increasing reproductive output. The basic tools (Stage 1) are already standard practice in most farms in Australia and New Zealand; most of those in Stage 2 and some in Stage 3 are becoming common in the more advanced enterprises, although they often require local interpretation. The Stage 4 tools, on the other hand, are still subject to a basic research and development.

Where Does High-End Reproductive Technology Fit?

Although we do not see reproductive technology as playing a direct role in extensive management systems, it is important to state clearly that our perspective is not "anti-technology." On the contrary, reproductive technology has two major roles in the development of our CGE framework:

(a) It is at the core of much of the scientific experimentation that has been done to date and that will be needed in the future; indeed, the whole CGE concept (as outlined in Fig. 9.3), can be seen as a distillation of decades of research that depended on reproductive and other technologies.
(b) It greatly accelerates genetic progress and we need massive progress in some areas.

The use of CGE management of our animals can be cost-effective and improve productivity, at the same time greatly improving the image of our industries in society and the marketplace. All we need is a little more research and development in genetics, physiology, nutrition, and behavior.

References

Banchero GE, Perez Clariget R, Bencini R, Lindsay DR, Milton JTB, Martin GB (2006) Endocrine and metabolic factors involved in the effect of nutrition on the production of colostrum in female sheep. Reprod Nutr Dev 46:447–460

Ben Salem I, Rekik M, Ben Hamouda M, Lassoued N, Blache D (2009) Live weight and metabolic changes and the associated reproductive performance in maiden ewes. Small Rumin Res 81:70–74

Blache D, Zhang S, Martin GB (2003) Fertility in males: modulators of the acute effects of nutrition on the reproductive axis of male sheep. In: Campbell BK, Webb R, Dobson H, Doberska C (eds) Reproduction in domestic ruminants V. Society for Reproduction and Fertility, Cambridge, pp 387–402

Bloomfield FH, Oliver MH, Hawkins P, Campbell M, Phillips DJ, Gluckman PD et al (2003) A periconceptional nutritional origin for noninfectious preterm birth. Science 300:606

Boukhliq R, Martin GB, White CL, Blackberry MA, Murray PJ (1997) Role of glucose, fatty acids and protein in the regulation of testicular growth and the secretion of gonadotrophin, prolactin, somatotrophin and insulin in the mature ram. Reprod Fertil Dev 9:515–524

Chemineau P, Pellicer-Rubio M-T, Lassoued N, Khaldi G, Monniaux D (2006) Male-induced short oestrous and ovarian cycles in sheep and goats: a working hypothesis. Reprod Nutr Dev 46:417–429

de Brosses C (1756) Histoire des navigations aux terres australes, contenant ce que l'on sait des moeurs et des productions des contrées découvertes jusqu'à ce jour. Durand, Paris

Delgadillo JA, Flores JA, Véliz FG, Hernández HF, Duarte G, Vielma J et al (2002) Induction of sexual activity of lactating anovulatory female goats using male goats treated only with artificial long days. J Anim Sci 80:2780–2786

Digby SN, Blache D, Masters DG, Revell DK (2010) Responses to saline drinking water in offspring born to ewes fed high salt during pregnancy. Small Rumin Res 91:87–92

Digby SN, Chadwick MA, Blache D (2011) Salt intake and reproductive function in sheep. Animal 5:1207–1216

Durmic Z, Blache D (2012) Bioactive plants and plant products: effects on animal function, health and welfare. Anim Feed Sci Technol 176:150–162

Durmic Z, Hutton P, Revell DK, Emms J, Hughes S, Vercoe PE (2010) In vitro fermentative traits of Australian woody perennial plant species that may be

considered as potential sources of feed for grazing ruminants. Anim Feed Sci Technol 160:98–109
Fleming TP, Lucas ES, Watkins AJ, Eckert JJ (2012) Adaptive responses of the embryo to maternal diet and consequences for post-implantation development. Reprod Fertil Dev 24:35–44
González de Bulnes A, Santiago Moreno J, López Sebastián A (1998) Estimation of fetal development in Manchega dairy ewes by transrectal ultrasonographic measurements. Small Rumin Res 27:243–250
Goodwin N, Norton BW (2004) Improving doe nutrition immediately prior to kidding increases kid survival. Proc Aust Soc Anim Prod 25:233
Goursaud AP, Nowak R (1999) Colostrum mediates the development of mother preference by the new born lamb. Physiol Behav 67:49–56
Hawken PAR, Beard AP, Esmaili T, Kadokawa H, Evans ACO, Blache D, Martin GB (2007) The introduction of rams induces an increase in pulsatile LH secretion in cyclic ewes during the breeding season. Theriogenology 68:56–66
Hawken PAR, Evans ACO, Beard AP (2008) Short term, repeated exposure to rams during the transition into the breeding season improves the synchrony of mating in the breeding season. Anim Reprod Sci 106:333–344
Hawken PAR, Esmaili T, Jorre de St Jorre T, Martin GB (2009) Do cyclic female goats respond to males with an increase in LH secretion during the breeding season? Anim Reprod Sci 112:384–389
Hernandez CE, Matthews LR, Oliver MH, Bloomfield FH, Harding JE (2010) Effects of sex, litter size and periconceptional ewe nutrition on offspring behavioural and physiological response to isolation. Physiol Behav 101:588–594
Hersom MJ (2008) Opportunities to enhance performance and efficiency through nutrient synchrony in forage-fed ruminants. J Anim Sci 86:E306–E317
Kenyon PR, Morris ST, Perkins NR, West DM (2004) Hogget mating use in New Zealand—a survey. Proc N Z Soc Anim Prod 64:217–222
Kenyon PR, Morel PCH, Morris ST, Burnham DL, West DM (2006) The effect of length of use of teaser rams prior to mating and individual liveweight on the reproductive performance of ewe hoggets. N Z Vet J 54:91–95
Lassoued N, Naouali M, Khaldi G, Rekik M (2004a) Influence of the permanent presence of rams on the resumption of sexual activity in postpartum Barbarine ewes. Small Rumin Res 54:25–31
Lassoued N, Rekik M, Mahouachi M, Ben HM (2004b) The effect of nutrition prior to and during mating on ovulation rate, reproductive wastage, and lambing rate in three sheep breeds. Small Rumin Res 52:117–125
Makkar HPS, Francis G, Becker K (2007) Bioactivity of phytochemicals in some lesser-known plants and their effects and potential applications in livestock and aquaculture production systems. Animal 1:1371–1391
Martin GB (1995) Reproductive research on farm animals for Australia—some long-distance goals. Reprod Fertil Dev 7:967–982
Martin GB, Greeff JC (2011) Genetic frontiers in the development of 'clean, green and ethical' management systems for the extensive sheep industry. Proc Assoc Adv Anim Breed Genet 19:143–150
Martin GB, Kadokawa H (2006) "Clean, green and ethical" animal production. Case study: reproductive efficiency in small ruminants. J Reprod Dev 52:145–152
Martin GB, Scaramuzzi RJ (1983) The induction of oestrus and ovulation in seasonally anovular ewes by exposure to rams. J Steroid Biochem 19:869–875
Martin GB, Oldham CM, Cognié Y, Pearce DT (1986) The physiological responses of anovulatory ewes to the introduction of rams—a review. Livest Prod Sci 15:219–247
Martin GB, Milton JTB, Davidson RH, Banchero Hunzicker GE, Lindsay DR, Blache D (2004) Natural methods of increasing reproductive efficiency in sheep and goats. Anim Reprod Sci 82–83:231–246
Martin GB, Blache D, Williams IH (2008) The costs of reproduction. In: Rauw WM (ed) Resource allocation theory applied to farm animals. CABI Publishing, Oxford, pp 169–191
Martin GB, Durmic Z, Kenyon PR, Vercoe PE (2009) Landcorp lecture: 'clean, green and ethical' animal reproduction: extension to sheep and dairy systems in New Zealand. Proc N Z Soc Anim Prod 69:140–147
Martin C, Morgavi DP, Doreau M (2010) Methane mitigation in ruminants: from microbe to the farm scale. Animal 4:351–365
Martin GB, Blache D, Miller DW, Vercoe PE (2010) Interactions between nutrition and reproduction in the management of the mature male ruminant. Animal 4:1214–1226
Min BR, Barry TN, Attwood GT, McNabb WC (2003) The effect of condensed tannins on the nutrition and health of ruminants fed fresh temperate forages: a review. Anim Feed Sci Technol 103:3–19
Nowak R (1996) Neonatal survival: contributions from behavioural studies in sheep. Appl Anim Behav Sci 49:61–72
Parr RA (1992) Nutrition-progesterone interactions during early pregnancy in sheep. Reprod Fertil Develop 4:297–300
Parr RA, Davis IF, Miles MA, Squires TJ (1993) Feed intake affects metabolic clearance rate of progesterone in sheep. Res Vet Sci 55:306–310
Rekik M, Ben Salem H, Lassoued N, Chalouati H, Ben SI (2010) Supplementation of Barbarine ewes with spineless cactus (*Opuntia ficus-indica f. inermis*) cladodes during late gestation-early suckling: effects on mammary secretions, blood metabolites, lamb growth and postpartum ovarian activity. Small Rumin Res 90:53–57
Rekik M, Gonzalez-Bulnes AM, Lassoued N, Ben Salem H, Tounsi A, Ben SI (2012) The cactus effect: an alternative to the lupin effect for increasing ovulation rate in sheep reared in semi-arid regions? J Anim Physiol Anim Nutr 96:242–249
Revell DK, Durmic Z, Bennell M, Sweeney GC, Vercoe PE (2008) The *in situ* use of plant mixtures including native shrubs in Australian grazing systems: the potential to capitalise on plant diversity for livestock

health and productivity. In: Skaife JF, Vercoe PE (eds) Harvesting knowledge, pharming opportunities. Cambridge University Press, Cambridge, pp 36–49

Rosa HJD, Bryant MJ (2002) The 'ram effect' as a way of modifying the reproductive activity in the ewe. Small Rumin Res 45:1–16

Rosales Nieto CA, Ferguson MB, Macleay CA, Briegel JR, Martin GB, Thompson AN (2013) Selection for superior growth advances the onset of puberty and increases reproductive performance in Merino ewe lambs. Animal 7:990–997

Scaramuzzi RJ, Baird DT, Campbell BK, Driancourt M-A, Dupont J, Fortune JE et al (2011) Regulation of folliculogenesis and the determination of ovulation rate in ruminants. Reprod Fertil Dev 23:444–467

Sneddon JN (2009) Identifying and exploiting opportunities for 'clean, green and ethical' animal production. Agrociencia 13:51–58

Thompson JG (2006) The impact of nutrition of the cumulus oocyte complex and embryo on subsequent development in ruminants. J Reprod Dev 52:169–175

Ungerfeld R (2007) Socio-sexual signalling and gonadal function: opportunities for reproductive management in domestic ruminants. In: Juengel JI, Murray JF, Smith MF (eds) Reproduction in domestic ruminants VI. Nottingham University Press, Nottingham, pp 207–221

Viñoles C, Forsberg M, Martin GB, Cajarville C, Repetto J, Meikle A (2005) Short-term nutritional supplementation of ewes in low body condition affects follicle development due to an increase in glucose and metabolic hormones. Reproduction 129:299–309

Viñoles C, Meikle A, Martin GB (2009) Short-term nutritional treatments grazing legumes or feeding concentrates increase prolificacy in Corriedale ewes. Anim Reprod Sci 113:82–92

Viñoles C, Glover KMM, Paganoni BL, Milton JTB, Martin GB (2012) Embryo losses in sheep during short-term nutritional supplementation. Reprod Fertil Dev 24:1040–1047

Zarazaga LA, Celi I, Guzmán JL, Malpaux B (2011) Enhancement of the male effect on reproductive performance in female Mediterranean goats with long day and/or melatonin treatment. Vet J 192: 441–444

A Perspective on the Impact of Reproductive Technologies on Food Production in Africa

10

Esté van Marle-Köster and Edward C. Webb

Abstract

Africa for the largest part is still regarded as part of the developing world and has a history of political instability, natural disasters, floods and droughts that all had an effect on the development of livestock production systems and the potential application of biotechnologies. It is expected that the human population in sub Saharan Africa will experience a growth of 1.2 % per year over the next 30 years. There is therefore pressure to increase sustainable productivity of livestock. Reproductive technologies such as Artificial Insemination in Africa were driven primarily by the need to control or prevent venereal diseases like Trichomoniases and *Campylobacter fetus* in cattle. Reproductive biotechnology had a limited impact in Africa due to several factors including a lack of infrastructure and animal recording systems, clear breeding objectives and continuously changing production systems and markets. Africa has a large variety of genetic resources adapted to the diverse environment and production systems and biotechnology should be applied within this context for an increase in food production.

Keywords

Reproductive biotechnology • Artificial insemination • Embryo transfer • Indigenous • Cattle sheep • Goats • Animal recording • Animal identification • Adaptation

---

E. van Marle-Köster (✉) • E.C. Webb (✉)
Department of Animal and Wildlife Sciences,
University of Pretoria, Private bag X20, Hatfield,
South Africa
e-mail: Evm.koster@up.ac.za; Edward.webb@up.ac.za

G.C. Lamb and N. DiLorenzo (eds.), *Current and Future Reproductive Technologies and World Food Production*, Advances in Experimental Medicine and Biology 752,
DOI 10.1007/978-1-4614-8887-3_10, © Springer Science+Business Media New York 2014

Introduction

Africa is the second largest continent in the world with an estimated population size of approximately one billion people. This continent boasts a range of diverse climatic regions including desert, semi-deserts regions, tropical rain forests and some of the most beautiful open savannas in the world. African people have always been dependent on some form of agriculture for their livelihood and through many centuries of discovery, colonization and independence, livestock has been kept under a range of production systems. Africa for the most part is still regarded as part of the developing world and has a history of political instability, natural disasters, floods and droughts that all had an adverse effect on the development of livestock production systems and potential application of new technologies. Traditional livestock systems have always favoured cattle, sheep and goats as the major species and each contributed to the needs of the household in different ways. Cattle are very much associated with status in the community and serve as an indication of prosperity, while sheep and goats are also used in rituals, cultural ceremonies and for consumption. Poultry is kept by most households for eggs and meat and is a major contributor of animal protein in deep rural areas. Swine are less popular due to religious reasons, as well as the intensive nature of production system and more complex feed requirements (Faustin and Kyvsgaard 2003).

The world population is expected to grow to an estimated 9.5 billion people by 2050 and most of the growth will occur in the developing world (Thornton 2010). It is expected that the human population in sub Saharan Africa will experience a growth of 1.2 % per year. Projected trends for meat and milk consumption in the developing world by 2015 are 32 and 55 kg, respectively, while current per capita consumption is approximately 28 kg of meat and 44 kg of milk per year. These trends indicate an ever increasing demand for protein and it is expected that this protein will be supplied by livestock (Delgado et al. 1999; Kahi and Rewe 2008; Pica-Ciamarra et al. 2010).

There is therefore pressure to increase the productivity of livestock production for both dairy and meat from the respective species on a commercial level. This increase in productivity and efficiency will have to take into account the improvement of sustainability of livestock production in Africa (Cunningham 1999; Scholtz et al. 2011). Due consideration is required for reduction of livestock emissions under natural grazing systems and welfare issues with regard to the technologies employed (Koehler-Rollefson 2012). This increase in productivity has to take place in environments subjected to more frequent droughts, with limited available pastures for grazing and high costs of intensive feeding. In the developed world biotechnology has been adopted on several levels for increasing productivity and lowering production costs by selection and genetic improvement in most livestock species (Van Arendonk 2011). Biotechnology has found application not only in reproduction and genetics, but also in improvements in health (Ruxandra 2010) and nutrition (Rode et al. 2010). Several papers have been published on the potential impact of biotechnology in African livestock production (Kahi and Rewe 2008; Rege et al. 2011). It is clear that the implementation of biotechnology, including reproductive technology has been complex with varying degrees of success in different African countries. In this chapter reproductive technologies and its associated effect on genetic improvement have been reviewed in context with the livestock production systems used in Africa with a perspective on the potential impact on food production.

Livestock Breeds in Africa

Africa has a large number of different local cattle, sheep and goat breeds and a number of exotic breeds that have been introduced to several countries over the past few centuries. Several surveys have been conducted to determine population sizes and breed trends but the statistics remain inadequate. Statistics on population sizes and different breeds reported by Scherf (2000) indicated that in Africa there are 251 cattle, 147 sheep and 89 goat breeds and from these 23 and 8 cattle and sheep breeds, respectively, are already extinct.

Table 10.1 Estimated population sizes for cattle, sheep and goats sub Saharan Africa (FAO STAT 2012)

Country	Cattle	Goats	Sheep
Algeria	1,650,000	3,800,000	20,000,000
Angola	5,143,000	2,571,000	355,000
Burkina Faso	9,845,000	12,377,500	8,050,000
Cameroon	5,700,000	4,450,000	3,850,000
Central African Republic	3,893,000	4,862,000	388,000
Chad	7,419,100	6,751,300	3,026,500
Egypt	4,524,950	4,200,000	5,591,580
Ethiopia	*50,884,000*	21,960,700	25,979,900
Kenya	17,862,900	13,291,700	9,899,300
Libya	195,000	2,700,000	7,000,000
Madagascar	9,900,000	1,280,000	730,000
Mali	9,163,280	16,522,500	11,865,300
Mauritania	1,677,630	5,500,000	8,860,000
Morocco	2,895,800	5,685,700	18,023,200
Niger	9,817,340	13,673,100	10,917,200
Nigeria	16,578,000	*56,524,100*	35,519,800
Somalia	5,350,000	12,700,000	13,100,000
South Africa	13,731,000	6,274,850	24,501,000
Sudan	41,726,700	43,441,000	*52,014,100*
Tunisia	670,900	1,295,940	7,234,070
Uganda	7,650,000	8,800,000	1,850,000
United Republic of Tanzania	19,500,000	12,900,000	4,200,000

http://faostat3.fao.org/home/index.html

In Table 10.1 the top 15 countries in Africa are shown with the most cattle, sheep and goats as provided by the FAO STAT (2012) and there is no doubt that these species have a major role to play in African agriculture. The focus for genetic improvement should be on these species and appropriate biotechnology and genetic tools should be employed to ensure success under diverse and challenging systems.

Since the late 1990s there has been awareness regarding conservation of farm animal genetic resources and several surveys and studies have followed on livestock breeds in Africa (Rege 1999). Studies were aimed to define breeds on common ancestry as livestock breeds were often "named" according to the geographical or ecological region where they were kept. Phenotypic and genetic characterization of a large number of African livestock breeds has been conducted that contributes to the knowledge and utilization of these breeds in the different countries. Phenotypic characterization, include studies based on performance and body measurements and type/breed traits (Mwacharo et al. 2006; Amare et al. 2012). Since the development of molecular techniques and DNA marker discoveries, most studies after the 1990s have focused on genetic diversity and population structure of the various indigenous and local breeds (Hanotte and Jianlin 2005; Lenstra et al. 2011). Studies such as these have provided valuable insight to the dangers of uncontrolled cross-breeding and decrease of effective population sizes for some of these breeds, for example, the studies by Ibeagha-Awemu and Erhardt (2004) on 12 African cattle breeds, Gizaw et al. (2007) on indigenous sheep in Ethiopia and Qwabe et al. (2012) on Namakwa Afrikaner sheep in South Africa.

Livestock breeds in Africa have evolved over many centuries in different regions that are characterized by large seasonal variations that affected the available quantity and quality of the grazing and the ecto- and endoparasite infestations. Production systems vary from smallholders with small herd or flock sizes to pastoralists that need to migrate on a seasonal basis and require different species and breeds to cope within these systems (Mirkena et al. 2010). Most of these local African breeds, however, are associated with lower reproduction and growth compared to exotic breeds. There are however sufficient evidence and examples for a number of African breeds that have unique characteristics with regard to adaptations to heat and disease susceptibility (Bonsma 1980). Cattle breeds such as the Ndama has resistance against tryponosomasis (Roberts and Gray 1973; Mattiolo et al. 2000) and Sanga breeds such as the Nguni of South Africa has been shown to be more adapted to subtropical areas with high temperature, humidity and feeding requirements and tick resistance (Spickett et al. 1989; Mapiye et al. 2007). Research has indicated that sheep breeds such as the Red Maasai has a higher resistance against endoparasites compared to Dorper sheep (Barker 1988), while West African Djallanke sheep showed resistance against tryponosomasis and endoparasite infections (Goosens et al. 1999).

Due to the poorer production potential of local breed types compared to exotics breeds, there have been several attempts to select for increased growth and higher outputs in these breeds. Reproductive biotechnologies were applied with the aim to genetically improve the breeds and enhance production. Local breeds were crossed with exotic breeds, but the outcomes were often disappointing due to the lack of defining breeding objectives and appropriate selection criteria (Van Arendonk 2011). Cross-breeding further has the danger of decreasing adaptability to both the specific climatic conditions and the type of production system that these breeds have become adapted to (Taberlet et al. 2008). The use of and impact of reproductive technologies to improve the genetic potential and increase food production in Africa should therefore be evaluated by keeping these factors in mind.

Reproductive Biotechnologies

Reproductive biotechnologies are employed for a variety of purposes in different species and these technologies are used worldwide. Reproductive technologies such as AI have the potential to increase genetic gain through the use of superior and proven sires. Through AI there is selection pressure on certain traits that need to be improved and this can be effectively done via sires with large numbers of offspring. The semen of a proven bull can be used to sire thousands of calves while the bull can only sire about 30 calves per natural mating in a breeding season. In addition, the expenses associated with the transportation of animals for breeding purposes as well as the constraints associated with the quarantine of animals, no longer limits the use of the best genetic material worldwide. AI has the advantages of importation of semen of the elite sires and used in local breeds, where a structured seed stock industry does not exist, for example, in some African countries. It also holds the advantage for use in small herds where farmers cannot afford to keep seed stock bulls and do not need to maintain a bull. Farmers can use different AI bulls in their programs to satisfy the breeding objective of the enterprise.

The initial commercialization of artificial insemination (AI) of livestock as reproductive biotechnology in Africa was driven primarily by the need to control or prevent venereal diseases like Trichomoniases and *Campylobacter fetus* (previously known as Vibriosis) in cattle. The real potential of AI only became apparent after the development of the recto-vaginal AI technique and dilution, processing and cryopreservation of bull semen in easily transportable liquid nitrogen (Dewar) flasks. The use of frozen and stored germplasm also has benefits in terms of the conservation of genetic resources and a number of scarce species already benefit from this technology (Holt and Pickard 1999).

Many scarce African mammals are deserving of conservation efforts, but the reproduction of only a few of these have been studied in sufficient detail to warrant the successful application of reproductive biotechnologies as conservation tool. Generally the use of in vitro fertilization and embryo transfer in wildlife has been limited to date (Holt and Pickard 1999). In South Africa these technologies have been used to breed disease-free buffalo (free from foot-and-mouth disease and tuberculosis) and other scarce

African ungulates like Roan and Sable antelope, due to the increasing monetary value of these species for the game and hunting industry.

Other reproductive biotechnologies that are used include the synchronization of estrus in female livestock to improve the practical management of AI within breeding programs. Synchronization of estrus makes it possible to artificially inseminate a large number of females on a predetermined date, which eliminates the necessity of daily estrus observations, followed by artificial insemination. The use of synchronization of estrus as management strategy in breeding programs has resulted in conception rates that are comparable to natural mating. The use of AI in extensive production systems in South Africa has yielded mixed results; estrus observation (heat detection) has been identified as the single most important factor that detracts from breeding success in such systems.

AI is beneficial when the genetic merit of the male is high, but the genetic contribution of sought after females in a herd can be increased by means of embryo transfer (ET) to recipient females (surrogate mothers). This technology has also been employed in a number of livestock species and wildlife on the African continent. The use of multiple ovulation and embryo transfer (MOET) has become an accepted practice in the breeding of stud cattle, sheep, goats and some wild ungulates. The success rates have not been exceedingly high (about 30 %) but the adoption of better embryo sorting methods and transfer of two embryos per recipient female have increased the conception rates to above 50 %.

Impact of Artificial insemination

It is widely agreed that AI of animals is the single most successful reproductive biotechnology ever employed for the genetic improvement of livestock, but the exact impact has been questioned. Thibier and Wagner (2002) conducted a world survey of the artificial insemination industry and also estimated the impact of AI on different continents. The data shows that already in 1998 there was about 648 semen collection centres and 1,635 semen banks worldwide. It is estimated that more than 40,000 bulls were housed at semen collection centres and about 264 million doses of semen were produced per annum towards the end of the 1990s. The largest proportion of semen processed (>95 %) was frozen in liquid nitrogen and more than 75 % of the semen dosses were from the *Bos taurus*-type dairy breeds.

Table 10.2 Breed of bulls used for semen collection purposes in Africa in 1998 (adapted from Thibier and Wagner 2002)

Breed of bull	Number of bulls used	Expressed as a proportion of the global pool of bulls used for semen collection (%)
Dairy breeds (*Bos taurus*)	553	1.83
Beef breeds (*Bos taurus*)	45	0.64
Beef breeds (*Bos indicus*)	16	2.07
Dual purpose breeds	7	0.49
Buffalo	25	4.92

Despite the technological developments and international acceptance of this reproductive biotechnology, Africa was relatively slow to adopt these biotechnologies. Africa produced less than 1 % of the total number of semen doses in 1998, from about 650 bulls per annum, which represent about 1.57 % of the global pool of bulls used for semen collection. Based on the data obtained by Thibier and Wagner (2002), South Africa was the most active semen producing country in Africa in 1998. Breeds of bulls used for semen collection purposes in Africa are summarized in Table 10.2.

Semen was initially stored in glass ampules which were kept at −196 °C in liquid nitrogen flasks, but plastic straws that contain between 0.25 and 0.5 mL of diluted semen became more popular in South Africa towards the end of the 1970s. Other countries in Africa adopted the same technology and the storage of semen straws in liquid nitrogen flasks and use of frozen–thawed semen in AI programs were adopted almost overnight. Similarly, the initial artificial insemination technique (vaginal deposition) that required a speculum and light source to deposit semen in the antrum of the cervix was soon replaced with the recto-vaginal technique (also known as the cervical fixation).

The recto-vaginal AI-technique was adopted with relative ease by cattle managers and technicians in Africa, after completion of a short course in artificial insemination at tertiary institutions or AI-cooperatives. Most countries in Africa have rules and regulations in place to control the use of these reproductive biotechnologies, but the aims are to guard against the exploitation of genetic resources, rather than the biotechnologies per se. The adoption rate of artificial insemination of cattle and buffalo in Africa is estimated at about 1.68 %, compared to the world average of 20.32 % (Thibier and Wagner 2002), which leaves room for further improvements in this regard.

There is a vast amount of literature that confirms the rapid uptake of AI, due to the technical, genetic and economic advantages in the commercial dairy industry in the developed countries around the world (Cunningham 1999; Corrigan and Parnell 2006). In Africa, about 85.6 % of semen produced is also from dairy type bulls, which generally agrees with the trends for semen production elsewhere in the world. The main reason for this trend is that AI is most practical and has the greatest impact on dairy production systems.

It has been indicated by Pollak (2005) that AI is less used in commercial beef cattle due to the fragmented structure of the beef industry compared to dairy, but it still makes a significant contribution with regard to genetic impact. Difficulties associated with the management of an AI program for extensively kept beef cattle and the generally lower success rates of such programs are reasons for the beef industry also being more reluctant to adopt AI as a reproductive biotechnology. The advent of synchronization of estrus has spurred new interest in the use of AI in extensive beef herds, and such programs appear to be much more lucrative. In South Africa, the use of AI in beef cattle is most popular in stud cattle breeding, although AI is used on a limited scale in three-way cross-breeding systems to improve the management of the breeding program. Smaller cattle operations often used AI in two-way cross-breeding systems due to the unavailability of good quality bulls at an affordable price. Despite these well-documented advantages of a relative simple technique African livestock farmers still face a number of limitations for successful implementation of AI. Several factors are discussed in literature ranging from a lack of infrastructure and animal recording systems (Cunningham 1999; Kahi and Rewe 2008; Rege et al. 2011; Mirkena et al. 2010) to cost of the technology, clear breeding objectives and insufficient capacity to adapt to continuously changing production systems and markets.

In small stock production systems in Africa, the purpose of AI is focused more on the exploitation of specific sires in a breeding flock, because a single ram can only impregnate 30–40 ewes with ease during the breeding season. A single ejaculate from such a ram can be diluted and used to impregnate 100 or more ewes. Ewes are thus synchronized by means of vaginal sponges containing progesterone and are inseminated with fresh-diluted semen from a sought after ram on the farm. Similar methodologies are employed in goats, especially in dairy goats where milk production has to be managed over a 12-month period. Synchronization of estrus of ewe or does plays an important part in the management of such breeding programs. In both sheep and goats, these technologies are also employed as part of accelerated breeding programs, by including flushing (flush feeding) with specific nutrients to increase twinning and improve conception rates.

Artificial insemination is primarily practiced in intensive pig production systems and seed stock operations. The technique is easy to learn and the conception rates in such systems are high. In an African context South Africa is the major producer of pork. In a number of other African countries, pigs as a species face several challenges with regard to the intensive nature of production as well as religious and social constraints.

Animal Recording Systems: A Prerequisite for Successful Reproductive Biotechnology

Animal recording is probably the most important tool or technology in the long process of genetic improvement of livestock. Recording begins with the identification of the animal and may then vary from a simple recording system consisting of pedigrees, calving dates and number of offspring to complex recording of all possible fitness, production, quality and behavioural traits (Peters and Zumbach 2004). Reproductive biotechnology as a technique may be simple enough to implement, but without a recording system it is unlikely that it will have the desired intended effect on genetic improvement.

In general, in African countries there is a lack of animal identification systems and pedigree and or performance recording. In countries such as Ethiopia and Kenya centralized programs were implemented for small stock (sheep and goats) improvement with varying degrees of success. Different approaches have been suggested for genetic improvement in tropical regions such as within breed selection and or cross-breeding. In the case of within breed selection artificial insemination is an option for producing offspring of improved quality for the desired traits. In Kenya within breed selection failed largely due to the absence of animal identification and performance and pedigree recording systems (Kosgey et al. 2006). In addition to these basic requirements for genetic improvement there are other constraints such as small flock sizes, single sire flocks and pastoralist systems, low level if literacy and cultural values. In several papers nucleus schemes have been recommended for genetic improvement in developing countries (Kosgey et al. 2006). Nucleus schemes have the advantage of all recording and selection to be done within the nucleus herds or flocks. This provides more opportunity for effective application of reproductive biotechnologies. The best animals are kept within the nucleus and selected for the desired traits from where good quality animals can be provided to the smallholder farmers who do not have the infrastructure or support to perform detailed animal recording. In this system native breeds can be used effectively without indiscriminate use of cross-breeding. In many of the African countries centralized programs were initiated, but failed on the long term due to the absence of technical support and participation (Rege et al. 2011).

Kenya had a recording scheme for beef and dairy cattle since 1963 that was supported by government with regard to animal health, AI schemes, tick control and livestock extension training (Kosgey et al. 2011). Since 1992, several organizations have been responsible for animal recording in Kenya providing the services for registration of stud animals (KSB, Kenya Studbook) and a milk recording (Dairy recording services of Kenya). The cattle and dairy breed societies and their breeders form part of the Kenyan Livestock Breeders Organization (KLBO) and record information is provided to the Livestock Recording Centre (LRC) with links to a central AI Station (CAIS). Despite a well-designed structure Kosgey et al. (2011) notes that participation remains low as breeders fail to see the benefits of the recording scheme. Even among Boran breeders taking part in the recording scheme, pedigree information for both dams and sires are often incomplete. Kios et al. (2011) found that misidentification of sires varied between 4.3 and 80 % in a study where DNA markers were applied to test for parentage in four Boran stud herds. The integration of the different structures with regard to AI services and recording is not at a level that promotes the potential genetic improvement that can be realized from these systems. Cattle and dairy breeders in Kenya have access to recording and the use thereof has had certain benefits. Boran breeders, for example, use reproductive technology and also export embryos to Southern Africa (http//www.embryoplus.com). Small holder dairy farmers in Kenya do use AI, although on a small scale (Bebe et al. 2003). There may be constraints and challenges to overcome but in Kenya a recording system is in place with potential for using reproductive technology and producing animals of high genetic merit for improved production.

In South Africa and Namibia animal recording systems have been successfully implemented making genetic evaluations of a number of cattle and small stock breeds possible. Estimated Breeding Values (EBVs) are calculated and applied in selection programs in combination with reproductive technologies (AI, Embryo Transfer). National performance schemes have been established in South Africa for dairy (1917), beef (1959) and small stock (1965) (Bergh 2010), where these schemes formed an integral part of the genetic improvement of livestock breeds in this country. Most breeds are involved in performance recording at least with regard to data collection for reproduction and weaning weight. Since the mid-1990s BREEDPLAN has also been available for stud breeders in South Africa and Namibia. In 2011 SA Studbook has extended their services for production recording and genetic evaluations for all livestock species. In Botswana a most effective system is used namely the "Livestock Identification and Trace-back system" (LITS) that is based on a reticular bolus containing a microchip in the middle. Animals can be traced at all times and extension officers can collect information at a herd level. The majority of beef produced in Botswana is exported to European countries where the EU regulations (EU 1760/2000) require traceability from all animals (Department of Animal Health and Production, Botswana 2000).

Managing Reproduction Within a Resource Limiting Environment

Different technologies are used in Africa to manage reproduction in livestock, most of which are based on improving conception rates, decreasing inter-calving periods and reducing postpartum anoestrus. Some of these principles are based on indigenous knowledge, but African farmers have also adopted new reproductive biotechnologies and management methods. One of the major challenges is to manage livestock production in a resource limited environment, which requires careful synchronization of the production level of livestock with the production potential of the environment, based on the most limiting climatological factors.

Environmental temperature and humidity, annual precipitation, quality and quantity of natural feed resources and endo-and ectoparasites are the most limiting environmental constraints for livestock production in Africa. Environmental temperature and humidity have major effects on reproduction of livestock and these effects are well documented. Bonsma's classic work on heat tolerance showed that non-adapted cattle breeds in the subtropics demonstrate severely stunted growth and reproductive development (Bonsma 1980). This condition was referred to as tropical degeneration and it is well known that heifers of the British beef breeds in Africa often have infantile reproductive systems at breeding age. Crossbreeding with indigenous cattle breeds was recommended to improve beef production in arid regions.

In Friesland bulls at an AI-centre in South Africa, better sperm morphology was observed in spring and winter than in summer and autumn (Vilakazi and Webb 2004). Percentage normal sperm decreased and major sperm defects increased with increasing environmental temperature. It was recommended that semen collection in Friesland bulls should be done at 36–48 months of age and preferably during spring or winter, to ensure best semen quality. Friesland bulls kept for semen collection at AI-centres in South Africa are often kept in cool and dry conditions and in some instances the bulls are sprayed and cooled down with fans to ensure good quality semen.

There are many examples of reproductive failure or poor reproduction due to nutrient deficiencies. Energy and protein content are the major nutritional causes of poor reproduction, especially in a resource limited environments. The maintenance of body energy reserves is considered to be the basis of any reproductive management strategy (Houghton et al. 1990; Morrison et al. 1999; Flores et al. 2008). It is well known that the Gonadotropin releasing hormone (GnRH) pulse generator system and subsequent secretion of GnRH from the hypothalamus are inhibited by under nutrition. Protein and energy supplementation have been an effective strategy to improve reproductive performance of livestock in Africa.

Protein supplementation in feeds is quite costly in modern beef and small stock production systems in South Africa, so supplementation is only beneficial if there are improvements in terms of production and reproductive rates. Dietary manipulation of reproduction has been investigated in small stock by focusing on the effects of high protein intake. The effects of different protein sources on ovulation rates were also studied in sheep (Webb et al. 2010) based on the concept of a lower level of protein degradation in the rumen and subsequent availability of protein in the lower digestive system. The results indicate that dietary protein supplementation in ewes has a positive effect on follicle development, but only a marginal effect on the number of lambs born per ewe mated (Webb et al. 2010).

In the early 1900s poor reproduction in cattle in southern Africa was linked to aphosphorosis in phosphorus-deficient grazing (Theiler et al. 1927). Large parts of South Africa, especially the grassland areas are considered phosphorus-deficient (Du Toit et al. 1940). Phosphorus supplementation of cattle in the form of salt-phosphate licks remedied this problem and this legendary strategy is now widely used for cattle production (Read and Engels 1986). Supplementary feeding has become an accepted practice in livestock production and a large proportion of this is based on salt licks with molasses and urea or other sources of non-protein nitrogen. In addition to calcium and phosphorus, a variety of other macro- and micro-minerals and vitamins are also supplemented to improve reproductive rate. Flush-feeding is a popular practice in small-stock production systems in South Africa. The energy level of the diet is increased resulting in higher ovulation and conception rates. Although this technique is used in cattle, it is less effective and therefore less popular.

Reproductive efficiency of postpartum dairy and beef cows has been improved by careful management of body condition and energy balance at parturition. Body condition scores exceeding 2.5 (scale 1–5) and an increasing energy balance at the beginning of the breeding season improved conception rates from 60 % to as high as 90.5 % in extensively kept Brahman-type cows in Moçambique (Escrivao et al. 2012). It is not exactly clear how suckling extends postpartum anoestrus in cattle, but suckling frequency correlates best with the duration and extent of postpartum anoestrus (Stewart et al. 1993). Results show that temporary calf separation at night improved the energy balance and conception rates of postpartum cows without adversely affecting calf-weaning weights (Escrivao et al. 2009). Temporary calf separation has long been practiced by African herdsmen to protect their calves from predators, but it is uncertain if they appreciated the benefits of this practice in terms of improvements in reproductive efficiency.

In some countries there is a preference for increased frame size in cattle due to the favourable correlation with growth rate (Du Plessis et al. 2005), but such animals are not favoured in extensive systems in Africa where grazing quality is frequently poor. Frame size significantly influences reproductive efficiency of beef cows in Africa. A study in Namibia found that calving rate was higher in small and medium frame cows compared to large frame cows (Taylor et al. 2008). Weaning rates of large frame cows were less than those of small and medium frame cows, although weaning weights of calves were higher for large and medium frame cows. Moreover, the total kilograms of calf produced per cow bred were higher for medium and small frame cows compared to large frame cows.

Breeding seasons is another strategy employed in sedentary African livestock production systems to combat the negative effects of extreme environmental temperatures and humidity. The use of breeding seasons allows the opportunity to breed cows in spring or early summer, so that cows calve down when the quality of the grazing starts to improve. This strategy has been employed for decades in the subtropical parts of Africa with great success. Breeding seasons are not used in nomadic livestock production systems, as in these systems the cattle are continuously moved in search of better grazing, which negates the use of breeding seasons. In intensive or semi-intensive dairy production, the "temperature-humidity Index" (THI) was developed to manage dairy cows in hot and humid environments (Du Preez et al. 1991). THI correlated significantly with conception rates in dairy cows in South Africa and this system provides an accurate index to calculate the risk of dairy cows for thermal stress during hot summer months.

An overall breeding soundness examination (OBE) is a quick, reliable and cost-effective method of screening and classifying bulls in terms of breeding potential (Ellis et al. 2005). The latter as well as fertility tests before the onset of the breeding season contribute significantly towards better conception rates in beef herds. However, reproductive (semen quality) and growth data does not appear to be reliable predictors of libido in extensively managed bulls (Scheepers et al. 2010). Scrotal circumference is used more effectively to select beef bulls for fertility and semen production potential.

Reproductive Technology and the Impact on Food Production

The aim of reproductive biotechnologies is to accelerate genetic gain with greater selection pressure on the males (AI) and females (ET). The point of departure for the effective implementation of these technologies is some form of animal identification and performance recording. From the literature reviewed here and elsewhere in World Bank and FAO reports, it is clear that Africa needs to overcome a number of obstacles before individual farmers and countries may reap the genetic and production benefits of reproductive technologies. The solutions to the limitations facing farmers, livestock owners and governments are unfortunately not all of a scientific nature. Some major drivers of change according to Rege et al. (2011) include population growth, urbanization and changes in food preferences; climate change that may lead to more frequent drought or flooding with a direct influence on available grazing and water for livestock and crops. Furthermore, Africa has little choice to become part of the global village and this may be seen as a threat or an opportunity to be able to produce for both local and international markets.

Apart from the non-genetic challenges that are beyond the scope of this discussion it may be time for Africa to adapt the available technology to suit the needs of African livestock producers. Machuka (2001) so rightfully stated with regard to biotechnology in crop production: "Biotechnology for Africa should mostly be done in Africa and by African themselves now. It can be done if there is consensus and goodwill."

There is a good consensus with regard to livestock breeds for Africa among scientists and clearly a need for understanding the role of adapted genotypes for different environments and production systems. The livestock production challenges of small holders and pastoralists need to be addressed with regard to adapted breeds although production may be lower compared to exotic breeds. For small holders and communal farmers and pastoralists, alternative practices for performance recording should be considered. Identification should be a relative simple procedure such as ear notching. In the absence of scales, alternatives measures for weight such as girth depth, chest depth and body length could be considered (Alderson 1999). Despite potential human error in using linear measurements, it can be used by the small holder farmer to establish norms. Animal recording should be combined with management practices where farmers can see the benefits on a short-term basis. New and innovative tools for recording by using handheld electronic data capturing systems could be considered with support from livestock officers and farmers cooperatives to collect and store data of identified animals.

Reproductive technology will find application within structures, private or via research institutions from where the improved genetics may be distributed to small holders keeping in mind that the genotype will fit the system. In countries such as Kenya, Zimbabwe, Zambia Botswana, Namibia and South Africa, animal recording and reproductive technologies had made an impact and it is expected that it will expand in these countries and across their borders. In some countries exotic breeds may be the obvious choice for intensive and semi-intensive production systems as successfully done in South Africa.

Animal scientists and agricultural policy makers should be aware of the diversity of climatic regions and production systems to ensure sustainability. Rege et al. (2011) suggested different breed options for different production systems ranging from pastoral systems using mostly indigenous breeds, mixed systems in semiarid/semi humid regions using indigenous with some cross-breds to highland regions where 50 % exotic breeds can be used in cross-breeding as well as pure bred dairy cattle. In both the mixed and highland regions biotechnology such as AI may find application with an expected medium to high impact. In smallholder systems in East Africa it is more likely that cross-breds will be used in cattle and goats with the development of composite breeds.

Strategies for increasing livestock production in Africa using any form of biotechnology needs careful planning and due consideration of the available indigenous livestock species (cattle, sheep, goats and poultry), indigenous knowledge and production systems.

Conclusion

Reproductive technology is an important tool for accelerating genetic progress through both AI and ET. For reproductive technology to be effective, it is of the utmost importance that breeding objectives will be well defined with regard to the breed choice and the specific traits with due consideration of the environmental challenges, production system and the use of some form of animal recording. In Africa reproductive technology generally had limited success, except for a few countries where animal recording schemes have been in place for a number of decades. The technology is available and Africa has a sufficient number of livestock breeds to comply with the increasing demand for proteins from animal origin. There is a need for improved management of non-genetic interventions for biotechnology to succeed on the African continent and contribute substantially to food production and security.

References

Alderson GLH (1999) The development of a system of linear measurements to provide an assessment of type and function of beef cattle. Anim Genet Resour 25:45–55

Amare B, Kefyalew A, Zeleke M (2012) Typical features, characterization and breeding objectives of Begait sheep in Ethiopia. Anim Genet Resour 50:1–7

Barker RL (1988) Genetic resistance to endoparasites in sheep and goats. A Review of genetic resistance to gastrointestinal nematode parasites in sheep and goats in the tropics and evidence for resistance in some sheep and goat breeds in sub-humid coastal Kenya. Anim Genet Resour 24:13–30

Bebe BO, Udo HMJ, Rowlands GJ, Thorpe W (2003) Small holder dairy systems in the Kenay highlands: breed preferences and breeding practices. Livestock Prod Sci 82:117–127

Bergh L (2010) The National beef recording and improvement scheme, chap 5. In: Scholtz MM (ed) Beef breeding in South Africa. Agricultural Research Council. www. Agric.za

Bonsma JC (1980) Livestock production: a global approach. Tafelberg, Cape Town

Corrigan L, Parnell PF (2006) Application of genetics technology in the temperate Australian beef seedstock industry. In: Proceedings of Australian Beef—The Leader! Conference. University of New England, Armidale, 7–8 March 2006

Cunningham EP (1999) The application of biotechnologies to enhance animal production in different farming systems. Livestock Prod Sci 58:1–24

Delgado C, Rosengrat M, Steinfeld H, Ehui S, Courbis C (1999) Livestock 2020: the next food revolution, food, agriculture and the Environment discussion paper 28 IFPRI/FAO/ILRI. IFPRI, Washington, DC

Department of Animal Health and Production, Botswana (2000) Omang Wa Dikgomo. Livestock identification and trace-back system. Department of Veterinary Services, Gabarone, Botswana

Du Plessis I, Hoffman LC, Calitz FJ (2005) Influence of reproduction traits and pre-weaning growth rate on herd efficiency of different beef breed types in an arid sub-tropical environment. S Afr J Anim Sci 35:89–98

Du Preez JH, Terblanche SJ, Giesecke WH, Maree C, Welding MC (1991) Effect of heat stress on a dairy herd model under South African conditions. Theriogenology 35(5):1039–1049

Du Toit PJ, Louw JA, Malan AI (1940) A study of the mineral content and feeding value of natural pastures in the Union of South Africa. Onderstepoort J Vet Res 14:123

Ellis RW, Rupp GP, Chenoweth PJ, Cundiff LV, Lunstra DD (2005) Fertility of yearling beef bulls during mating. Theriogenology 64:657–678

Escrivao RJA, Webb EC, Garces APJT (2009) Effects of 12 hour calf withdrawal on conception rate and calf performance of Bos indicus cattle under extensive conditions. Trop Anim Health Prod 41:135–139

Escrivao RJA, Webb EC, Garces APJT, Grimbeek RJ (2012) Effects of 48 hour calf withdrawal on conception rates of Bos indicus cows and calf weaning weights in extensive production systems. Trop Anim Health Prod 44:1779–1782

FAO Statistical yearbook (2012) Part 1 Livestock cattle 46–49. FAO Rome

Faustin PL, Kyvsgaard NC (2003) Improving pig husbandry in tropical resource-poor communities and its potential to reduce risk of porcine cysticercosis. Acta Trop 87:111–117

Flores R, Looper ML, Rorie RW, Hallford DM, Rosenkrans CF Jr (2008) Endocrine factors and ovarian follicles are influenced by body condition and somatotropin in postpartum beef cows. J Anim Sci 86:1335–1344

Gizaw S, Van Arendonk JAM, Komen H, Windig JJ, Hanotte O (2007) Population structure, genetic variation and morphological diversity in indigenous sheep of Ethiopia. Anim Genet 38:621–628

Goosens B, Osaer S, Ndao M, Van Winghem J, Geerts S (1999) The susceptibility of Djanllonke and Dkallonke-Sahelian crossbred sheep tp Trypanosoma congolense and helminth infection under different diet level. Vet Parasitol 85(1):25–41

Hanotte O, Jianlin H (2005) Genetic characterization of livestock populations and its use in conservation decision-making. In: The role of biotechology, Villa Gualino, Italy, 5–7 March 2005

Holt W, Pickard AR (1999) Role of reproductive technologies and genetic resource banks in animal conservation. Rev Reprod 4:143–150

Houghton PL, Lemeneger RP, Moss GE, Hendrix KS (1990) Prediction of postpartum beef cow body composition using weight to height ratio and visual body condition score. J Anim Sci 68:1428–1437

Ibeagha-Awemu EM, Erhardt G (2004) Genetic structure and differentiation of 12 African Bos Indicus and Bos Taurus cattle breeds, inferred from protein and microsatellite polymorphisms. J Anim Breed Genet 122:12–20

Kahi AK, Rewe TO (2008) Biotechnology in livestock production: overview of possibilities for Africa. Afr J Biotechnol 7(25):4984–4991

Kios D, Van Marle-Köster E, Visser C (2011) Application of DNA markers in parentage verification of Boran cattle in Kenya. Tropl Anim Health Prod 41(3):471–476

Koehler-Rollefson I (2012) Sustainable solutions need smallholder systems. Global donor platform for rural development. http://wwwdonorplatformorg/livestock-and-pastoralism

Kosgey IS, Baker RL, Udo HMJ, Van Arendonk JAM (2006) Successes and failures of small ruminant breeding programmes in the tropics: a review. Small Rumin Res 61:13–28

Kosgey IS, Mbuki SM, Okeyo AM, Amimo J, Philipsosson J, Ojango JM (2011) Institutional and organizational frameworks for dairy and beef cattle recording in Kenya: a review and opportunities for improvement. Anim Genet Resour 48:1–11

Lenstra JA, Groeneveld LF, Eding H, Kantanen J, Williams JL, Taberlet P et al (2011) Molecular tools and analytical approaches for the characterization of farm animal genetic diversity. Anim Genet 41:1–20

Machuka J (2001) Agricultural biotechnology for Africa African scientists and farmers must feed their own people. Plant Physiol 126:16–19

Mapiye C, Chimomyo M, Muchenje V, Dzama K, Mumyaradzi CM, Raats JG (2007) Potential for value—addition of Nguni cattle products in the communal areas of South Africa: a review. Afr J Agric Res 2(10):488–495

Mattiolo RC, Pandey VS, Murray M, Fitzpatrick JL (2000) Immunogenetic influences on tick resistance in African cattle with particular reference to trypanotolerant N'Dama (Bos taurus) and trypanosusceptible Gobra zebu (Bos indicus) cattle. Acta Trop 75(3):263–277

Mirkena T, Duguma G, Haile A, Tibbo M, Okeyo AM, Wurzinger M, Solkner J (2010) Genetics of adaptation in domestic farm animals: a review. Livestock Sci 132:1–12

Morrison DG, Spitzer JC, Perkins JL (1999) Influence of prepartum body condition score change on reproduction in multiparous beef cows calving in moderate body condition score. J Anim Sci 77:1048–1053

Mwacharo JM, Okeyo AM, Kamande GK, Rege JEO (2006) The small East African shorthorn zebu cows in Kenya. I: linear body measurements. Trop Anim Health Prod 38:65–74

Peters KJ, Zumbach B (2004) Needs for research and development in livestock recording systems (LRS) in transition and developing countries. Technical series no 9. In: Pauw R, Mack S, Maki-Hokkonen J (eds) Development of animal identification and recording systems for the developing countries. ICAR technical series are published by the International committee for animal recording. pp 152–174

Pica-Ciamarra U, Baker D, Bedane B, Emwanu T, Morgan N (2010) Intergrating livestock into agricultural statistics Joint paper of the World Bank, FAO, October 2010

Pollak EJ (2005) Application and impact of new genetic technologies on beef cattle breeding: a "real world perspective". Aust J Exp Agric 45:739–748

Qwabe SO, Van Marle-Köster E, Visser C (2012) Genetic diversity and population structure of the endangered Namaqua Afrikander sheep. Trop Health Prod. doi:101007/s 11250-012-0250 online

Read MVP, Engels EAN (1986) Phosphorus and the grazing ruminant. 2. The effects of supplementary P on cattle at Glen and Armoedsvlakte. S Afr J Anim Sci 16:7–12

Rege JEO (1999) The state of African cattle genetic resources. I. Classification framework and identification of threatened and extinct breeds. Anim Genet Resour 25:1–25

Rege JEO, Marshall K, Notenbaert A, Ojango JMK, Okeyo AM (2011) Pro-poor animal improvement and breeding—what can science do? Livest Sci 136:15–28

Roberts CJ, Gray AR (1973) Studies on trypanosome-resistant cattle. II. The effect of trypanosomiasis on

N'dama, Muturu and Zebu cattle. Trop Anim Health Prod 5(4):220–233

Rode LM, McAllister TA, Beauchemin KA, Morgavi DP, Nsereko VL, Yang WZ, Iwaasa AD, Wang Y (2010) Enzymes as direct-feed additives for ruminants. In: Renaville R, Burny A (eds) Biotechnology in animal husbandry. Kluwer Academic, Dodrecht

Ruxandra D-AR (2010) Gene therapeutic enhancement of animal health and performances. In: Renaville R, Burny A (eds) Biotechnology in animal husbandry. Kluwer Academic, Dodrecht

Scheepers SM, Annandale CH, Webb EC (2010) Relationship between production characteristics and breeding potential of 25-month old extensively managed Bonsmara bulls. S Afr J Anim Sci 40(3):163–173

Scherf BD (ed) (2000) World Watch list for domestic animal diversity, 3rd edn. FAO/UNEP, Rome

Scholtz MM, McManus C, Okeyo AM, Theunissen A (2011) Opportunities for beef production in developing countries of the southern hemisphere. Livestock Sci 142:195–202

Spickett AM, De Klerk D, Enslin CB, Scholtz MM (1989) Resistance of Nguni, Bonsmara and Hereford cattle to ticks in a Bushveld region of South Africa Onderstepoort. J Vet Res 56:245–250

Stewart IB, Louw BP, Lishman AW (1993) Suckling behaviour and fertility in beef cows on pasture, 1. Suckling behaviour. S Afr J Anim Sci 23:176–179

Taberlet P, Valentini H, Rezaei R, Naderi S, Pompanon F, Negrini R (2008) Are cattle, sheep and goats endangered species? Mol Ecol 17:275–284

Taylor G, Swanepoel FJC, Webb EC, Stroebel A (2008) Effect of heifer frame size on their subsequent reproductive performance and preweaning performance of their calves. Aust J Exp Agric 48:945–949

Theiler A, Green HH, Du Toit PJ (1927) Minimum mineral requirements in cattle. J Agric Sci 17(3):291–314

Thibier M, Wagner H-G (2002) World statistics for artificial insemination in cattle. Livestock Prod Sci 74:203–212

Thornton PK (2010) Livestock production: recent trends, future prospects. Philos Trans R Soc 365: 2853–2867

Van Arendonk JAM (2011) The role of reproductive technologies in breeding schemes for livestock populations in developing countries. Livestock Sci 136:29–37

Vilakazi DM, Webb EC (2004) Effect of age and season on sperm morphology of Friesland bulls at an artificial insemination centre in South Africa. S Afr J Anim Sci 34(1):62–69

Webb EC, Van Niekerk WA, Lee K, Marais WJ (2010) Reproductive performance of semi-extensively kept Döhne Merino ewes fed with different protein sources. S Afr J Anim Sci 40(5):451–454

11 International Perspectives on the Impacts of Reproductive Technologies on Food Production in Asia

Takeshi Osawa

Abstract

The greatest numbers of domestic animals are raised in Asia. The Asian animal industry is characterized by the involvement of a high percentage of the population, mostly smallholders, of which 95 % rear domestic animals. In exploring the best ways to formulate sustainable society, it is essential to make the most of livestock products by applying appropriate reproductive technologies. There is no doubt that reproductive technologies such as AI and ET have made a great contribution to increase the number of excellent animals. Although more advanced cutting edge reproductive technologies have become available and some of them have indeed a potential of revolutionary changes in livestock industry, the most important problem for increasing productivity concerns the maintenance of optimum nutrition and prevention of heat stress to support reproductive performance and increased supply of animal proteins. International societies should be involved in binding together developed and newly developing countries in the construction of a novel model for future livestock rearing management that suits diverse environmental circumstances.

Keywords

Asia • OPU • Smallholders • Wagyu

Introduction

In 2011, the human population of Asia was approximately 4.1 billion, representing about 60 % of the world's people. Population increase during the last 3 decades, combined with rapid economic development and rising per capita income, has led to increasing demand for livestock foods. However, the particular circumstances and conditions within different Asian countries are diverse; while some countries are located in temperate zones, where livestock industries are relatively well-developed and stable, others are located in subtropical and tropical zones, where livestock industries are still growing rapidly.

T. Osawa, D.V.M., M.Sc., M.Phil., Ph.D. (✉)
Faculty of Agriculture, Laboratory of Theriogenology, Department of Veterinary Sciences, University of Miyazaki, Miyazaki 889-2192, Japan
e-mail: osawa@cc.miyazaki-u.ac.jp

G.C. Lamb and N. DiLorenzo (eds.), *Current and Future Reproductive Technologies and World Food Production*, Advances in Experimental Medicine and Biology 752,
DOI 10.1007/978-1-4614-8887-3_11,

The total milk production in India and China (the second and third highest producers in the world, respectively) exceeded that of the USA in 2010. The very high growth rates of these so-called BRIC countries (Brazil, Russia, India, and China) are exerting an enormous impact on world animal production. The Asian animal industry is characterized by the involvement of a high percentage of the population, mostly smallholders, of which 95 % rear domestic animals. For the less prosperous smallholders, domestic animals represent an important source of household income. However, their production systems typically rely on traditional methods employing small numbers of locally bred animals with low productivity. Such producers have been unable to meet the rapidly increasing demand for livestock products or the changes in distribution systems in recent years. To avoid future risk of food shortage, many countries have responded by importing exotic breeds to cross with local breeds to increase domestic animal production levels. In addition, efforts have been made to establish high quantity production systems by introducing modern animal industry technologies. However, the crossing of local breeds with European breeds could lead to reduced diversity of genetic resources, and this has become a social issue associated with such economic development. Currently there is inadequate infrastructure to meet the increasing demand for feedstuff, so ensuring adequate feedstuff production to support the increasing animal population is another important concern. Consumption of livestock products has an impact on grain markets and economic trends at a global level. Importation from animal products, particularly beef from the USA and Australia, may satisfy the demand caused by insufficient local production, but could disturb the balance of trade in such merchandise. In order to increase domestic production of animal products and to become internationally competitive, improvement of local breeds through introduction of reproductive technologies and realization of intensive management are vital in a country that has limited amount of land. At the same time it is also important to remember the adaptability of "improved" breeds to local climate and environmental consciousness.

This chapter summarizes and discusses statistics related to the livestock industry in some Asian countries, animal reproductive technologies in Japan, and the challenges associated with a decline in reproductive performance. A perspective for future developments is provided.

Domestic Animals in Asia

The greatest numbers of domestic animals are raised in Asia, with numbers continuing to increase at high rates in China and India. The world population of the main domestic animals is about 4.1 billion, being composed of 1.36 billion cattle, 0.94 billion pigs, 1.03 billion sheep, and 0.75 billion goats. Of the 1.36 billion cattle in the world, one-third (472 million) are present in Asia. This represents the greatest population of cattle in the world, followed by South America (350 million). India (210.2 million), Brazil (209.5 million), the USA (94 million), and China (83 million) are the four countries with the largest cattle populations. Of the 965 million pigs in the world, 60 % (583 million) are present in Asia. This represents the greatest population of pigs in the world, followed by Europe (189 million). China (476 million), the USA (65 million), Brazil (39 million), Vietnam (27 million), and Germany (26.5 million) are the five countries with the largest pig populations. Including other domestic animals (e.g., water buffaloes, horses, donkeys, and camels), the total is 4.37 billion (FAO 2002). Zebu cattle, water buffaloes, Bali cattle, and Yaks play significant roles in livestock industry in Asia (Abeygunawardena and Dematawewa 2004; Zi et al. 2006; Perera 2011; Martojo 2012).

Thus, animal production (both in terms of animal numbers and animal products) in Asia is increasing at an accelerating rate. Animal production and animal industries are prevalent in many countries of Asia and animal rearing technologies adapted to their particular environments are now beginning to be established in this part of the world.

Livestock Production in Asia

Beef production in the world was 66.6 million tonnes (Mt) in 2011 (FAO). The five countries with the greatest beef production in 2012 were the USA (11.9 Mt), Brazil (9.3 Mt), China (5.5 Mt), India (3.5 Mt), and Argentina (2.6 Mt) (USDA 2013).

Pork production in the world was 108.8 Mt in 2011 (FAO). The two countries with the greatest pork production in 2012 were China (49.5 Mt) and the USA (10.3 Mt) (USDA). In Asia, other than China, Vietnam (1.96 Mt), Japan (1.27 Mt), and the Philippines (1.25 Mt) have high production. Japan, however, imports 125 Mt besides domestic production and is the biggest pork importing country, having 20 % of the total pork import in the world.

Egg production in the world was 64 Mt in 2010, and Asia is the largest area (37 Mt, 59 % of the total), followed by Europe (10 Mt) and North America (8.8 Mt). By country, China (24 Mt), the USA (5.4 Mt), India (3.4 Mt), Japan (2.5 Mt), and Mexico (2.4 Mt) are the five biggest countries.

The demand for milk and milk products in Asia, a region that includes the enormous populations of the two BRIC countries China and India, is expected to maintain growth into the future. The growth rate of milk production in China has been particularly marked in recent years. World milk production increased by 51 Mt between 2000 and 2006 (to 489 Mt), 24 Mt of which was attributed to China.

World milk production in 2010 was 599 Mt, and Asia (158 Mt) had the second largest output after Europe (207 Mt), followed by North America (112 Mt). The USA (87 Mt), India (50 Mt), China (36 Mt), Russia (31.9 Mt), and Brazil (31.7 Mt) were the five countries with the greatest production (FAO 2010).

Internationally, the most competitive regions for milk production are Southeast Asia, Oceania, South America, and several Eastern European countries. The lowest production costs were recorded in farms having 1,400 head in Argentina (US$0.1 per kg of milk produced), 10 head in Pakistan (US$0.11), and 50 head in Poland (US$0.12). Despite their small farm sizes, animal industries in many Asian countries may thrive because of the low labor cost. However, South American countries have the advantage of lowest production costs and favorable exchange rates, and are expected to increase their export share of the world dairy product market.

Country Reports

China

In the past, Chinese people had virtually no tradition for drinking milk except in specific parts of the country. Until a few decades ago, milk was consumed only by sick people. However, the demand for milk and dairy products has increased as the living standards in China have improved. The promotion of the livestock industry is an important national policy that could vitalize the rural economy. Although the scale is different, China is following a similar path to that followed by Japan in earlier times. To meet the demand for better quality milk, it will be necessary to refine the technologies for improving breeds and feeding systems.

Consumption of livestock products has increased in both urban and rural areas (Kitakura and Li 2007). Milk production in China tripled between 2002 and 2008. As a national policy, the Chinese Government is reinforcing the dairy industry from the viewpoint of improving the health status of the nation's people. In 1997, the State Council published the "National Nutrition Improvement Plan," which emphasized increasing the number of dairy cattle and development of the dairy industry. In August 2000, the Ministry of Agriculture, the National Development and Reform Commission, the Ministry of Education, the Ministry of Finance, and other sectors of the Government proposed the "Drink Milk Plan for Students" under the slogan of "a cup of milk strengthens the people," and started to provide school students in the major cities with milk at lunchtime. This movement is similar to one proposed by the Japanese Government, which planned to change the eating habits of their children to enhance their health and physical development through provision of school lunches.

Thus, milk consumption is expected to steadily increase as a result of the widespread habit of drinking milk among children in China.

In 2007, the consumption of milk and dairy products in China was 18 kg per capita in urban areas and 3 kg per capita in rural areas, much lower than their consumption in Japan (32 kg per capita per year). Given the traditional dietary culture in Asia, it seems unlikely that milk and dairy product consumption in China will catch up with that in North America or in European countries, such as Germany or France, where the annual consumption is 80–100 kg of milk and dairy products per capita. However, it should be possible to equal that of Japan. Nevertheless, the impact of such an increase in China would be huge because of the size of the nation. A substantial increase in milk production would be required to meet the expected increase in demand.

The total population of cattle (both beef and dairy) in China was 117 million in 1997 and 138 million in 2004 (a 1.2-fold increase). In contrast, dairy cattle increased from 1.5 million in 1997 to 12 million in 2005 (an eightfold increase) (Kitakura and Li 2007). However, the percentage of Holstein-Friesian (HF) breeds, grass production, and the quality of feedstuff are still relatively low and are associated with low productivity.

Milk production in China was 22.6 Mt in 2004 and increased by 16 Mt by 2010. This increase accounted for 41 % of the total increase in the world during the same period. In 2010, Chinese milk production ranked third in the world after the USA and India. Besides the rapid increases in the number of animals and milk yield per animal, the spread of artificial insemination (AI) technologies using frozen semen and prevention of reproductive disorders have also contributed to the increase in milk production in the country.

In 2004, the proportion of farms with between 5 and 19 head of cattle was 87 %, as a proportion of animals it was 52 %, and in terms of milk production it was 47 %. The average number of head per farm was 10.6, a value much smaller than that in Japan (59 head in 2004) (Kitakura and Li 2007). At the same time, 180 farms had more than 1,000 animals. Most of these farms were run by dairy companies or national companies. In some, as many as 10,000–20,000 head of animals were reared on farms equipped with modern milking machinery, and cattle of Holstein breed with high milk yield were imported from Australia or New Zealand.

India

Since the late 1990s, India has become the biggest producer of fresh milk (including cow's and buffalo's milk) in the world, with a production that meets its domestic demand (114.9 Mt in India, followed by the USA with 89.0 Mt). With respect to production of whole fresh cow's milk, the country is the second largest producer (37 Mt) after the USA (77 Mt) (FAOSTAT Agriculture 2013). Raw milk is India's principal agricultural product and accounts for about 70 % of the total livestock production.

Development of the dairy industry has involved national policies supporting smallholders in rural areas. In addition, the organization of a network supporting farmers has played an important role. Therefore, unlike other newly developing countries, fundamental structural changes and expansion of scale are not applicable in India. The dairy sector, which comprised 5.5 % of the national work force in 1999, was underpinned not by genetic improvement programs for cattle and buffaloes but by the importation of HF and Jersey bulls and semen and by a few institutional herd-based small-scale progeny testing programs. No significant progress has been made in the genetics of indigenous cattle breeds, partly because of the absence of rigorous selection, unplanned breeding and poor monitoring, and partly the legal ban on slaughtering cows and young bulls in most states of India. Many of farmers still possess only one or two animals on less than 2 ha of land. More than half of all cattle and buffaloes, almost all sheep and goats, 77 % of pigs, and 64 % of poultry in India belong to smallholders who own 1 ha or less land, or who are landless (Nimbkar 2011).

The Indian Government has adopted a package of policies to support and stimulate smallholder dairy production since the 1970s. These include the establishment of dairying infrastructure and marketing, making India the largest milk-producing country in the world. The fourfold increase in milk production between 1963 and 2003, more than half of which was produced by buffaloes, was primarily not a result of genetic improvement (Nimbkar 2011).

India has several national species-specific and veterinary research institutes with highly trained scientists, sophisticated facilities, and equipment for advanced genomic analyses. However, practical objectives, animal identification, and field recording of performance are lacking. Reproductive technologies such as multiple ovulation and embryo transfer (MOET) and in vitro fertilization (IVF) have been established but are not being applied to multiply animals with superior genetics. Accurate performance records on large numbers of animals from a reference population are necessary for extensive validation of the association between genotypes and phenotypes. Simply genotyping animals without such validation or efficient data analysis will be fruitless.

Vietnam

The total value of agriculture products (agroforestry and aquaculture) amounts to 20.6 % of GDP, and livestock products constitute 21.6 % of the total agriculture value in 2010 (General Statistics Office of Vietnam 2013).

In 2002, there were 55,848 dairy cows, and the average growth rate of the population was 35.7 % per year. The total population of dairy cows was about 95,000 in 2004.

Of the present population of 95,000 dairy cows, more than 82,500 are HF crossbred cows (HF × improved cows) with different levels of genetic purity in F_1, F_2, F_3, and F_4 generations. There are about 12,000 pure HF and Jersey cows (12.5 % of the total dairy cow population), and about 0.5 % of other breeds. A total of 94 % of dairy cows are raised in households and private companies and 6 % are raised in State companies.

During the period 1999–2003, milk production/305 days of milking period increased from 2,200 to 3,400 kg in crossbred herds and from 3,200 to 4,600 kg in pure herds.

The average herd size in the North of Vietnam is 4–5 cows/household, and in the South of Vietnam it is 7–10 cows/household. However, herd sizes vary greatly with some households keeping as many as 130–150 cows. The cost of feed is about 68–70 % of the cost of milk. The selling price of milk was about 24 % higher than the cost.

By 2005, the total herd size had reached 100,000 dairy cows, producing 165 kilo tonnes of milk, which satisfied 20 % of milk consumption in the country. In 2009, there were 115,000 cows producing 278 kilo tonnes of milk and satisfying only 21 % of milk consumption. The Vietnamese Government sets up a plan to achieve a half million dairy cows producing 1 Mt of milk in 2020.

In May 2010, a groundbreaking venture was launched for the construction of a very large dairy factory. According to Voice of Vietnam Radio, the dairy factory will have the latest technologies in its 22 ha site in Nghê An. The factory, which represents the epitome of high-tech agriculture in Vietnam, is owned by a private company; it will have a production capacity of 500 tonnes of milk per day and 8,000 dairy cows will be imported from New Zealand.

Japan

The Japanese dairy industry is similar to that in many other developed countries. The average milk yield per cow per year was 8,011 kg in 2008. The total number of dairy cattle was 1.29 million in 1965 and 1.64 million in 2006. The average number of cows per farm was 3.4 in 1965 and this increased to 61.5 in 2006. Conversely, the number of farms dropped sharply from 417.6 thousand in 1963 to 26.6 thousand in 2006. Annual per capita milk consumption was 22 kg in 1960 and 93 kg in 1996. Japan is fully self-sufficient with respect to domestic milk consumption.

Of the four domestic beef breeds in Japan (Japanese Black, Japanese Brown, Japanese Shorthorn, and Japanese Poll), Japanese Black is most common, with a population of 1.77 million cows (MAFF 2013). The breed is characterized by prominent intramuscular fat deposition (marbling) and has recently gained greater interest and reputation from around the world for the unique texture, tenderness, and premium quality of Wagyu beef.

After the establishment of the Japanese Black breed in 1944, most of the breeding programs were carried out within prefectures, and little migration of sires to other prefectures occurred. As a result of the intensive breeding from a few "top-class" sires with high marbling, to produce offspring of excellent quality and monetary value, the decline of genetic diversity has become an increasing problem. The number of sires started to decline after 1991, the year when beef imports became liberalized in Japan, and the variance of progeny number per sire increased sharply to become 2.5–3 times greater during the decade from the mid-1980s to the mid-1990s (Nomura et al. 2001). This increase in the variance was due primarily to the intensive use of a few popular sires. Domestic production of high-quality beef has received increased emphasis in Japan since 1991. Since 1991, the percentage of registered animals sired by the five bulls most intensively used in each year has increased rapidly and, in 1998, more than 42 % of all registered animals were the progeny of only five sires (Nomura et al. 2001). Semen units collected from a small number of prominent sires are intensively used.

The effective population size of the Japanese Black breed in the late 1990s was around 17. Previous studies suggested that 40–50 is the minimum viable effective size for dairy breeding or for conservation of endangered animal species (Goddard and Smith 1990; Lande and Barrowclough 1987). Current genetic evaluation of Japanese Black was limited to carcass traits, but evaluations of other economically important traits such as reproductive performance and growth rate have been planned so that the effective population size should increase in future.

At present, there are potential uses for new reproductive techniques in the beef industry. However, such new techniques do not always result in the enhancement of overall economic efficiency because of low reproductive performance, high mortality, and high production costs, and therefore should be carefully evaluated before implementation (Herd et al. 1993).

A previous study indicated the importance of the ratio of live male to live female calf prices in the use of sex control in Japanese cow/calf production systems. An increase of less than 2 % in the ratio can lead the production system with male sexing to have higher annualized net revenue than the base production system without sex control and twinning techniques. The ratio also affects the choice between male or female sexing; the threshold value for the ratio was lower than 1.04 in the situation where the female calf price was nearly equal to the carcass price of culled cows (950 yen per kg) (Oishi and Hirooka 2012). The ratio was set at 1.134 (13.4 % higher price for male calves) in this study and was reported as about 1.073 in the USA (Stockton et al. 2010) and 1.067 in Ireland (Mc Hugh et al. 2010). Therefore, it can be concluded that male sexing is usually more economically valuable than female sexing in beef cow/calf production when the herd size is fixed. However, calf market conditions may be highly changeable among seasons and regions.

Under the present conditions of Japanese beef production, introduction of sex control does not have economically appreciable effects, but twinning was found to be economically beneficial (Oishi and Hirooka 2012). For sex-controlled production, improvements in the conception rate per mating and/or reduction of technical costs were required for this technology to be profitable. In addition, improvements of the calving rate and preweaning survivability in the production system, combined with the twinning technique, could greatly increase the economic benefit. However, it should be noted that, if the cost of the twinning technique is extremely high, its beneficial effect is negated.

Animal Reproductive Technology in Japan

Livestock improvement policies should be implemented steadily so that the country can produce and store excellent animals through breeding and selection of sires, indexed with respect to pedigree, performance, and traits. Then, animal industry should reliably provide consumers with good quality animal products as well as contributing to increased income for producers, and management stability.

In relation to the dissemination of reproductive technologies for increasing animal production, it is important to emphasize the control of the storage and distribution of semen. First, no semen from bulls other than sires that have passed breeding soundness examinations should be distributed and used for AI. Second, pedigrees (excellent breeds) should be strictly controlled via a complete sire registration system. The quality and quantity of animal products have increased as a result of continuous efforts in promoting livestock improvement. The Act for "Improvement and Increased Production of Livestock" has played an important role in this process.

An example of quality improvement is the increase in the proportion of Wagyu beef of fourth grade or higher from 39.4 % in 1995 to 48.1 % in 2007. Likewise in dairy industry, milk fat content increased from 3.44 % in 1975 to 3.99 % in 2005, and nonfat solids increased from 8.18 % in 1975 to 8.79 % in 2005. These quality improvements have added value to livestock products. Consequently, the profitability for producers has increased and, in turn, this has raised their motivation toward increasing product quality, and has provided animal products that better meet the requirements of consumers.

In terms of quantitative improvements, the average daily gain in fattening Japanese Black cattle increased from 0.60 kg in 1975 to 0.73 kg in 2005. For HF cows, the average 305-day milk production per cow increased from 5,718 kg in 1975 to 9,315 kg in 2008. As a result of these accomplishments, the production cost per kg decreased, farm businesses improved, and consumers are now able to purchase livestock products at reasonable prices. Reliable provision of these products should improve consumers' willingness to buy.

Artificial Insemination (AI)

The first AI of domestic animals in Japan was implemented in horses in 1896. In Japan, AI was introduced in dairy cows in 1928, in rabbits in 1930, chickens in 1936, pigs and goats in 1938, sheep in 1939, and Japanese Black cows in 1940.

AI of cattle using frozen semen started in Japan in 1954. In 1965, the Livestock Improvement Association of Japan was established and took charge of the production and distribution of semen straws. Frozen semen straws then rapidly became popular in the country. Imported semen from North America has been used since 1984.

Animal reproductive technologies began with AI and advanced to the transfer of in vivo-developed embryos and, later, of in vitro-developed embryos. All of these technologies have now been widely utilized in the field. Now, sexing semen and ovum pick-up and in vitro embryo production (OPU-IVP) technologies are prevalent in many developed countries, including some Asian countries.

The livestock industry in Japan briskly introduced AI, which does not require space for keeping a bull, to overcome the difficulties of efficient management on limited land. At the same time, the introduction and dissemination of the technique has contributed to increased productivity of livestock through improvement of breeds and prevention of infectious diseases.

The AI technique in Japan began in the 1920s using fresh semen and propagation of the technique accelerated as the Act for "Improvement and Increased Production of Livestock" became effective in 1950. The law was revised in 1961 for frozen semen, in 1983 for in vivo-cultured ET, and in 1992 for in vitro-cultured ET. The framework for use of these technologies in the field was clarified by these revisions, and therefore these reproductive technologies gradually prevailed. Now, breeding of cattle is carried out almost exclusively within Japan.

The number of beef cows in Japan is 2.6 million and number of bulls is 2,000. Currently, the proportion of AI procedures using frozen

Japanese Black semen to HF cows is more than 30 %. Breeding of cattle is implemented almost 99 % by AI and 1 % by ET. Therefore, natural breeding of cattle is now very rare in Japan although it occurs exceptionally, at a rate of much less than 1 %, in grazing beef cattle.

Importation of semen for AI of domestic animals, and of embryos, is permitted only from countries that have established a bilateral agreement with Japan in relation to animal health. In addition, the use of imported semen or embryos in Japan requires the germ cells or embryos to be certified by organizations designated by Minister of Agriculture, Forestry and Fishery or by governmental organizations in the exporting country.

As of August 2008, the countries permitted to export semen to Japan for AI of domestic animals are the USA, Canada, Australia, France, The Netherlands, and Germany for cattle semen; the USA for pig semen; and Australia and New Zealand for sheep semen. Those that can export embryos for domestic animals are the USA and Canada. The number of imported cattle semen straws was 540,000 in 2002, and increased to 650,000 in 2006. The number of imported cattle embryos was 1,217 in 2002 and doubled to 2,445 in 2006.

Frozen semen is produced at a national scale in AI centers owned by incorporated associations or private companies, such as the Livestock Improvement Association of Japan (for dairy and beef cattle), and also prefectural AI centers (for beef cattle).

Use of Sexed Semen

If producers do not intend to increase their herd size and their cows produce a calf every 13 months, more than half of the calves born within the farm can be sold, because the number of heifers required to maintain the herd size is approximately 30 % of all milking cows (Sasaki et al. 2011). The Japanese beef industry places great emphasis on carcass quality, and Japanese Black cattle produce very high-quality meat. It is common for Japanese dairy farmers to use Japanese Black semen for heifers, partly because the price of a Japanese Black × Holstein crossbreed (F_1) calf 7–10 days old is nearly 1,000 dollars higher than that of a male Holstein calf, and partly because the crossbreeds are smaller than a pure Holstein, which reduces the risk of dystocia. A computer simulation has shown that the dairy producer's revenue could be increased by using sexed semen to produce calves of more valuable breed and sex (Sasaki et al. 2011).

ET

Research on ET began in the 1940s. Viable offspring of cattle derived from fresh embryos were produced in 1964 and those derived from frozen embryos were produced in 1979. Sugie (1965) was the first in the world to successfully transfer embryos by nonsurgical methods, and he also elaborated the techniques of collecting embryos.

In recent years, more than 70,000 cows have been used annually as recipients for ET (Fig. 11.1). A total of 22,666 calves were produced by ET in 2009. Currently, approximately 10–15 % of all the calves born in Japan are derived from in vivo-cultured ET in Japan.

In 1989, the biotechnology center of the Livestock Improvement Association was opened, and this center has provided dairy cooperatives and private organizations throughout the country with about 10,000 in vitro-cultured embryos each year. As a consequence, 2,308 calves derived from IVF were born in 2006.

The performance of bulls can be demonstrated by a progeny test, but "excellent" cows tend to be selected for the "good response" of multiple ovulations to hormonal treatment, and heritability is not necessarily high.

Production of calves by transfer of Japanese Black embryos to HF cows provides an additional income for dairy farmers because Japanese Black calves can be sold for better prices. Ensuring successor heifers has become easier than before, thanks to sexing technologies, such as increased utilization of XX sorted semen, and transfer of female embryos.

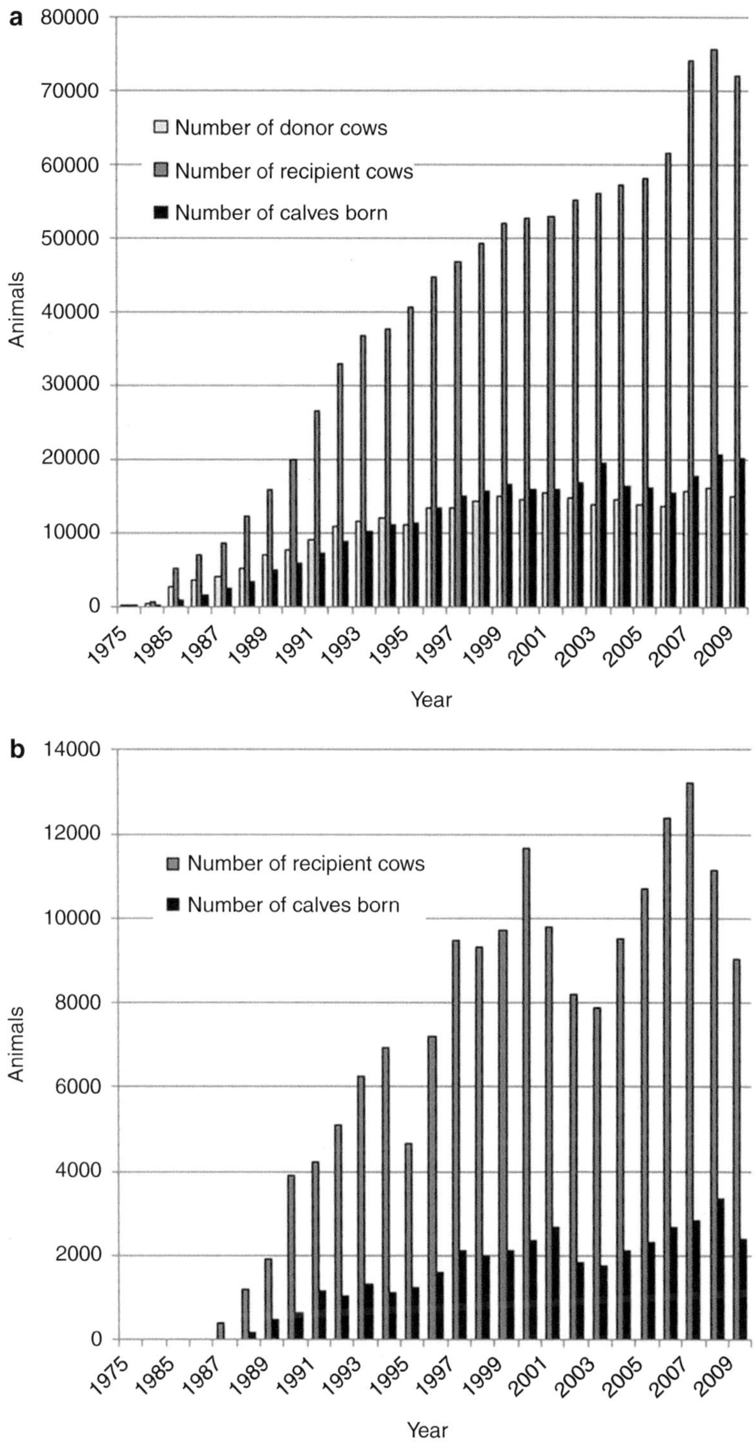

Fig. 11.1 Change in numbers of donor cows, recipient cows, and calves born by transfer of in vivo- (**a**) and in vitro- (**b**) cultured embryos in Japan

Ovum Pick-up and In Vitro Production

Conventionally, multiple ovulation treatment using FSH injections has been recognized as a means for collecting embryos for ET. Subsequently, techniques were established to produce embryos in vitro by collecting immature ova from the ovaries in the slaughterhouse, and then maturing, fertilizing, and culturing them in the dishes. Several fertilized ova and embryos may be produced in vitro and utilized for producing calves. Later, the collection (aspiration) of immature ova from the ovaries of live cows using ultrasonography became available. Such ova are also able to be matured, fertilized with selected sperm, and grown to the blastocyst stage before transfer to recipients. This series of techniques is termed "ovum pick-up and in vitro production" (OPU-IVP). The advantages of OPU-IVP are as follows:

1. It is possible to produce greater numbers of embryos with repeated collection from the same donor cows (once a week, for example).
2. It does not require labor and cost for hormonal treatment.
3. Variation in the number of ova is lower than that of the number of embryos collected from different donor cows.
4. Timing of implementation of the treatment is more flexible (practically, collection of ova can be started at any time regardless of the stage of the ovarian cycle, unlike the FSH injection protocols).
5. The range of applications is wider, i.e., the OPU-IVP method can be applied to old cows, pregnant cows, fattening cows, and even cows with certain reproductive disorders that are not suitable for the conventional method of FSH treatment, multiple ovulations (MO), and embryo flushing.

For example, with the OPU-IVP method, an average of 11.8 embryos were obtained, which was significantly higher than the 6.4 embryos obtained with the conventional method (Imai et al. 2007). Moreover, no transferable embryo was obtained from only 1.7 % of the total donor cows, compared with 27.8 % of the animals treated with MO. OPU-IVP is therefore more time- and cost-efficient.

Compared with embryo production from slaughterhouse ovaries, OPU-IVP has advantages. First, pedigree registration is always possible, and excellent pedigree cows can be used as donors. Second, collected ova are fresher so that percentage developing to blastocysts is expected to be higher.

An approximately 15-year-old cow provided 253 premature ova in 38 OPU operations, at about 10-day intervals, during 1 year. In addition, a 17-year-old cow provided 53 premature ova in two OPUs (Hirata et al. unpublished data). These results suggest that old cows still can be used as donors for OPU.

Typically, 20–30 % of donors exhibit a weak response to FSH. For example, using a cow from which only 13 embryos were collected in 5 years, and a cow from which only 10 embryos were collected in 4 years, 175 ova were collected from each animal (a total 350 ova from the two animals) by 15 OPUs in 6 months. Although the numbers of transferable embryos were only 18 and 20, the yield still surpassed that achieved by FSH treatment, and in a shorter period, indicating the advantage that OPU has over the conventional method (Hirata et al. unpublished data).

Recommendations for the use of OPU-IVP

1. Apply to pregnant cows after AI: Even though the number of follicles aspirated and collected ova is less, the quality of the collected ova tends to be better and the development rate to blastocyst is higher than in those from non-pregnant cows. OPU does not have adverse effects on the calving rate in pregnant donor cows.
2. OPU and timed AI in Japanese Black beef cattle in the early postpartum period (<40 days postpartum): Timed AI protocols combined with OPU-IVP have been reported. Immediately after OPU, the progesterone-based Ovsynch protocol was implemented. This trial showed that the protocol was effective in early postpartum suckled beef cattle, that the conception and ovulation synchronization rates were satisfactory, and that the

high quality of embryos cultured in vitro was maintained, even after several consecutive OPUs in Japanese Black cows weaned during early postpartum (Hirata et al. 2007, 2008, 2011).

3. Timed AI in early postpartum suckled and non-suckled cows and subsequent OPU-IVP: Three OPUs in inseminated cows produce numerous in vitro-cultured embryos without adverse effects on the pregnancy status of donor animals. Non-suckled cows yielded more ova and produce more embryos in vitro, and the conception rates of recipient cows compare well with suckled cows.

For these reasons, it is expected that OPU-IVP is a technique that could be widely adopted in the field, particularly for Japanese Black cattle, whose value per head and per embryo is very high.

Work to do in OPU-IVP:

1. Education of workers with the technical skills to carry out the series of manipulations required for collection and culture of ova.
2. Investment in relatively expensive special facilities and kits required for ova collection, including ultrasonography, cell culture, and microscopy.
3. Minimizing the variability of the growth rate from ovum to blastocyst among donor animals.
4. Improvement of the efficiency of production of transferable embryos cultured in vivo and improvement of the conception rate in recipient cows.

In 2005, 1,552 cows were used for OPU and 3,846 embryos were produced by OPU-IVP (c.f. 48,939 embryos produced in vitro).

Cloning Technology

The current situation (as of December 2012) with respect to the cloning of domestic animals in Japan is summarized below.

1. Embryonic cloning of cattle
   The first birth in Japan of a calf from a cloned blastomere of an embryo was in 1990. Between 1999 and March 2012, 731 embryonically cloned cattle were born in Japan, of which 620 were by normal calving, 76 (10 %) were stillborn, and 35 (5 %) died immediately after birth. Of the 620 normal births, 105 (14 %) died from illness, 20 (3 %) died by accident, and 98 (13 %) were used for research and other purposes. From the 731 calving episodes, 334 cattle (46 %) entered the market for consumption.

   The number of embryonically cloned calves born up to 1998 was 461 and in 1999, 2000, 2001, and 2002 there were 65, 52, 51, and 49 calvings, respectively. The number went down to nine in 2003 and became none by 2012.
2. Somatic cell cloning of cattle
   In 1998, the first calf cloned from a somatic cell derived from a cultured fibroblast was born in Japan (Kato et al. 1998). By the end of March 2012, 594 cattle were born in Japan, of which 411 (69 %) were by normal calving, 88 (15 %) were stillborn, and 95 (16 %) died immediately after birth. Of the calves born normally, 112 (19 %) died within 6 months of birth, 37 (6 %) died 6 or more months after birth, 9 (1 %) died accidentally, and 253 (43 %) were used for research and other purposes. None of these cattle entered the market for consumption.

   The mortality rates in newborn calves produced by various techniques were as follows: inseminated dams (5.3 %); transfer of in vivo-produced embryo (4.6 %); transfer of in vitro-produced embryo (7.5 %); embryonically cloned cattle (716; 15 %); and somatic cell-derived cloned cattle 535 (31 %).

   Japan is probably the country in which the highest number of cloned cattle have been born, and also the country with the highest number of research institutes (47) where cloned cattle have been reared. The average market price of a Japanese Black calf is about half a million yen. Clearly, a calf born from an excellent sire can be sold for a high price and calves derived from an excellent cow also command a good price. However, cloned cattle did not dominate the market. As mentioned above, the calving rate is low in cloned animals and the problems of high stillbirth and mortality rates remain to be solved. Nevertheless, cloned cows that had calved normally had higher growth rates than normal cows, and were able to get pregnant and calve repeatedly. Lactation occurred normally and their offspring were all normal with normal blood tests. Furthermore, it has been reported that cloned fattening animals grew normally, and that the variability in weight and rib-eye area among cloned animals from the same donor was lower (National Livestock Breeding Center).

The first risk assessment of the food safety of somatic cell nuclear transfer (SCNT) animals was released by the Food and Drug Administration (FDA) of the USA in January, 2008 (FDA 2008). It concluded that meat products derived from cloned food animals, such as cattle, goats, and pigs, are as safe for human consumption as conventional food animals. A similar report was issued in Europe (European Food Safety Authority, EFSA).

In Japan, a report of the Food Safety Commission (FSC) of the Cabinet Office in March 2009, which was the first risk assessment report concerning foods derived from SCNT animals and their progeny in Asian/Australasian countries, concluded that food products derived from SCNT cattle, pigs, and their progeny are substantially equivalent to cattle and pigs produced by conventional assisted reproduction technologies, such as AI and ET (FSC 2009). Thus, the safety of food products derived from SCNT has been upheld scientifically. This information is available to the general public in the form of printed and online documents. However, in August 2009, the Ministry of Agriculture, Forestry and Fisheries (MAFF) of Japan issued a notification demanding a continuation of the voluntary moratorium in place since 1999, because of the low efficiency of animal SCNT and the lack of consumer acceptance of clone animals (Watanabe and Nagai 2011). It appears that scientific data are unable to eliminate consumer anxiety concerning SCNT-derived food products. It has been reported that most Japanese consumers did not change their attitudes toward cloned beef (cattle cloned from either embryonic or somatic cells) after receiving technological information about nuclear transfer (e.g., the cells of cloned cattle have identical genes to those of the cells used in the nuclear transfer), bovine embryo and somatic cell cloning processes, the relationship between donors and their cloned calves, and the expected merits of animal cloning (Aizaki et al. 2011).

A nationwide survey concerning ETs using SCNT embryos, and the calves produced, was carried out in 2009 (Watanabe and Nagai 2011). The data referred to 301 cloned calves that were born between 1998 and 2007. The survey revealed that survival of transferred bovine embryos and calves derived from SCNT had not improved over a decade. The survival rates of transferred bovine embryos and calves derived from SCNT were lower than those of MOET and IVP because of the conception rate (21.6–32.9 %), which was lower than for MOET (50–52 %) and IVP (37–46 %), and because of the high frequency of abortions (8.3–27.1 %) that occurred during 100 days of gestation (Watanabe and Nagai 2011). Moreover, the incidences of stillbirths, neonatal deaths, and fatal diseases were high in the calves derived from SCNT (Watanabe and Nagai 2009).

How to Tackle the Decline in Reproductive Performance of Cattle

The conception rate by AI of dairy and beef cattle has been declining in Japan and many other nations. A questionnaire to farmers indicated that, in their perception, the best approaches to increasing pregnancy rates in cattle were, in order of importance, feeding and hygiene management, improvements to the detection of estrous signs, and improvement of AI techniques.

Strategies to Mitigate the Adverse Effects of Heat Stress on Reproduction

A decrease in the conception rate of dairy cows is one of the major problems in reproductive management in Asia as well as other parts of the world. It has been suggested that one of the causes of the fall in conception rate of lactating dairy cows during the summer season in Japan is increased core body temperature, with the maximum temperature–humidity index of 70 or higher on the day before AI (Nabenishi et al. 2011).

Heat shock to early stage embryos (day 0–2) has been reported to delay development to blastocyst stage, and reduce cell proliferation (Sakatani et al. 2004). Therefore, ET could be an option to improve the conception rate in dairy cattle in summer. Transfer of embryos (day 7 blastocysts) could avoid the risk of heat stress that oocytes or embryos at earlier stages may face when AI is implemented, and thus improve conception rates in dairy cows during the summer season (Sartori et al. 2002).

The genetic effects of heat stress on fertility in Holstein-crossbred cows have been studied in Thailand, and better reproductive performance (fewer days open) could be achieved through selective breeding of cows with less than 87.5 % Holstein genetics (Boonkum et al. 2011).

Improving Conception Rate in Repeat Breeders

Significantly higher pregnancy rates ($P<0.05$) were achieved in heifers by ET following AI (49.2 %) than by ET alone (29.5 %). A similar result was observed in cows (41.5 % and 20.4 % for ET following AI and ET alone, respectively), suggesting that ET may be useful in conjunction with AI during repeat breeding (Dochi et al. 2008). ET following AI was reported to overcome low conception rate in dairy cows during summer (Ambrose et al. 1999; Rutledge 2001; Tani et al. 2010).

Future Perspectives

The populations of domestic animals are increasing throughout the world. In 1950, the total number of cattle was 720 million and this had doubled to 1,530 million by 2001. Likewise, the number of sheep and goats increased from 1,040 million in 1950 to 1,750 million in 2001. Although grazing by herbivores could be considered effective utilization of land that is unsuitable for cultivation, damage by overgrazing, methane gas production, and desertification are becoming worldwide problems.

Domestic milk production falls far below the total quantities milk consumed locally in many Asian countries, so a substantial proportion must be imported from other countries (e.g., about 80 % in Vietnam). Some governments have definite plans to develop their dairy farming now and in the future, which means that there is considerable potential for further development of animal industries.

It is necessary to maintain a balance between increasing productivity and preserving local breeds that have adapted to local climates. Crossbred cows (e.g., F_1, F_2, and F_3 generations of European breed bulls × improved local breeds) should therefore be developed in many different ecological areas within the region.

It is important to establish systems for collection of accurate data from individual farms on animal identification, pedigree, reproductive performance, and lactation. Overall planning for dairy cow development should be carried out at both central and local levels. Specifically, it is vital to set up systems for ear-tagging of cows for identification. This will facilitate the recording and updating of notes for monitoring data on milk production and reproduction, and will help to promote and implement productivity testing of breed bulls through their offspring.

Eventually, the development of human resources will be the key to success of reproductive technologies and livestock production. People who are actively working in the field in the regions are essential, and it may be necessary to introduce a system of monetary or non-monetary incentives to encourage such people.

In countries where rapid development is ongoing, the desire for consumption of animal protein is hard to control, but efforts could be made to reduce the pressures on grazing land. For example, utilization of rice straw, wheat straw, and corn stalks for domestic animals should be considered. In fact, Japan has a tradition of circulating animal husbandry that utilizes rice straw left after rice cultivation for rearing cattle in the same farm. In exploring the best ways to formulate a sustainable society, it is essential to make the most of livestock products by applying appropriate reproductive technologies. The raising of dairy cows should be integrated with eco-environmental conditions and take into account waste treatment, environmental protection, and animal welfare issues. Developed and developing countries must share a common awareness and concern for these issues. Dealing with these problems is a further hurdle in the pursuit of productive efficiency by developing countries. International societies should be involved in binding together developed and newly developing countries in the construction of a novel model for future livestock rearing management that suits their diverse environmental circumstances.

Major technical breakthroughs in reproductive biology in the last century are cryopreservation of gamete and embryos, nonsurgical embryo transfer, in vitro maturation of eggs, assisted fertilization such as IVF and intracytoplasmic sperm injection, sperm-sexing, and preimplantation genomic analysis of embryos (Yanagimachi 2011). Some of future research areas related to the improvement of animal reproduction should be permanent preservation of spermatozoa at ambient temperature, mass production of mature gametes in vitro, conversion of somatic cells to germ cells, and genomic manipulation of gametes for the production of superior animals.

In fish species, some important breakthroughs in germ cell development have been made and a method to create masu salmon (*Oncorhynchus masou*) broodstock that produce both eggs and sperm of rainbow trout (*Oncorhynchus mykiss*) succeeded in deriving offspring trout populations through fertilization using these eggs and sperm (Okutsu et al. 2007). This technology has potential to be applied into other areas including the transplantation of spermatogonia of the Pacific bluefin tuna (*Thunnus orientalis*), one of the most commercially valuable fish, taking 3–5 years to reach sexual maturity at which time it reaches a body weight of more than 100 kg, into the closely related chub mackerel (*Scomber japonicus*), which reaches sexual maturity in 1 year at a body weight of less than 500 g, so that the Pacific bluefin tuna gametes could be produced more easily and rapidly, even in a land-based small fish tank (Yoshizaki et al. 2012). A series of these technologies could have a fundamental change in the way of fish production systems in future.

Conclusions

The history of animal reproductive technology follows the progress of efforts to meet the increasing demand for animal proteins in the developing countries of Asia and elsewhere in the world. AI, ET, and related scientific technologies in the livestock industry have enabled increases in animal numbers and productivity per animal, together with the development of lines having excellent quality.

With the availability of these cutting edge reproductive technologies in the livestock industry, the most important problem for increasing productivity in developing countries now concerns the maintenance of optimum nutrition to support reproductive performance and increased supply of animal proteins.

Such developments will not only generate employment and improve the health of local people but also help to make a more secure and safe society.

References

Abeygunawardena H, Dematawewa CMB (2004) Prepubertal and postpartum anestrus in tropical Zebu cattle. Anim Reprod Sci 82–83:373–387

Aizaki H, Sawada M, Sato K (2011) Consumers' attitudes toward consumption of cloned beef. The impact of exposure to technological information about animal cloning. Appetite 57:459–466

Ambrose JD, Drost M, Monson RL, Rutledge JJ, Leibfried-Rutledge ML, Thatcher MJ et al (1999) Efficacy of timed embryo transfer with fresh and frozen in vitro produced embryos to increase pregnancy rates in heat stressed dairy cattle. J Dairy Sci 82:2369–2376

Boonkum W, Misztal I, Duangjinda M, Pattarajinda V, Tumwasorn S, Buaban S (2011) Genetic effects of heat stress on days open for Thai Holstein crossbreds. J Dairy Sci 94:1592–1596

Dochi O, Takahashi K, Hirai T, Hayakawa H, Tanisawa M, Yamamoto Y et al (2008) The use of embryo transfer to produce pregnancies in repeat-breeding dairy cattle. Theriogenology 69:124–128

FAO (Food and Agriculture Organization of the United Nations) (2002) FAOSTAT database. FAO, Rome

FAO (Food and Agriculture Organization of the United Nations) (2010) FAOSTAT database. FAO, Rome

FAOSTAT Agriculture (2013) Livestock primary. Production of top 5 producers. Milk, whole fresh cow. Average 1992–2011. http://faostat.fao.org/site/339/default.aspx

Food and Drug Administration (FDA) (2008) Animal cloning: a risk assessment [cited 14 Jan 2008]. FDA, Silver Spring, MD. Available from http://www.fda.gov/cvm/Documents/Cloning_Risk_Assessment.pdf

Food Safety Commission (FSC) of Japan (2009) Safety assessment of novel foods. Foods derived from cloned cattle and pigs produced by somatic cell nuclear transfer (SCNT) and their offspring [cited 25 June 2009]. FSC, Tokyo, Japan. Available from http://www.fsc.go.jp/english/evaluationreports/hy_detail_clone.pdf

General Statistics Office of Vietnam (2013) Monthly statistical information, Statistical data. http://www.gso.gov.vn

Goddard MG, Smith C (1990) Optimum number of bull sires in dairy cattle breeding. J Dairy Sci 73:1113–1122

Herd RM, Bootle BW, Parfett DC (1993) An economic evaluation of traditional, twinning and sex-controlled systems of beef production in Southern Australia. Aust J Agric Res 44:1541–1556

Hirata T, Hoshina T, Sasaki S, Sasaki O, Osawa T (2007) Applicability of a progesterone-based timed artificial insemination protocol after follicular fluid aspiration using the ovum pick-up technique in suckled beef cows. J Reprod Dev 53:171–177

Hirata T, Sato M, Sasaki S, Sasaki O, Osawa T (2008) Effect of suckling on embryo production by repeated ovum pick-up before and after timed artificial insemination in early postpartum Japanese black cows. | J Reprod Dev 54:346–351

Hirata T, Kon N, Sugiyama A, Sato M, Osawa T (2011) Effect of follicular aspiration at the onset of progesterone-based timed artificial insemination on the follicular dynamics and fertility of early postpartum Japanese black cows. J Reprod Dev 57:613–619

Imai K, Inaba Y, Yoshioka H, Aikawa Y, Ohtake M, Suzuki M et al (2007) Effect of follicular wavesynchronization and super stimulation on *in vitro* embryo production. Reprod Fertil Dev 20:182

Kato Y, Tani T, Sotomaru Y, Kurokawa K, Kato J, Doguchi H, Yasue H, Tsunoda Y (1998) Eight calves cloned from somatic cells of a single adult. Science 282:2095–2098

Kitakura T, Li K (2007) A study of situation and promotion policy of dairy farming and dairy industry in China. Hokkai-Gakuen University. J Econ Q 54:31–50 (in Japanese)

Lande R, Barrowclough GF (1987) Effective population size, genetic variation, and their use in population management. In: Soulé ME (ed) Viable populations for conservation. Cambridge University Press, Cambridge, MA, p 87

Martojo H (2012) Indigenous Bali cattle is most suitable for sustainable small farming in Indonesia. Reprod Domest Anim 47(suppl 1):10–14

Mc Hugh N, Fahey AG, Evans RD, Berry DP (2010) Factors associated with selling price of cattle at livestock marts. Animal 8:1378–1389

Ministry of Agriculture, Forestry and Fishery (MAFF) (2013) Statistics of animal husbandry. Ministry of Agriculture, Forestry and Fisheries, Tokyo

Ministry of Agriculture, Forestry and Fishery (2013) e-Stat http://www.e-stat.go.jp/SG1/estat/List.do?lid=000001105713

Nabenishi H, Ohta H, Nishimoto T, Morita T, Ashizawa K, Tsuzuki Y (2011) Effect of the temperature-humidity index on body temperature and conception rate of lactating dairy cows in southwestern Japan. J Reprod Dev 57:450–456

Nimbkar C (2011) Animal breeding in India—a time for reflection and action. J Anim Breed Genet 128:161–162

Nomura T, Honda T, Mukai F (2001) Inbreeding and effective population size of Japanese Black cattle. J Anim Sci 79:366–370

Oishi K, Hirooka H (2012) Effects of sex control and twinning on economic optimization of culling cows in Japanese Black cow-calf production systems. Theriogenology 77:320–330

Okutsu T, Shikina S, Kanno M, Takeuchi Y, Yoshizaki G (2007) Production of trout offspring from triploid salmon parents. Science 317:1517

Perera BMAO (2011) Reproductive cycles of buffalo. Anim Reprod Sci 124:194–199

Rutledge JJ (2001) Use of embryo transfer and IVF to bypass effects of heat stress. Theriogenology 55:105–111

Sakatani M, Kobayashi S, Takahashi M (2004) Effects of heat shock on in vitro development and intracellular oxidative state of bovine preimplantation embryos. Mol Reprod Dev 67:77–82

Sartori R, Sartor-Bergfelt R, Mertens SA, Guenther JN, Parrish JJ, Wiltbank MC (2002) Fertilization and early embryonic development in heifers and lactating cows in summer and lactating and dry cows in winter. J Dairy Sci 85:2803–2812

Sasaki O, Kimura H, Ishii K, Satoh M, Nagamine Y, Yokouchi K (2011) Economic effects of using sexed semen in Japanese dairy herds. Anim Sci J 82: 486–493

Stockton MC, Besseler DA, Wilson RK (2010) Price discovery in Nebraska cattle markets. J Agric Appl Econ 42:1–14

Sugie T (1965) Successful transfer of a fertilized bovine egg by non-surgical techniques. J Reprod Fertil 10:197–201

Tani M, Hayashida T, Tomokawa K, Mito Y, Funakoshi D, Tani C et al (2010) Effect of embryo transfer following artificial insemination (ETFAI) on reproductive performance in dairy cows in South-Western Japan. J Vet Med Sci 72:627–629

USDA (United States Department of Agriculture) (2013) Data. FAS Databases. Production, Supply and Distribution. Reports. Livestock. Beef and veal summary selected countries. http://www.fas.usda.gov/psdonline/

Watanabe S, Nagai T (2009) Death losses due to stillbirth, neonatal death and diseases in cloned cattle derived from somatic cell nuclear transfer and their progeny: a result of nationwide survey in Japan. Anim Sci J 80:233–238

Watanabe S, Nagai T (2011) Survival of embryos and calves derived from somatic cell nuclear transfer in cattle: a nationwide survey in Japan. Anim Sci J 82:360–365

Yanagimachi R (2011) Fertilization studies and assisted fertilization in mammals: their development and future. In: International meeting for evolution of reproductive biology and tasks of frontiers: trajectory and prospects of IVF, stem cell and epigenetic studies. Society for Reproduction and Development, Morioka, Japan, 13–15 Sept 2011, pp 20–30

Yoshizaki G, Okutsu T, Morita T, Terasawa M, Yazawa R, Takeuchi Y (2012) Biological characteristics of fish germ cells and their application to developmental biotechnology. Reprod Domest Anim 47(suppl 4):187–192

Zi X-D, He S-M, Lu H, Feng J-A, Lu J-Y, Chang S et al (2006) Induction of estrus in suckled female yaks (*Bos grunniens*) and synchronization of ovulation in the non-sucklers for timed artificial insemination using progesterone treatments and Co-synch regimens. Anim Reprod Sci 92:183–192

12 International Perspectives on Impacts of Reproductive Technologies for World Food Production in Asia Associated with Poultry Production

Vishwajit S. Chowdhury, Halima Sultana, and Mitsuhiro Furuse

Abstract

Poultry meat and eggs are valuable sources of dietary protein in almost every country in the world. A number of breeding techniques, methods, and technology have been applied to obtain maximum production under different environmental and economic conditions. Indigenous and local breeds share 90 % of the total population of poultry in many developing countries in Asia. However, indigenous chickens are low in productivity. Many studies have found that crossbreeding of exotic with indigenous chickens resulted in birds that performed better, even superior to pure exotic chickens, with respect to body weight, egg production, survivability, fertility, hatchability, and egg quality. There are some other technologies for efficient use of male genetic resource and conservation of rare genetic make-up, namely artificial insemination and chimeric chicken, respectively. It was reported that 25 % of the world's meat supply is derived from poultry, and the proportion is increasing rapidly.

V.S. Chowdhury, Ph.D. (✉)
Division for Arts and Science, Faculty of Arts and Science, Kyushu University, Fukuoka 819-0395, Japan
e-mail: vc-sur@artsci.kyushu-u.ac.jp

H. Sultana, Ph.D.
Laboratory of Animal Science, North Florida Research and Education Center (NFREC), University of Florida, 3925 HWY 71, Marianna, FL, USA

M. Furuse, Ph.D.
Faculty of Agriculture, Laboratory of Regulation in Metabolism and Behavior, Department of Bioresource Sciences, Kyushu University, Fukuoka 812-8581, Japan

G.C. Lamb and N. DiLorenzo (eds.), *Current and Future Reproductive Technologies and World Food Production*, Advances in Experimental Medicine and Biology 752,
DOI 10.1007/978-1-4614-8887-3_12, © Springer Science+Business Media New York 2014

The continent of Asia produces almost one third of the world's eggs. However, there are still many scopes to improve the production of poultry in many developing countries in Asia. Therefore, continuous research works would be essential to determine the suitable technologies for more poultry production to feed the increasing habitants on earth.

Keywords
Poultry • Reproductive technology • Poultry production • World food production

Introduction

The original site of domestication of chicken in the area inhabited by a single subspecies of the red jungle fowl (*Gallus gallus gallus*) is Indonesia (Fumihito et al. 1994). Commercial genetic selection for chickens is mainly done in western countries and the products can be efficiently obtained. However, the center of the poultry industry is shifting to subtropical countries, and this trend is likely to continue. The poultry industry in Asian region will grow even faster if maximum benefits can be derived from modern broiler stocks (Singh 1999). In many developing countries in Asia, commercial poultry farming has recently emerged in different areas of the country to satisfy market demands for poultry meat and eggs by the mostly urban and municipal populations. Despite rapid growth of commercial poultry farming, eggs and meat are still at small holder farms under traditional scavenging system in many developing countries (Das et al. 2008). The traditional "backyard" operations, universally known as "family poultry production," have a long historical background in many Asian countries that makes it popular with the rural and certain urban populations. Scientists are concerned to obtain maximum production, involving landless or marginal farmers with a minimum land, investment, and time (Islam and Nishibori 2010). A number of breeding techniques, methods, and technologies have been applied to obtain this goal. Some poultry industries have imported high-yielding exotic varieties of chickens in some tropical and subtropical countries which originated in temperate countries where they produce well. In tropical climate like Bangladesh, the production performance of these improved chickens is often below the standard of the breeder company because of their genetic make-up inherent for temperate region, genotype × environment interaction, unsophisticated management, inadequate nutrition and harsh environment, as well as high susceptibility to disease and lack of availability of good quality vaccines and therapeutics (Hutt 1958; Al-Soudi and Al-Azzawi 1974; Okoye and Aba-Adulugba 1996; Tadelle et al. 2000; Singh et al. 2004; Islam and Nishibori 2010). Besbes (2009) showed that indigenous and local breeds share 90 % of the total population of poultry in developing countries. Therefore, some nongovernment organizations (NGOs), foreign aid projects, and governmental organizations have concentrated on adopting a model for using crossbred chickens in scavenging or semi-intensive systems to increase family poultry production (Das et al. 2008). Under extensive or traditional systems of poultry rearing, indigenous chickens performed better with respect to survivability, fertility, and hatchability, although they have poor productivity (Huque and Haque 1990; Barua et al. 1998; Islam 2000, 2006) which does not encourage farmers to extend the present level of their poultry operations. Indigenous chickens are low in productivity due to their inherent genetic characteristics, poor husbandry practices, seasonal effects, low level of nutrition, and broodiness (Sarkar and Bell 2006; Besbes 2009). Many studies have found that crossbreeding of exotic with indigenous chickens resulted in birds that performed better, even superior to pure exotic chickens, with respect to body weight, egg production, survivability, fertility, hatchability, and egg quality (Islam et al. 1981; Barua and Howlider 1990; Khondoker et al. 1996; Rahman et al. 1998). Besbes (2009) reported improved body weight, egg production, and survivability in crossbred chickens. Although it is very complex to maintain a crossbreeding program under village conditions, in most cases the cock has been used to utilize high-yielding breeds to upgrade local chicken (Besbes 2009). Considering the above facts, we would try to review the potential of using crossbreds for poultry production in hot-humid climate which is a typical environment in many Asian countries. Artificial insemination (AI) was developed in the middle of the twentieth century. With the use of AI, a single male animal can have a much larger influence on a population than with the use of natural mating. Therefore, AI has had a very significant effect in terms of improving the genetic make-up of poultry. Both AI and natural mating are used in the breeding of broiler chickens. In turkey breeding, AI has become indispensable, because the toms are so large and heavy that they would cause serious harm to the hens in natural mating. The turkey industry initially was hesitant to use AI, but now is dependent on this reproductive biotechnology to maintain flocks; it had no alternative (Singh 1999). As with turkeys, broiler breeder males have also difficulty of floor mating. Therefore, broiler breeder farms also practice AI in many countries in Asia. Chimeric chicken would be another reproductive technology which has been practiced in some countries in Asia for conservation of genetic resources of birds and the production of transgenic chickens (Furuta 2011).

Breeding Systems in Poultry

In poultry breeding, several breeding systems are being practiced. There is no evidence, however, to indicate that any one system is best for all purposes. Although there are a number of available breeding systems (outcrossing, grading, line breeding, top crossing, etc.), crossbreeding is a very popular means to increase the production of poultry in many Asian countries. Thus, here we will review the potential of using crossbreds and the development of suitable genetic make-up for future poultry production in hot-humid climate in many Asian countries. The methods of mating used will have a marked influence on fertility and consequently on the number of offspring used.

Performance of Exotic, Indigenous, and Crossbred Chickens

Growth Performance

Table 12.1 shows the growth performance of indigenous, exotic, and crossbred chickens. D. Nana (improved naked neck chicken), and D. nana × Rhode Island Red (RIR) crossbred chickens showed better growth and feed conversion efficiency compared with pure exotic and other crosses at 16–17 weeks of age (Table 12.1). Singh et al. (1996) observed better growth performance in Nana chickens compared to their full feathered counterparts in India. Fayoumi × RIR are found to be the best performer among the crossbreds in terms of body weight gain at 20–24 weeks of age (Table 12.1). Azharul et al. (2005) reported that Fayoumi × RIR crossbred ("Sonali") performed better in terms of growth performance in intensive systems under rural condition of Bangladesh compared to pure breed Fayoumi. Therefore, crossbred chickens are widely used in many developing countries where hot-humid climate is prevailing like Bangladesh.

Table 12.1 Growth performance of indigenous, exotic, and crossbred chickens

Genetic resource	Age (week)	BW (g/bird)	FI (g/bird)	FCR
D. nana	16–17	1213.8	5831.0	5.40
WLH	16–17	987.7	5093.2	5.40
Fayoumi	16–17	975.8	6080.9	6.20
D. nana × RIR	16–17	1142.4	5854.0	5.10
D. nana × WLH	16–17	1082.9	5640.6	5.20
D. nana × Fayoumi	16–17	1094.8	5712.0	5.25
Significance level (CRD)		*	**	**
Fayoumi	20–24	1200.0	5844.0	7.46
RIR	20–24	1560.0	6386.0	5.46
WLH	20–24	1278.0	5806.0	6.87
Fayoumi × WLH	20–24	1309.0	6072.0	7.04
Fayoumi × RIR	20–24	1462.0	6060.0	6.15
Significance level (CRD)		***	NS	**

Source: Islam and Nishibori (2010). Journal of Poultry Science. 47:271–279

BW body weight, *FI* feed intake, *FCR* feed conversion ratio, *WLH* White Leghorn, *RIR* Rhode Island Red, *CRD* complete randomized design, $*P<0.05$; $**P<0.01$; $***P<0.001$; *NS* Non significant

Table 12.2 Egg production performance of indigenous, exotic, and crossbred chickens

Genetic resource	Laying period (year)	Hen day egg production (%)	Egg weight (g)
D. Nana	1	26.86	42.55
Hilly	1	25.10	42.00
Aseel	1	9.40	45.00
WLH × Fayoumi	1	61.63	49.45
WLH × RIR	1	58.71	55.10
Fayoumi × D. Nana	1	41.0	46.0
Fayoumi × Hilly	1	35.00	45.00

Source: Islam and Nishibori (2010). Journal of Poultry Science. 47:271–279

D. Nana indigenous naked neck, *WLH* White Leghorn, *RIR* Rhode Island Red

Egg Production Performance

White Leghorn (WLH) × Fayoumi was found to be the highest egg producing chicken compared to other chickens when reared in intensive management systems, but not in scavenging or semi-scavenging systems (Table 12.2). Haque and Howlider (2000) showed better egg production in D. nana × Fayoumi and D. nana × RIR under scavenging condition compared to exotic chickens. In developing countries, crossbreds from indigenous with exotic chickens perform better under scavenging or semi-scavenging condition (Das et al. 2008). Horst (1991) observed improved egg production in crossbreds of Nana × Dahlem Red (German strain) and WLH × Kadaknath (Indian) chickens. Egyptian local chicken (Dandarawi and Fayoumi) were lower for egg production (202.2 and 212.5 eggs/year, respectively) in temperate countries compared to Dahlem Red (279 eggs/year) but, in warm climates, egg production was depressed in both. However, the depression was lower in Egyptian chickens (137.9 and 161.0 eggs, respectively) than in Dahlem Red chickens (181.7 eggs) (Valle Zarate et al. 1988). In India, a crossbred hen from Aseel × CARI Red, called "CARI Nirbheek," has been developed for improving egg production under traditional condition. This crossbred was able to produce 163 eggs annually with a survival rate of 90–95 % (Singh et al. 2004). Therefore, some crossbreds under scavenging or semi-intensive system in tropical climate may be worthy combinations for increasing the production of egg.

Artificial Insemination (AI)

The impact of AI in broiler breeder industry is well recognized. Economic benefits of AI over natural mating coupled with spiking of males in floor groups have been extensively studied (Reddy and Sadjadi 1990; Reddy 1995, 1996). One of the main advantages of AI is that poultry raisers need to use less male compared to floor mating. The cost of labor is less in many Asian countries compared to some developed countries which is an additional advantage of conducting AI in many broiler farms in Asia. In India, production costs of broiler chicks and hatching eggs will be lower in flocks using AI than natural mating (Singh 1999). It is possible to house roosters in cages, collect semen for AI, and inseminate broiler hens housed in floor pens. Regardless of the exact system, with proper implementation of AI, number of chicks hatched per valuable male, or even per group of hens, can be increased vs. floor mating (Singh 1999). Quinn and Burrows were the first to describe a practical method of manual semen collection and AI of the fowl (see Sexton 1984). With some modification this technique became the basis of the current method of semen collection and insemination of chickens and turkeys. For the success of any AI program, both males and females must be given proper care to ensure the uninterrupted production of semen and eggs.

Production of Chimeric Chickens

Germline chimeric chickens are considered effective means for the conservation of endogenous birds and for the production of transgenic chickens.

Blastodermal Cells

Somatic cell chimeras have been produced by transferring blastoderm. Donor blastodermal cells from freshly ovipositioned fertilized eggs were transferred to a recipient blastoderm, and, subsequently, the donor-derived cells developed and differentiated in the recipient gonads (Kagami and Hanada 1997; Trefil et al. 2002).

Implantation and Immigration of Primordial Germ Cells

Germline chimeric chickens can be produced using primordial germ cells (PGCs). Fertilized eggs are incubated until embryonic developmental stages 12–15 (Furuta and Fujihara 1999). PGCs are collected from the veins of the developing donor embryos and injected into the blood vessels of developing recipient embryos. Donor-derived offspring were obtained from germline chimeric chickens (Naito et al. 1994; Furuta et al. 2001; Kuwana et al. 2006). When the Japanese quail was used, chicken and quail hybrid chimeras were produced (Nakamura et al. 1992; Ono et al. 1996, 1998a, b). However, when PGCs were introduced into the germinal crescent of recipient embryos at stage 9, the donor PGCs immigrated to the recipient gonads via embryonic blood vessels (Furuta and Fujihara 1999). In Japan and China there are many endeavors for producing chimeric chickens.

Poultry in World Food Production

Poultry furnish an immense supply of food for the world. Yamada (1988) estimated that 25 % of the world's meat supply is derived from poultry, and the proportion is increasing rapidly. Eggs also provide a great contribution to animal protein in the human diet. Estimates of poultry numbers throughout Asia are shown in Table 12.3 (FAO 2012). There are very few cultures in Asia and the world that do not consume eggs or poultry meat or both in large quantities (Crawford 1995). Millions of people in the world are today suffering from starvation or malnutrition and the number will increase with the inevitable rise in the world population (Hunton 1995). This leads an unalterable challenge to all scientists involved in problems underlying the production of human food.

Chickens: Wherever there is human habitation there are chickens. When measured by contributions to the human diet, they are probably the most important of all domesticated birds and mammals. Stocks kept primarily for food production are three kinds, namely indigenous stock, crossbred stock, and industrial stock. Indigenous chickens are characteristic of developing countries everywhere. Their status and importance have been reported by Mukherjee (1990) and Gunawan (in FAO 1990).

Local stocks in some countries have received a lot of study and are recognized as distinct breeds, for instance Kadaknath of India, Deshi of Bangladesh, Fayoumi in Egypt, and Segur of

Table 12.3 Estimates in millions of poultry population in some Asian countries (FAO 2012)

	Chickens	Ducks	Geese and guinea fowl	Turkey
Afghanistan	0.1	–	–	–
Armenia	0.05	–	–	0.002
Azerbaijan	0.2	–	–	0.007
Bahrain	0.005	–	–	0.0006
Bangladesh	20.1	2.2	–	0.009
Bhutan	0.003	–	–	–
Cambodia	0.2	0.07	–	–
China	46.5	7.7	2.9	0.002
India	6.2	0.3	–	–
Indonesia	13.2	0.4	–	–
Iran	4.3	0.002	0.001	0.002
Israel	0.4	0.0002	0.001	0.005
Japan	31.3	–	–	0.00003
Jordan	0.3	0.00004	0.00001	0.0001
Malaysia	2	0.4	–	–
Myanmar	1	0.1	0.007	0.00002
Nepal	0.2	0.004	–	–
Pakistan	2.4	0.04	–	–
Philippines	1.5	0.1	0.004	0.006
Republic of Korea	1.3	0.1	0.0001	0.0002
Saudi Arabia	1.5	–	–	–
Thailand	2.5	0.2	0.0003	–
Vietnam	1.8	0.8	–	–

Indonesia. Indigenous chickens contribute immensely to local food supplies. Because the performance of crossbred chickens is superior to indigenous lines as described by their potentiality above, crossbred chickens are popular in many developing countries. Industrial stocks comprise nearly all the chickens in developed countries and those in urban areas of developing countries.

Turkeys: Turkeys are major meat producers in developed countries; however, many countries in Asia are also producing turkeys (Table 12.3). Turkeys also contribute to meat supply, with leading consumers being Israel and the USA.

Ducks: Ducks are very important in Asia, especially in China and in Southeast Asian countries. Gunawan (in FAO 1990) and Mukherjee (1990) have described the Asian ducks and their remarkable productivity. Although there are no industrial layer ducks in developing countries, industrial ducks are produced for meat.

Geese: Major consumers of geese meat are China, Russia, and eastern European countries. Geese are a major poultry meat product in developed countries. Very few are raised in developing countries (Table 12.3).

Poultry Meat and Meat Products in Some Asian Countries

Processed poultry meat and their fabricated products are consumed worldwide. However, socio-political and economic circumstances modify the scope of commercialization. Here some trends are presented in various population groups in Asia to demonstrate the importance and prospects of poultry as human food.

India: Broiler meat production grew at over 15 % per annum in recent years. Due to wide acceptance and affordable prices, poultry meat is widely consumed in India. High mutton prices, religious restrictions on beef and pork, and availability of fish in coastal regions made poultry the preferred meat (Guerrero-Legarreta and Hui 2010). Indians typically prefer fresh chicken meat, and thus the processed poultry meat market is limited. However, processing of poultry meat is performed in some southern Indian industries to export in Middle East and Southeastern Asian countries.

Thailand: Thailand imports cooked chicken from the EU and Japan; China is reducing its export to Thailand for their increasing domestic consumption (Guerrero-Legarreta and Hui 2010). Thai chicken boneless leg and skinless and boneless breast are being exported to Japan.

Japan: Broiler meat makes up about 90 % of the Japanese poultry meat market, including both domestic production and imports (Guerrero-Legarreta and Hui 2010). Spent laying hens account for about 10 % of the poultry meat market; Japanese people consume less duck, turkeys, and other poultry meat. Japan imports cut meat mainly from Brazil, also some US bone-in leg cuts are utilized. Japan also imports cooked products from Thailand and China to prepare yakitori (grilled chicken).

China: Broiler production in China in 2007 was 12.5 million metric tons (Guerrero-Legarreta and Hui 2010). Poultry meat is traditional in the Chinese culture. China is exporting poultry in some Asian countries including Japan.

Malaysia: Malaysia has one of the highest per capita consumption rates in the world for chicken (Guerrero-Legarreta and Hui 2010). Chicken meat is the most popular and cheapest source of meat protein among Malaysians, due to fact that there is no dietary prohibition or religious restrictions against chicken consumption. Some chicken parts are imported from Netherlands and Denmark. All turkey meat is imported mainly from the USA.

Taiwan: The USA is the main supplier of poultry in Taiwan (Guerrero-Legarreta and Hui 2010). They have domestic production of non-broiler chickens, ducks, geese, and turkeys.

Saudi Arabia: Broiler chicken is the main source of poultry meat in Saudi Arabia (97 %). Brazil has been the leading frozen broiler meat supplier to Saudi Arabia. Consumption of whole turkeys is seasonal, while duck is consumed primarily in Chinese restaurants and some Arabic restaurants.

Conclusions

In conclusion, poultry has been playing very important role for the human food supply throughout the globe. Although traditional breeding and management are being practiced in many Asian countries, there are some advanced biotechnologies and breeding policies which are playing significant roles for the human food production. However, more advanced reproductive biotechnologies would be needed to cope with the progressing food demand.

Acknowledgements The authors express cordial thanks to Professor G. Cliff Lamb, Assistant Director and Professor, North Florida Research and Education Center, University of Florida, for his invitation of this chapter to the book entitled "World Food Production: The Impact of Current and Future Reproductive Technologies."

References

Al-Soudi KA, Al-Azzawi I (1974) Protein requirements of two breeds of laying hens in subtropical climate. Indian J Anim Sci 44:703

Azharul IM, Ranvig H, Howlider MAR (2005) Comparison of growth rate and meat yield characteristics of cockerels between Fayoumi and Sonali under village conditions in Bangladesh. Livestock Research for Rural Development, p 17. http://www.lrrd.org/lrrd17/2/azha17021.htm

Barua A, Howlider MAR (1990) Prospects of native chicken in Bangladesh. Poult Adviser 22:57–61

Barua A, Howlider MAR, Yoshimura Y (1998) Indigenous naked neck fowl of Bangladesh. Worlds Poult Sci J 54:279–286

Besbes B (2009) Genotype evaluation and breeding of poultry for performance under sub-optimal village conditions. Worlds Poult Sci J 65:260–271

Crawford RD (1995) Chapter 1. Origin, and distribution of commercial poultry. In: Hunton P (ed) Poultry production. Elsevier, Amsterdam

Das SC, Chowdhury SD, Khatun MA, Nishibori M, Isobe N, Yoshimura Y (2008) Poultry production profile and

expected future projection in Bangladesh. Worlds Poult Sci J 64:99–118
FAO (1990) Animal genetics resources. A global programme for sustainable development. Proceedings of an FAO expert consultation, Rome, September 1989. FAO Animal Production and Health Paper 80. Food and Agriculture Organization of the United Nations, Rome
FAO (2012) Statistical year book. http://faostat3.fao.org/home/index.html#VISUALIZE_BY_DOMAIN
Fumihito A, Miyake T, Sumi S, Takada M, Ohno S, Kondo N (1994) One subspecies of the red junglefowl (*Gallus gallus gallus*) suffices as the matriarchic ancestor of all domestic breeds. Proc Natl Acad Sci U S A 91:12505–12509
Furuta H (2011) Establishing germline chimeric chickens using primordial germ cells. J Poult Sci 49:1–4
Furuta H, Fujihara N (1999) Proliferation of exogenously injected primordial germ cells (PGCs) into the busulfan-treated chicken embryos. Asian J Androl 1:187–190
Furuta H, Kinoshita K, Maeda Y, Fujihara N (2001) Restoration of genetic resources from Ehime native chicken via transferred primordial germ cells (PGCs). J Poult Sci 38:302–307
Guerrero-Legarreta I, Hui YH (eds) (2010) Processed poultry products: a primer. In: Handbook of poultry science and technology, vol 2: Secondary processing, John Wiley & Sons, pp 3–11
Haque ME, Howlider MAR (2000) Growth and meat yield in native naked neck, exotic chicken and their crossbreds; F2 generation. Indian J Anim Sci 70:501–503
Horst P (1991) Native fowl as a reservoir for genomes and major genes with direct and indirect effects on the adaptability and their potential for tropically oriented breeding plans—a review. Anim Res Dev 33:63–79
Hunton P (1995) Poultry production. Elsevier, Amsterdam
Huque QME, Haque ME (1990) The onset of lay in indigenous hens following hatching of chicks. Poult Adviser 12:57–60
Hutt FB (1958) Genetic resistance to disease in domestic animals. Cornell University Press, Ithaca, NY
Islam MA (2000) Effect of local and exotic strains of chicken for broiler production at hot-humid climate. Ph.D. Thesis, Institute of Animal Science, Faculty of Agriculture and Horticulture, Humboldt University of Berlin, Germany
Islam MA (2006) Comparative egg production and egg quality of indigenous full feathered and naked neck chicken at hot-humid climate. Bangladesh J Anim Sci 35:99–105
Islam M, Nishibori M (2010) Crossbred chicken for poultry production in the tropics. J Poult Sci 47:271–279
Islam ABMM, Hoque MM, Rahim QME (1981) Reproductive performance of upgraded indigenous chicken. Poult Adviser 14:33–36
Kagami H, Hanada H (1997) Current knowledge of sexual differentiation in domestic fowl. Worlds Poult Sci J 53:111–123
Khondoker MAMY, Faruque MO, Howlider MAR, Ali A (1996) Performance of upgraded indigenous desi chicken under farm condition. Bangladesh J Anim Sci 25:85–89
Kuwana T, Kawashima T, Naito M, Yamashita H, Matsuzaki M, Takano T (2006) Conservation of a threatened indigenous fowl (Kureko Dori) using the germline chimeras transplanted from primordial germ cells. J Poult Sci 43:60–66
Mukherjee TK (1990) Chapter 42. Breeding and selection programs in developing countries. In: Craford RD (ed) Poultry breeding and genetics. Elsevier, Amsterdam
Naito M, Tajima A, Yasuda Y, Kuwana T (1994) Production of germline chimeric chickens, with high transmission rate of donor-derived gametes, produced by transfer of primordial germ cells. Mol Reprod Dev 39:153–161
Nakamura M, Yoshinaga K, Fujimoto T (1992) Histochemical identification and behavior of quail primordial germ cells injected into chick embryos by the intravascular route. J Exp Zool 261:479–483
Okoye JOA, Aba-Adulugba EP (1996) Comparative study of the resistance or susceptibility of local Nigerian and exotic chickens to infectious bursal disease. In: Proceedings of the XX World's poultry congress, New Delhi: [World's Poultry Science Association] p 11
Ono T, Yokoi R, Aoyama H (1996) Transfer of male or female primordial germ cells of quail into chick embryonic gonads. Exp Anim 45:347–352
Ono T, Yokoi R, Maeda S, Nishida T, Aoyama H (1998a) Settlement of quail primordial germ cells in chicken gonads. Anim Sci Technol 69:546–555
Ono T, Yokoi R, Maeda S, Nishida T, Aoyama H (1998b) Transfusion of chick primordial germ cells into quail embryos and their settlement in gonads. Anim Sci Technol 69:911–915
Rahman M, Islam MN, Sarker NR, Islam MM (1998) Effect of supplementary feeding on production performance of RIR, Fayoumi and their crossbred chicken in rural Bangladesh. Bangladesh J Livestock Res 1:184–193
Reddy RP (1995) Artificial insemination of broilers, economic and managemental implications. In: Bakst MR, Wishart GJ (eds) Proceedings first international symposium on the artificial insemination of poultry. Poultry Science Association, Savoy, IL, pp 73–89
Reddy RP (1996) Use of artificial insemination in broilers, under various production, management, and marketing conditions. Proc World Poult Congr 20:519–529
Reddy RP, Sadjadi M (1990) Selection for growth and semen traits in the poultry industry. What can we expect in the future. In: Control of fertility in domestic birds. Institut National de la Recherches Agronomiques, Tours, France, pp 47–60
Sarkar K, Bell JG (2006) Potentialities of the indigenous chicken and its role in poverty alleviation and nutrition

security for rural households. INFPD Newsl 16: 15–26

Sexton TJ (1984) Breeding by artificial insemination. In: Cunningham FJ, Lake PE, Hewitt D (eds) Reproductive biology of poultry, Harlow: British Poultry Science. pp 175–182

Singh H (1999) Optimizing delivery of genetic merit in subtropical climates through advanced reproductive technologies. Poult Sci 78:453–458

Singh DP, Johari DC, Mohapatra SC and Sharma RD. 1996. Efficiency of growth and production performance of four Indian native breeds of chicken. In: Proceedings of the XX World's Poultry Congress, New Delhi, vol 4, p 18

Singh DP, Johri, TS, Singh UB, Narayan R, Singh D (2004) Impact of constraints minimization on productivity and popularity of traditional backyard poultry production. In: Proceedings of the XXII world's poultry congress, Istanbul

Tadelle D, Alemu Y, Peters KJ (2000) Indigenous chicken in Ethiopia: their genetic potential, attempts at improvement. Worlds Poult Sci J 56:45–54

Trefil P, Kotrbova A, Mikus T, Poplstein M, Ruzkova A (2002) The fate of female blastodermal donor cells in chimeric cockerels. Czech J Anim Sci 47:8–14

Valle Zarate A, Horst P, Harren-kiso AV, Rahman A (1988) Comparative performance of Egyptian local breeds and high yielding German medium-heavy brown layers under controlled temperature and warm environmental conditions. In: Proceedings of the XVIII world's poultry congress Tsukuba: Japan Poultry Science Association, pp 389–391

Yamada Y (1988) The contribution of poultry science to society. Worlds Poult Sci J 44:172–178

Beef Cattle in the Year 2050

13

George E. Seidel Jr.

Abstract

In 2050, beef likely will be produced much as occurs currently, as (1) a by-product of dairying—cull cows and calves not needed as replacements; (2) intensively managed cattle in environments rich in feed resources; or (3) extensively managed cattle grazing land unsuitable for tillage, with calves often moving to richer feed environments. Genetic progress will continue to depend on information such as weaning weights, but in addition, genetic, epigenetic, and phenotypic information will be obtained from blood, hair, semen, and/or biopsies of embryos.

Most cattle will be genetically modified for efficient growth, desirable carcass traits, and management traits such as disease resistance. Some strains of cattle will have Y-chromosome-dependent terminal cross traits; sexed semen thus will automatically result in males with terminal cross characteristics and females with maternally desirable traits. In most cases, mother cows will have shorter gestations and smaller frame sizes than currently to decrease nutrient requirements for maintenance. The cow herd may disappear with some intensively managed systems; with sexed semen, each female can replace herself with a female calf and then be fattened for slaughter. The flexibility of being a ruminant will continue to be exploited by using a variety of feedstuffs, some of which are otherwise of little value.

Keywords

Beef cattle • Genetic selection • Food production • Ruminants • Meat • Dairy

G.E. Seidel Jr., B.S., M.S., Ph.D. (✉)
Animal Reproduction and Biotechnology Laboratory, Department of Biomedical Sciences, Colorado State University, Fort Collins, CO 80523, USA
e-mail: george.seidel@colostate.edu

G.C. Lamb and N. DiLorenzo (eds.), *Current and Future Reproductive Technologies and World Food Production*, Advances in Experimental Medicine and Biology 752,
DOI 10.1007/978-1-4614-8887-3_13,

Large ruminants are enormously adaptable, and can even be managed primarily as a monogastric if grain or by-products are available. While all cattle are multipurpose, most have been selected for specific purposes, ranging from the high-producing dairy cow to animals that survive in essentially feral environments for beef and hide production. With current genetic selection tools and ten generations of selection, I contend that it would be possible to make a population of almost feral beef cattle into a respectable population of high-producing dairy cows or the converse.

Currently, there are a few billion head of cattle on the earth. Similarly, there are hundreds of millions of water buffaloes, bison, yaks, and a few other species with broadly similar characteristics and uses as cattle. These large ruminants vary enormously in genotypes/phenotypes and have a myriad of uses, predominantly to provide milk, meat, power, and hides, in addition to dozens of other uses including entertainment, a means of accumulating/storing/dispersing of capital, and providing symbols of prestige and religion.

Raw Materials and Limitations

Barring a worldwide discontinuity, such as nuclear war, there likely will be around nine billion people on earth in 2050, and well over a billion head of cattle probably will continue to be maintained; they likely will vary greatly in genotype/phenotype. In this chapter, I will concentrate on thinking about 2050 model cattle managed primarily for profitable beef production, mostly in herds of hundreds to thousands of head. Cattle provide higher value of product per animal than small ruminants, which justifies investing more time/resources in management per individual animal. For example, costs of artificial insemination with a dose of semen can be justified more often if producing $1,000 of product vs. $200 of product. While cattle will be in a spectrum of environments, I will concentrate on three: (1) reasonably intensive management in environments relatively rich in feed resources; (2) fairly extensive management in environments such as semi-arid grasslands, in which calves often will move to locations with richer feed resources; and (3) dairies, where beef is a by-product of culling and making dairy cows pregnant to induce lactation.

It is unclear how bovine genetics will be managed and controlled. It is inevitable that companies will continue to develop genetic products such as semen and embryos that have desired genetics for convenient and efficient production of high quality meat. What is less clear is the extent of vertical integration such as has already occurred with poultry and swine. Due to capital requirements with cattle and other factors, vertical integration is likely to be limited, but not by any means absent with beef production in 2050.

The main tool for cattle breeding currently is use of information, which requires records such as parentage, birth weights, and weaning weights. This will become even more important than at present; an increasing informational need is a tissue sample for genotyping such as semen, blood, hair follicles, or an embryo biopsy. Assisted reproductive technologies such as synchronization of ovulation, artificial insemination, embryo transfer, sexed semen, and cloning will continue to be available. One of the most valuable of these technologies will be transgenic technology to make genetic modifications. Despite current opposition to this technology from certain quarters, it will ultimately be morally indefensible not to use such technology for food production from animals; already over 85 % of corn and soybeans produced in North America has such genetic modifications. Management traits such as resistance to disease and parasites, the polled trait, and excretion of excess waste products such as methane, nitrogen, and phosphorous are examples of obvious targets for genetic modifications. Probably the main consideration will be selecting for traits that decrease labor needs such as easy calving and disease resistance, but efficiency traits such as high rates of converting feed to product will remain very important.

Characteristics of the 2050 Beef Cow

Many Beef Cows Will Continue to Be Dairy Cows

As alluded to earlier, many of the mothers of beef animals will be dairy cows, and, while there will be some selection of dairy cows for the beef value of some of their calves and for the salvage value of dairy cows themselves for beef, the overriding criterion for the ideal dairy cow herself will be for profitable milk production, since the average value of the milk she will produce will be worth about ten times the value of her calves and her carcass for beef. Selection of males to cross with dairy cows for beef calves will, however, be very important, especially due to changed dynamics with use of sexed semen for producing dairy female replacements, which will free up matings for beef crosses. To summarize this part, a huge amount of beef has its origin from a dairy cow mother, and producing milk profitably will be the main characteristic in selecting those females that produce meat as a by-product.

One caveat is that in future most dairy cows in some management systems may have only had one or two calves in their lifetime due to hormonal maintenance of lactation, thus avoiding the "roller coaster" cycle of positive energy balance during the dry period and negative energy balance during early lactation. This already occurs to some extent with use of bovine somatotrophin, and carried to extreme would decrease calf production.

Each Heifer Is Slaughtered After Producing a Heifer Calf

A second model of beef production, different from the current main approach in North America, also will not, strictly speaking, involve beef cows. For this model, each heifer is bred with sexed semen to produce a female replacement, and all are fattened and slaughtered after having their first calf. The calves likely would be weaned at around 3 months of age, and the resulting first-calf heifers fattened for another 2–3 months and slaughtered prior to 30 months of age. With such a system, there is no beef cow herd, and all cattle are in a growth phase. There are all sorts of management issues with this system because not all heifers will get pregnant, and a few will have male calves, but much more beef could be produced per unit of feed, manure, methane, etc. than the current system in which the majority of beef cows are simply being maintained for years without any net accretion of muscle other than the calves produced.

There are obvious disadvantages with producing nearly all beef from slaughtering heifers that have calved. The first problem that usually comes to mind is that all dams are primiparous. However, this is much less of an issue than in decades past because of effective selection for calving ease plus with this system most calves are heifers, weighing on average 2 kg less than males (Tubman et al. 2004). A second issue is that females accrete muscle less efficiently than steers or bulls, although this also is routinely partially circumvented with implants and/or feeding melengestrol acetate.

There also are obvious advantages in addition to dispensing with maintaining cows year round with no net accretion of muscle themselves. For example, the problems of low or delayed pregnancy rates with nursing first-calf heifers disappear. So do most of the problems as animals age, such as lameness, cancer eye, and problems with the mammary gland. In fact, selection for longevity and fertility traits such as returns to fertility postpartum would not be relevant. A completely different selection index from those now practiced would be in order.

For such an all-heifer, no-cow, no-male-calf system, there would be considerable value in shortening gestation length. This probably is occurring already as a correlation to selection for calving ease and related lower birth weight. Most such cattle likely will be crossbreds, but there would be intense selection for growth, selection for moderate milk production, and especially selection for desirable carcass traits that hold up in 30-month-old heifers previously exposed to the hormones of pregnancy; some information is currently available on the effects of pregnancy on carcasses of 30-month-old heifers (Waggoner et al. 1990).

Traditional Cow/Calf Beef Production

Despite the ascendance of the two beef production systems just mentioned, the majority of beef production in North America and many other areas will still likely originate from more traditional cow/calf operations. This is because of the requirement for extensive management for most beef cattle due to environmental and other considerations. Many cattle graze in arid and semi-arid landscapes, graze otherwise marginal pastures and resources such as cornstalks, or simply are in small units that do not justify having a critical mass of intensive management.

For these more traditional, extensive systems, I predict smaller mature size than currently is fashionable, simply to lower the costs of maintaining animals for years on end. Although it may not be sufficiently appreciated by many, there has been a major decrease in dystocia over the past 2 decades in all breeds, and this can be managed very effectively using artificial insemination of heifers with sires proven to have calves with low rates of dystocia, and further managed by selection for dams with low rates of dystocia. Superimposed on this can be use of sexed semen for heifers to have heifer calves, which as alluded to earlier average about 2 kg lower birth weights than bull calves with the added benefit of obtaining herd replacements from the youngest females in the herd, which should be genetically the best animals if one has a progressive breeding program.

It will be especially beneficial to select for shorter gestation in many situations so that cows have more time to get pregnant within a breeding season. A target of shortening by 5 days should be feasible without compromising calf health and vigor. This would be especially valuable for those breeds that currently have the longer gestations on average. Selection for fertility likely will be more practicable with the recently developed genomic tools (e.g., Seidel 2010) than was possible in the past.

There also will continue to be opportunities for intensively managed, traditional cow/calf systems. Simply optimizing the use of tools of reproductive management plus the tools of genetic selection will be profitable under some conditions. Superimposed could be the option of having a high percentage of twin pregnancies. Seedstock production will continue to require an element of intensive management, even if done in an extensive environment. There will be niches such as "organic" beef production, breeding rodeo stock, etc. Essentially all of these cattle will, however, end up as beef.

Heterosis and Genomics

For the foreseeable future, crossbreeding has such a huge performance advantage due to heterosis that the majority of beef cows will continue to be crossbreds. Currently there is a bit of a fad that straightbreds can perform as well as crossbreds; straightbred performance can be impressive with optimal use of tools such as EPDs and management of nutrition, particularly for selected traits such as some carcass characteristics. However, straightbreds do not even come close to crossbreds for traits such as pounds of beef per breeding cow or per unit of feed.

Crossbreeding does, however, suffer from a serious limitation: the need for straightbred cows to produce crossbreeds. There is also the problem of mating the F_1 for replacements. One can of course go to three-way and four-way crosses, but eventually one ends up with a composite or runs out of breeds of appropriate phenotype that have substantive genetic information obtained in a relevant environment. In other words, straightbreeding is much simpler than crossbreeding. Even so, crossbreds are markedly more "trouble-free," and thus will continue to predominate for decades. Furthermore, use of sexed semen can be a great advantage for systems of producing crossbred females.

However, genomic and other tools may eventually be used to identify and manipulate the alleles that produce beneficial heterotic performance. The concepts of crossbred and straightbred may eventually disappear, and selection will be based on manipulating alleles directly, as already occurs to a great extent in poultry, pig, and plant breeding. A transitional breeding scheme might mimic what is often done with sheep in Australia and New Zealand in which there are seedstock, multiplier, and terminal phases to the breeding system.

Power of Transgenic Approaches

There are a myriad of options with transgenic procedures. One of my favorites is to have terminal cross genes expressed on the Y-chromosome. With such a scheme, the females would have maternal traits including small size, but give birth to males that are born small but grow explosively due to growth genes (note genes, not alleles) on the non-pseudoautosomal portion of the Y-chromosome. One simply uses sexed semen to breed for female replacements or male terminal crosses. Of course this already can be done to some extent as males grow more efficiently than females in the feedlot, the only justification for the premium paid for steer vs. heifer feeder calves. One simply enhances this advantage by adding (possibly moving from autosomes in a few cases) genes to the Y-chromosome. A theoretical advantage would be the hemizygous nature of such genes; they automatically breed true without issues of heterozygosity if sexed semen is used.

Another transgenic example of interest, already demonstrated in mice (Herrmann et al. 1999), is to distort sex ratio transgenically so that 70–90 % of calves from a particular sire will be one sex or the other without sexing semen. I will not go through a list of theoretically desirable transgenic modifications, but point out that remarkable changes in the efficacy of transgenic procedure are evolving (Carlson et al. 2012), some of which result in homozygous modifications in the one-cell embryo. These procedures circumvent the generation interval problem of applying transgenics to cattle with stem cell/cloning approaches and breeding to homozygosity.

Less Conventional Tools of Animal Breeding

I have long maintained that the most powerful assisted reproduction tool available for cattle is artificial insemination coupled with use (and generation) of genetic information and tools such as estrus synchronization and possibly sexed semen. For reasons of simplicity, effectiveness, and low cost, artificial insemination is likely to remain in this preeminent position for decades for most production purposes. Tools such as superovulation, in vitro oocyte maturation and fertilization, cloning, biopsy of embryos for genotyping, and transgenics require the relatively expensive tool of embryo transfer, but will be important to create elite breeding stock, which then can be amplified by less costly tools like artificial insemination for the main production endpoints.

Note that there are dozens of other exotic tools that I have summarized elsewhere (Seidel 2011) such as creating two-father, no-mother (except for mitochondrial genetics) animals using two X-sperm or an X- and a Y-sperm for genetic parents. This approach is problematic due to the phenomenon of gametic imprinting, but two-father, no-mother mice have already been produced by using various genetic and epigenetic tricks (Deng et al. 2011).

Phenotypes

The 2050 beef cows will look remarkably like the 2000 models except for average smaller size. They will of course vary among environments, e.g., more skin surface when heat tolerance is required. Dark pigment will prevail in colder climates. Nearly all will be polled. Body condition scores likely will be moderate to thin on average, but of course vary due to season, stage of gestation, lactation, etc. Cattle within herds likely will be more uniform than at present. Despite selection for longevity, beef cattle likely will be younger on average, but that will depend on objectives of managing individual herds. Females will be younger at puberty than at present. However, cows will still look and act like cows.

References

Carlson DF, Tan W, Lillico SG, Stverakova D, Proudfoot C, Christian M, Voytas DF, Long CR, Whitelaw CBA, Fahrenkrug SC (2012) Efficient TALEN-mediated gene knockout in livestock. Proc Natl Acad Sci U S A 109:17382–17387

Deng JM, Satoh K, Wang H, Chang H, Zhang Z, Steward MD, Cooney AJ, Behringer RR (2011) Generation of viable male and female mice from two fathers. Biol Reprod 84:613–618

Herrmann BG, Koschorz B, Wertz K, McLaughlin KJ, Kispert A (1999) A protein kinase encoded by the t complex responder gene causes non-Mendelian inheritance. Nature 402:141–146

Seidel GE Jr (2010) Brief introduction to whole-genome selection in cattle using single nucleotide polymorphisms. Reprod Fertil Dev 22:138–144

Seidel GE Jr (2011) Future reproductive technology. In: Mckinnon AO, Squires EL, Vaala WE, Varner DD (eds) Equine reproduction, 2nd edn. Wiley-Blackwell, Chichester, pp 3051–3056

Tubman LM, Brink Z, Suh TK, Seidel GE Jr (2004) Characteristics of calves produced with sperm sexed by flow cytometry/cell sorting. J Anim Sci 82: 1029–1036

Waggoner AW, Dikeman ME, Brethour JR, Kemp KE (1990) Performance, carcass, cartilage calcium, sensory and collagen traits of longissimus muscles of open versus 30-month-old heifers that produced one calf. J Anim Sci 68:2380–2386

Index

G.C. Lamb and N. DiLorenzo (eds.), *Current and Future Reproductive Technologies and World Food Production*, Advances in Experimental Medicine and Biology 752,
DOI 10.1007/978-1-4614-8887-3, © Springer Science+Business Media New York 2014

Printed by Printforce, the Netherlands